Great and Desperate Cures

Steve

Science Studies Unit
University of Edinburgh

GREAT AND DESPERATE CURES

The Rise and Decline of Psychosurgery and Other Radical Treatments for Mental Illness

ELLIOT S. VALENSTEIN

Basic Books, Inc., Publishers / New York

Grateful acknowledgment is made for permission to reprint from the following sources:

W. Freeman and J. Watts, "Prefrontal Lobotomy in Agitated Depression: Report of a Case," *Medical Annals of the District of Columbia* 5(1936): 326–28.

"Explorers of the Brain," *New York Times,* 30 October 1949. Copyright © 1949 by The New York Times Company. Reprinted by permission.

F. R. Ewald, W. Freeman, and J. W. Watts, "Psychosurgery: The Nursing Problem," *American Journal of Nursing* 47(1947): 210–13. Copyright © 1947 by the American Journal of Nursing Company.

Letters from J. F. Fulton to Walter Freeman of 12 September 1936, 16 April 1948, 25 February 1935, and 3 January 1950, John F. Fulton Papers, Yale University Library.

Editorial, "The Surgical Treatment of Certain Psychoses," *New England Journal of Medicine* 215 (1936): 1088.

W. Freeman and J. W. Watts, *Psychosurgery: Intelligence, Emotion and Social Behavior Following Prefrontal Lobotomy for Mental Disorders,* 1st ed. (1942); and W. Freeman and J. W. Watts, *Psychosurgery: In the Treatment of Mental Disorders and Intractable Pain,* 2nd ed. (1950). Courtesy of Charles C Thomas, Publisher, Springfield, Illinois.

"Frontal Lobotomy," *Journal of the American Medical Association* 117(1941):534–35. Copyright © 1941, American Medical Association.

"New Brain Operations for Mental Illness," *Life,* 16 August, 1948, 57–60. Life Magazine, © 1948 by Time Inc. Reprinted by permission.

Library of Congress Cataloging-in-Publication Data

Valenstein, Elliot S.
 Great and desperate cures.

 Includes index.
  1. Psychosurgery—History. 2. Frontal lobotomy—History. 3. Mental illness—Treatment—history.
 I. Title. [DNLM: 1. Mental Disorders—therapy.
 2. Psychosurgery. WM 400 V153g]
 RD594.V35 1986   616.89'1   85-43104
 ISBN 0–465–02710–5 (cloth)
 ISBN 0–465–02711–3 (paper)

Copyright © 1986 by Basic Books, Inc.
Printed in the United States of America
Designed by Vincent Torre
87 88 89 90 RRD 9 8 7 6 5 4 3 2 1

To Thelma: I gave my love a story that had no end.

Physicians get neither name nor fame by pricking of wheals, or picking out thistles, or by laying of plasters to the scratch of a pin: every old woman can do this. But if they would have a name and a fame, if they will have it quickly, they must . . . do some great and desperate cures. Let them fetch one to life that was dead; let them recover one to his wits that was mad; let them make one that was born blind to see; or let them give ripe wits to a fool: these are notable cures, and he that can do thus, if he doth thus first, he shall have the name and fame he desires; he may lay abed till noon.

JOHN BUNYAN

The Jerusalem Sinner Saved;
or Good News for the Vilest of Men
(1668)

Contents

List of Illustrations

Preface and
Acknowledgments

THIS HISTORY of lobotomy began, in one sense, with an earlier book of mine, *Brain Control,* which was published in 1973, a time when concern over what was perceived to be the beginning of a "new wave" of lobotomy had led to a raging controversy. Although my book had been written for another purpose, a part of it critically evaluated the capacity of lobotomy and other brain operations to change behavior. I was thus soon drawn into the "psychosurgery controversy"—at least as a "regular" participant in the many symposia held to debate the scientific, ethical, and legal questions raised by these operations.

In 1976, as a result of this activity, the National Commission for the Protection of Human Subjects of Biomedical and Behavioral Research asked me to survey the extent of psychosurgery being performed around the world, to determine the results of these operations, and to comment on related ethical and social problems. In performing this task, I learned much more about the early history of lobotomy as well as about current psychosurgery. After writing my report for the commission, I edited *The Psychosurgery Debate* (1980), which presented a range of views on the subject.

By this time, I had talked to many of the major figures who had participated in the history of lobotomy. Much of the information I had collected about its early history and the personalities involved was either unknown or had been distorted in earlier accounts of these events. When I was invited to give the Kenneth Craik Lecture at St. John's College of Cambridge University in April 1981, I pulled together some of this material for my talk. It was then that I decided the story had to be written.

It had become evident that lobotomy was not an aberrant event but very much in the mainstream of psychiatry. There were many other examples, although perhaps none so dramatic, of uncritical enthusiasm running rampant and causing great harm to desperate patients. I began to see the history of lobotomy as casting light on current practice, not only in psychiatry but in all of medicine: that is, the events that made for its wide acceptance were not unique but are actually quite commonplace.

In researching this history, my starting point was the archival record—the clinical and experimental reports published in journals and books. Particularly valuable were speeches at medical meetings, as the comments from the audience—often published with the speeches—reflect the prevailing views at the time. Patients' records and the diaries, correspondence, memoirs, and autobiographies of the major participants in this history were also most informative. Especially noteworthy were the unpublished papers and correspondence of Walter Freeman, on file at the Himmelfarb Health Science Library of George Washington University School of Medicine. This collection includes case histories and medical records of the patients in the Freeman-Watts standard prefrontal lobotomy series and in Freeman's transorbital lobotomy series. The manuscript of Freeman's unpublished and unfinished "History of Psychosurgery" is also available in this library.

Walter Freeman's unpublished autobiography, written for his immediate family, proved to be a most valuable source of information about its author, the events he shaped, and those that shaped him. It is a remarkably honest and revealing document. Freeman's unpublished manuscripts are not generally available but are cited in the text to indicate the source of some information. In almost all cases, it was possible to check the accuracy of this information either against accounts published by others or against those told to me by people with firsthand experience of an event. To portray Freeman's personality and views accurately and vividly, I quote some of his brief remarks; but to avoid cluttering the text with too many notes, they are not referenced. In all such instances, where they are not otherwise referenced, they are from one of the two unpublished Freeman manuscripts. Freeman's son, Walter Freeman III, kindly granted me permission to quote from these manuscripts as well as from his father's letters and other papers.

John Fulton's forty-seven-volume diary and his extensive correspondence file at the Yale University Library also proved useful.

A great many people helped in my research. The reference librarians at the Taubman Medical Library of the University of Michigan were most cooperative in obtaining books and journals not in their collection, and numerous people at the National Library of Medicine and the Library of Congress willingly searched for some of the more obscure references. People who had firsthand knowledge of events in this story generously gave all the time I requested. They attempted to answer all my questions, shared their recollections, and searched through their files for photographs or correspondence that would help me to interpret these events. Literally no one whom I approached failed to help. I owe them all a great debt and can only hope that, where I have disagreed with them, they will understand that I seriously considered their views.

I gratefully acknowledge the cooperation of the following people. Walter Freeman III, in addition to granting permission, answered my questions and also made available to me a copy of his father's unpublished autobiography. The neurosurgeons Pedro Almeida Lima in Lisbon and James Watts in Washington, both pioneers in the development of lobotomy, spent several hours reminiscing about their experiences. Jonathan Williams, who was Walter Freeman's neurosurgical associate after Watts, shared his memories of Freeman and duplicated photographs from his personal collection. The neurosurgeon Paul Bucy helpfully talked with me about the introduction of lobotomy in the United States. Charles Jones, one of the many psychiatrists in state hospitals trained by Walter Freeman to perform transorbital lobotomies, described his experiences in detail. Eric Cunningham Dax, now residing in Hobart, Tasmania, searched through his files for information that might be useful in reconstructing the early history of lobotomy in England. J. Sydney Smith's hospitality during my week-long visit to the Prince Henry Hospital in Sydney, Australia, was truly memorable. He provided an opportunity for me to observe contemporary psychosurgery as practiced at the Neuropsychiatric Institute of that hospital, and helped me to understand the history of lobotomy in his country. David McKenzie Rioch, who had supported me in my early laboratory work at the Walter Reed Institute of Research, spent an afternoon and evening talking about John Fulton and Walter Freeman. I am indebted to Walter Freeman's former residents—Zigmond Lebensohn, Oscar Legault, and his office assistant, Paul Chodoff—for sharing their memories of Walter Freeman and their perspectives on this history. Max Fink, Harald Fodstad, Sadao Hirose, Leopold Hofstatter, Lothar B. Kalinowsky, Lauri Laitinen, Pavel Nádvornik, J. Laurence Pool, Karl Pribram, B. Ramamurthi, William Scoville, and Joseph Wortis provided information useful in answering specific questions. I am especially grateful to Allan F. Mirsky and Haldor Enger Rosvold for sharing with me information about lobotomized patients. Rosvold also described his experiences working with John Fulton on the Connecticut Lobotomy Studies. I also wish to acknowledge the assistance of Anita Lundmark of the Karolinska Institute, who provided me with parts of the records of the Medical Nobel Archives available for scholarly research.

Barahona Fernandes, Pedro Polonio, Juan Miller Guerra, Lobo Antunes, A. Castro-Caldas, and many members of the staff of the Centro de Estudos

Egas Moniz in Lisbon were most cooperative in helping me collect background material on Egas Moniz. It is difficult to express adequately my debt to the Portuguese neurologist Carlos Garcia, who spent the better part of a week arranging my schedule in Lisbon and traveling with me to Avanca to visit Egas Moniz's former home, now a museum. Mr. Boaventura Pereira Melo, director of the Egas Moniz Foundation, interrupted his vacation to open the museum for me.

Leo Bernucci and Carl Valenstein translated some of Moniz's papers from the Portuguese. Otto Sellinger and Susan Bachus helped with other translation. Roberta Y. Arminio, director of the Ossining Historical Society, and James Sullivan, superintendent of the Sing Sing Correctional Facility, helped me obtain background information related to the case of J.S., presented in chapter 12.

In 1982, the Rockefeller Foundation supported me at the Villa Serbelloni in Bellagio, Italy, when I turned from collecting information to working seriously on writing this book. In 1984, a fellowship from the National Humanities Center in the Research Triangle, North Carolina, provided six months in a congenial atmosphere with most helpful secretarial and library assistance. A substantial part of the first draft of the manuscript was completed during this period.

Anyone who has undertaken a task such as this can fully appreciate the importance of competent and reliable secretarial help. Judy Baughn not only deciphered my scribbling and somehow followed my cryptic arrows but returned clean manuscript copies to me with amazing speed and volunteered to work extra hours and weekends when necessary. I am truly grateful to her. I am also grateful to Kathe Davids for help in typing; to Guy Mittleman, who took time from his Ph.D dissertation to help with our sometimes obstreperous word processor; and to Buda Martonyi, for his photographic skill.

Judith Greissman and Phoebe Hoss, my editors at Basic, were both enormously helpful in suggesting ways, large and small, to improve the manuscript. Needless to say, had they not been "gentle" in making these valuable and essential suggestions, I should have been hard put to return to a task I had thought was just about completed. My thanks go, too, to Linda Carbone for her tender loving care of the manuscript through the final editorial process.

Lastly, a very special debt is owed to my wife, Thelma, who always manages to overcome my resistance by balancing her constructive criticisms with support and encouragement.

Great and Desperate Cures

1

The Treatment of Mental Illness: Organic versus Functional Approaches

> Perhaps the greatest danger that threatens neurology today is the possibility of the passage of the care of the psychoneuroses into the hands of the psychoanalytic psychologists and psychiatrists.
>
> —HENRY ALSOP RILEY (1933)

BETWEEN 1948 and 1952, tens of thousands of mutilating brain operations were performed on mentally ill men and women in countries around the world, from Portugal, where prefrontal leucotomy was introduced in 1935, to the United States, where under the name of "lobotomy" the procedure was widely used on patients from all walks of life. From our present perspective, these operations—referred to collectively as "psychosurgery"—seem unbelievably primitive and crude. After drilling two or more holes in a patient's skull, a surgeon inserted into the brain any of various instruments—some resembling an apple corer, a butter spreader, or an ice pick—and, often without being able to see what he was cutting, destroyed parts of the brain. In spite of the huge amount of psychosurgery done during the peak of its popularity, by 1960 this practice was drastically curtailed. Not only had chlorpromazine and other psychoactive drugs provided a simple and inexpensive alternative, but it had also been discovered that these operations were leaving in their wake many seriously brain-damaged people. Today lobotomy has largely fallen into disrepute and is now considered an evolutionary throwback, akin more to the early practice of trepanning the skull to allow the demons to escape than to modern medicine.

Why then pursue the history of a treatment at once bizarre and obsolete? The answer, it seems to me, is clear. Psychosurgery was not a medical aberration, spawned in ignorance. In a real sense, the history of psychosurgery is a cautionary tale: these operations were very much a part of the mainstream of medicine of their time, and the factors that fostered their development and made them flourish are still active today.

Psychosurgery was recommended by distinguished psychiatrists and neurologists and performed by equally prominent neurosurgeons, many of whom were affiliated with highly respected medical centers and universities. Other renowned medical and scientific figures, while not primarily involved in treating the mentally ill, nevertheless found the practice "interesting," "promising," and based on "sound theory" and "convincing evidence." Editorials in the most influential and prestigious medical journals praised these operations and their scientific foundations, even while at times suggesting the need for caution and more research. All this support from within the medical profession increased the demand for these operations, leading to their ultimate accolade when, in 1949, the man who had introduced psychosurgery was awarded the Nobel Prize in Medicine.

The patients on whom psychosurgery was performed were not only—as has been asserted—the indigent in publicly supported institutions. While the largest number of lobotomies were done in state hospitals, a substantial number were performed on wealthy and socially advantaged patients in private hospitals or in the psychiatry or the neurology departments of university medical schools. While some psychiatrists and neurologists were strongly opposed to psychosurgery from the outset, this opposition, for reasons I shall explore, was not effective until more than a decade had elapsed.

Psychosurgery was not an isolated development, but one of many drastic physical or, as they were called, "somatic" treatments for mental disorders. However drastic these treatments may seem today—and, indeed, were—it was commonly accepted that the more severe mental disorders were caused by some pathological, organic condition whose treatment required physical or somatic intervention. Most physicians were in agreement that, without effective treatment, the fate of seriously ill mental patients would be progressive deterioration, both mental and physical. In spite of the revolution Freud and his theories had effected in the field of psychology, few psychiatrists believed that psychotic conditions were susceptible to treatment by psychoanalysis. Even had they been, the one-to-one practice of psychotherapy made it totally impractical for coping with the magnitude of the problem in the many crowded, understaffed, and underfunded asylums for the mentally ill.

Psychosurgery also had a significant place in the politics of medicine. There had been, since the late nineteenth century, an ongoing jurisdictional dispute between psychiatrists and neurologists over the responsibility for treating the mentally ill. The differences were due not only to conflicting theories about mental illness—whether it was caused by organic pathology or life experience and how, therefore, it should be treated; but, as I shall describe, basic economic considerations played a significant role in the dispute between the two evolving medical specialties. Neurologists accused psychiatrists of abandoning medicine, and contrasted their own "scientific," somatic treatments of mental illness with the "ineffectual" and "metaphysical" search for intrapsychic conflict of the psychiatrists. Some psychiatrists were critical of the somatic therapies, refusing to accept the assertion that they had a solid scientific foundation or the claims that they provided effective treatment; but others—some of whom called themselves "neuropsychiatrists"—adopted many of the somatic therapies and were among their leading protagonists. Regardless of orientation, however, both neurologists and psychiatrists were subjected not only to competition from each other but also to pressure from desperate mental patients and their desperate families.

Equally desperate were the superintendents of the public institutions charged with the responsibility for treating mental patients. As these patients remained hospitalized for long periods, sometimes for life, and funding did not keep stride with the increasing patient population, economic considerations often became the major factor determining choice of treatment. Any suggested treatment that had a chance of decreasing the patient population, *and* was relatively inexpensive to administer, had great appeal.

Outside the medical profession itself, lobotomy was promoted by the popular press. Magazines and newspapers, whose readers numbered in the millions, popularized each new "miracle cure" with uncritical enthusiasm, while commonly overlooking its shortcomings and dangers. These popular accounts created an enormous interest in lobotomy among patients and their relatives, many of whom had abandoned hope, and they sought out the physicians mentioned in the articles with the desperation of a drowning person reaching for anything to stay afloat.

Lobotomy, along with the other somatic therapies, was ultimately created by physicians—in many instances, able men who had contributed significantly to medicine earlier in their careers—whose energy and determination in their work was matched by their ambition to carve a lasting niche for themselves in the annals of medicine. The two outstanding figures in the history of psychosurgery were Egas Moniz, the Portuguese neurologist who initiated prefrontal-lobe operations in 1935 and later, for this work, received the Nobel Prize; and Walter Freeman, the American neuropathologist and neuropsychiatrist who, more than anyone else, was responsible for the wide adoption of these operations around the world.

Thus, the story of lobotomy involves many factors: opposing theories of mental dysfunction; a long political struggle within medicine between psychiatrists and neurologists; a desperate human need and a procedure that offered to cure it; immediate enthusiasm in the popular press; uncritical acceptance by the medical profession, which not infrequently paid little attention to the validity of the claims of success; and determined and ambitious doctors. I shall begin this multifaceted story with the struggle within the medical establishment and go on, in chapter 2, to the variety of somatic treatments developed and tried on the mentally ill from the turn of the century up through the 1920s. In chapter 3, I shall discuss three radical somatic therapies introduced in the 1930s—insulin coma, metrazol shock, and electroshock. Chapters 4 to 6 cover the story of Egas Moniz, the development of lobotomy, and the theory he devised to justify its use. In the following chapters, I examine the dissemination of lobotomy throughout the world, but especially in the United States, where it was promoted principally by Walter Freeman; and finally, its eventual decline.

Throughout the book, I devote considerable attention to the personalities and lives of Moniz and Freeman and other leading figures in this drama because I believe this history was not the inevitable outcome of social forces. While much of the medical profession and the world at large was clearly ready to accept lobotomy, its practice was catalyzed and shaped by persons able to inspire others to follow their lead, and who, at critical moments, had an enormous impact on the course of events.

First, then, to the rivalry within the medical establishment between neurologists and psychiatrists—a rivalry dating back to the latter part of the nineteenth century. To a large extent, this rivalry was paralleled by two opposing views of the causes of mental illness—the somatic or organic versus the functional, which emphasized life experiences rather than biological determinants; but some neurologists were sympathetic to the functional view, and many psychiatrists accepted biological explanations. Indeed, the functional view is epitomized by Sigmund Freud, who started his medical career as a neurologist, while the somatic view was shaped to a large extent by German psychiatrists, particularly Emil Kraepelin.

Kraepelin was the foremost authority in psychiatry during much of the first half of this century. His descriptions and classifications of mental illnesses continued to have a great influence even after his death in 1926. While he recognized that the neuroses—the less severe mental disturbances—may be caused by life experiences, he was convinced, along with most psychiatrists, that the psychoses—that is, the serious mental disorders—have biological causes, which are often genetic in origin and essentially incurable. Although Kraepelin distinguished over twenty subtypes of psychosis, the two main categories were dementia praecox and the manic-depressive disorders. *Dementia praecox*—its name reflects the fact that "demented" thought processes often become evident during adolescence—was in 1911 given the name "schizophrenia"* by the Swiss psychiatrist Eugen Bleuler. Kraepelin divided dementia praecox into four types: simple, hebephrenic, catatonic, and paranoiac.† The manic-depressive psychoses were serious disorders of mood or emotion in which patients were so depressed or manic they could no longer function. Dementia praecox was thought to be a progressive and incurable disease; manic-depressive patients were known to improve, but because the disorder usually recurred, often with increasing severity, it was thought to spring from some biologically determined predisposition.[1]

Kraepelin's view that the psychoses have biological and probably genetic causes was accepted at the time by most psychiatrists as almost axiomatic. It is reflected in part in the works of Italian physician-criminologist Cesare Lombroso and German psychiatrist Ernest Kretschmer, both of whom argued that inherited physical characteristics predispose a person toward a particular behavior and personality. In 1911, Lombroso had written an extensive treatise purporting to demonstrate that criminals have identifiable external physical traits (and presumably the internal nervous system to go along with them); Kretschmer argued, during the 1920s and 1930s, that the biological factors predisposing individuals toward particular kinds of mental illness are evident in one's body type. Kretschmer argued that certain physiques are commonly associated with "cyclothymic" temperaments (individuals likely to develop manic-depressive disorders). Other body types are found among people with "schizothymic" or "schizoid" temperaments. These ideas continued to have a major influence into the 1940s: for example, in 1936, Walter Freeman was moved to study the body types of 1,400 former mental patients at St. Elizabeth's Hospital in Washington, D.C., and in the 1940s, American psychologist William Sheldon developed an elaborate quantitative method for denoting body type—expressed in terms of the relative contribution of the ectoderm, the

---

* *Schizophrenia* refers to a split or separation from reality and sometimes to a separation between affect (emotions) and thoughts; the term does not refer to split or "multiple" personalities, which are considered to be a hysterical disorder.

† Only *paranoid schizophrenia*—referring to the prevalence of delusional ideas—is used today. *Simple schizophrenia* designated profound withdrawal and lack of interest, initiative, and drive. *Hebephrenia* referred to patients whose behavior appeared "silly"—often giggling inappropriately; *catatonic schizophrenia* applied to individuals who were mute and often rigid, maintaining the same posture for hours. *Dementia* is now used mainly in cases of known brain pathology, such as *senile dementia,* and rarely to refer to schizophrenic thought processes.

endoderm, and the mesoderm*—relating each to "varieties of temperament."²

A contemporary of Kraepelin's, Sigmund Freud was, during this same period, developing an alternative theory of mental illness. Although it was not widely accepted initially, psychoanalysis, helped to a great extent by the attention given to it in the arts and popular media, eventually came to play a major role in strengthening the position of the functional psychiatrists. Prior to psychoanalysis, there was no coherent explanatory theory of mental illness, based on life experiences. Psychoanalysis never offered any possibility of helping significant numbers of institutionalized psychotic patients, but it had created an illusion of a growing body of knowledge that would eventually be able to attack that problem.

Freud started his career in Vienna as a neurologist with a strong interest in laboratory research. He seemed headed for a bright future as a "clinical investigator" searching for the causes of mental illness in the nervous system— that is, looking for the basis of psychopathology in neuropathology. First working with Theodor Meynert, generally considered one of Europe's leading brain anatomists, Freud switched to the Institute of Physiology headed by Ernst Brücke and, while there, published several anatomical papers, most of them on the nervous system.

Freud gradually became pessimistic, however, about the possibility of localizing psychological problems in pathological brain cells. In 1880, he began to spend long hours discussing with his friend, the physician and physiologist Josef Breuer, the latter's now-famous patient Anna O.† This woman was a textbook case of hysterical symptoms, including paralysis of three limbs and a multiple personality with two distinct states of consciousness. Breuer had observed that after she talked about her disturbing thoughts, her symptoms were alleviated. He began to recognize the therapeutic value of "talking out problems," calling it by such names as "the talking cure," "chimney sweeping," or the "cathartic treatment." Later Freud and Breuer would collaborate on a monograph about hysteria, although Freud eventually differed with his older friend about its cause.³

Freud's fascination with hysteria led him eventually to abandon his bio-

---

* Endoderm, mesoderm, and ectoderm are embryological layers from which different parts of the body develop. Simply put, the digestive system and other viscera develop from the endoderm; the muscles and bones, from the mesoderm; and the nervous system and skin, from the ectoderm. "Endomorphs" are soft and round, the viscera predominate, and they have low specific gravity, floating high in water; "mesomorphs" are strong and muscular; and "ectomorphs" are flat-chested and fragile.

† Breuer was older than Freud and at the time a highly respected medical scientist. As a physiologist, he worked with Ewald Hering and, with him, described the Hering-Breuer reflex controlling breathing. Breuer also did research on the semicircular canals of the ear and was elected to the Vienna Academy of Science. As a neurologist he was sought out for treatment by the medical faculty, including Professor Brücke, and also by the prime minister of Hungary.

psychoanalysis but to any of its derivative forms of psychotherapy. Even if psychotherapy proved effective for treating psychotics—a concession most psychiatrists would not make—it was totally impractical. Psychotics were generally institutionalized in large hospitals, where according to Hinsie, the physician-patient ratio in 1934 was more than 1 to 200 in New York State, better than in many areas of the country. As a full daily schedule for a psychoanalyst could not exceed eight patients, who were usually seen three to five days a week for an average of eighteen months, psychoanalytic therapy could be of little help to most mental patients.[7]

Partly because they were dealing with different patients in different settings, the organic and the functional approaches in psychiatry developed independently, each being strongly prejudiced against the ideas and practices of the other. Most of the early leaders in organic psychiatry were trained as neurologists, with strong research backgrounds in neuroanatomy and neuropathology. The eminent organic psychiatrists were often professors in major universities, heads of departments of neurology or of neurology and psychiatry—there being virtually no separate departments of psychiatry before the 1920s. The leading functional psychiatrists, on the other hand, often developed their own clinics, psychoanalytic institutes, and professional societies.

Most mental hospitals in the United States were isolated from the rest of medicine. Up to 1894, the professional organization concerned with institutionalized mental patients was the Association of Medical Superintendents of American Institutions for the Insane. This group had refused overtures to affiliate with the American Medical Association, partly out of fear of diluting its own privileged position. If allowance was made for all the fringe benefits, especially housing and food, the average income of psychiatrists—or "alienists," as they were still generally called until around the turn of the century—working in institutions exceeded that of most other physicians. Many newly graduated physicians were more than pleased to receive the regular salary that came with a position in a mental hospital, while the real income of the superintendents of the asylums—considering the fringe benefits—was greater than that of all but the most successful practitioners.

The staff in the mental hospitals rarely had any specialized training in psychiatry, as medical schools offered almost no instruction in the field—or, at best, a few lectures included as part of a neurology course. Lack of courses was not, however, a drawback, as there was little to learn. Often young, untrained physicians were put in charge of a ward within a few weeks after their appointment to the staff of an asylum; and the somatic treatments they used were a matter of personal preference and institutional convenience.

During the last quarter of the nineteenth century, open warfare broke out between the neurologists interested in mental disorders and the superintendents of asylums. The neurologists accused the superintendents and their staffs of being divorced from the rest of medicine, of doing nothing for the patients, of making no progress, and of running the hospitals mainly for their own benefit. At the first meeting of the American Neurological Association in 1875, a vote to bar the superintendents from membership was passed without opposition. The neurologists began publishing the *Journal of Nervous and Mental Disease* in competition with the older *American Journal of Insanity,* which represented the superintendents.[8]

In 1878, Edward C. Spitzka, a twenty-five-year-old neurologist recently turned psychiatrist, gave a speech entitled "The Study of Insanity Considered as a Branch of Neurology, and the Relations of the General Medical Body to This Branch." He attacked institutional psychiatry and the superintendents, stating that, after reading their reports, he had concluded

> that certain superintendents are experts in gardening and farming (although the farm account frequently comes out on the wrong side of the ledger), tin roofing (although the roof and cupola is usually leaky), drain-pipe laying (although the grounds are often moist and unhealthy), engineering (though the wards are either too hot or too cold), history (though their facts are incorrect, and their inferences beyond all measure so); in short, experts at everything except the diagnosis, pathology and treatment of insanity.[9]

Spitzka did not seem to want to reform the superintendents so much as to usurp them. He characterized the members of the Association of Superintendents as "deficient in anatomical and pathological training, without a genuine interest in their noble specialty, [and] untrustworthy as to their reported results."[10] Although he was exceptionally acerbic, his opinion of "institutional psychiatry" was shared by the vast majority of neurologists. Spitzka had received his medical training in New York City and then spent three years in Berlin, Leipzig, and Vienna, studying embryology, morphology, and psychiatry. When he returned to New York, he was completely convinced that, above all, psychiatry should be "scientific" and thus should confine itself to studying only the neurological manifestations of mental disorders. Spitzka was an influential figure in neuropsychiatry. As a young man, he wrote the essay "Somatic Etiology of Insanity," which won a prize in international competition. In 1883, when he was thirty-one, his successful textbook *Insanity: Its Classification, Diagnosis and Treatment* was published. Seven years later, he was elected president of the American Neurological Association, which he used as a platform to spearhead the attack on institutional psychiatry.

The superintendents had not, in 1878, simply ignored Spitzka; they had too much to lose, and counterattacked by arguing that he and his neurologist allies had a "hidden agenda" and were trying to take over the asylums for their own teaching and research. Indeed, Spitzka had emphasized in his 1878 speech that a change in administrative responsibility would make it possible for "mental pathologists" and general neurologists "to conduct [their] difficult and interesting researches"; he added that, as a result, "there is hardly a specialty in medicine which will not benefit." He noted that with such a change even the general practitioner would become familiar with insanity and better able to cope with it when it arises as a "complication of other diseases." Spitzka recommended reforming the asylums so that they could become teaching hospitals and a source of "rich material"—not "dead material." He also advised "uniting psychiatry with neurology, in our college courses."[11]

Spitzka's call for reform in the asylums, however, would have amounted to their "takeover" by the neurologists. In the inevitable power struggle, the charges and countercharges became increasingly severe, even to personal attacks on the ethical behavior of some of the protagonists. The superintendents, for example, accused William A. Hammond, a neurologist and former surgeon general of the United States, of testifying on any side of a legal suit as long as he was given a large fee—an accusation that led him to threaten a libel suit.[12]

A series of editorials during 1879 in the *New York Times* reflected the backlash against the neurologists, whose motives were becoming suspect and whose scientific claims, when examined closely, seemed exaggerated. The *Times* commented on a report written by members of the committee established to inspect the conditions in the asylums:

> The eloquent talk about the wonderful properties of nerve-cells and their laws of action have held the attention of medical men altogether too long, and the time has come to dismiss this high-sounding rigmarole from the literature of science, and to place insanity where it belongs—as a symptom of disease which may or may not have its primary seat in the nervous system.[13]

The lines were further drawn when a letter to the editor of the *New York Times*—signed only with the letter "D"—responded to this last editorial.* The writer acknowledged that "no specific lesion of the brain has always been found in insanity," but went on to state that the reason was only that present instruments were not sufficiently sensitive "to determine the fact pos-

---

* This same controversy was conducted, more temperately, in England, most leading psychiatrists there believing psychological explanations to be useless in therapy. Henry Maudsley, writing as early as the 1870s, had expressed the prevailing belief that while psychology might be useful for descriptive and diagnostic purposes, the treatment of mental disorders should emphasize existing pathology and use somatic therapies "attacking the mental through the bodily humours."[14] J. C. Bucknell and D. H. Tuke recommended, in their *Manual of Psychological Medicine* (1879), that psychiatric diagnostic categories be reformulated to conform to recent information about cerebral localization.[15]

itively one way or another." "D" then observed that for the committee to place insanity among the "non-cerebral diseases," would be to put it where it has been for the last forty or fifty years "with almost as little success as when it was placed in the misty regions of metaphysics."[16] The *Times* responded the next day, alluding to

> a clique of physicians who are radically opposed to everything in existing Asylum management because they are not themselves the managers at liberal salaries. . . . A combination, evidently formed for the purpose of obtaining possession of these institutions in the interest of a small society of so-called experts, holding monthly meetings of mutual compliment, has been signally defeated, [despite the efforts of] a single filibustering expert [William A. Hammond] and his hungry followers [the neurologists].[17]

In 1894, the Association of Medical Superintendents of American Institutions for the Insane celebrated its fiftieth anniversary. The association had changed its name that year to the American Medico-Psychological Association, a title that reflected a willingness to compromise with the medical establishment while maintaining a separation from the neurologists and organic psychiatry. (In 1921, the American Medico-Psychological Association would change its name once again—this time to its present title, the American Psychiatric Association.) Partly motivated by a spirit of compromise, S. Weir Mitchell, possibly the leading American neurologist of the day, was invited to give the keynote lecture at the 1894 meeting.

Weir Mitchell started by explaining that, in preparing his remarks, he had consulted thirty neurologists and other physicians, and that therefore the opinions he was going to express were not his alone. All that he had learned, he said, pointed to the devastating result of the isolation of "institutional psychiatry from the rest of the medical profession." He observed that the meeting was being held in Philadelphia where, in the late eighteenth century, Benjamin Rush had practiced in both psychiatry and general medicine. Noting that psychiatry had started to isolate itself after Rush's time, Weir Mitchell said that

> you have never come back into line. It is easy to see how this came about. You soon began to live apart, and you still do so. Your hospitals are not our hospitals; your ways are not our ways. You live out of range of critical shot; you are not preceded and followed in your ward by clever rivals, or watched by able residents fresh with the learning of the schools.[18]

Weir Mitchell was correct about the physical isolation of psychiatry: only four of the ninety-nine people in his audience had city addresses; and as late as 1930, nearly three-quarters of the members of the American Psychiatric Association worked in state hospitals.[19]

Warming to the subject, Mitchell resorted to theatrical hyperbole:

> The cloistral lives you lead gives rise, we think, to certain mental peculiarities. . . . One is the superstition (almost is it that) to the effect that an asylum is in itself curative. You hear the regret in every report that patients are not sent soon enough, as if you had ways of curing which we have not. Upon my word, I think asylum life is deadly to the insane.

He then inquired rhetorically, "Where, we ask, are your annual reports of scientific study, of the psychology and pathology of your patients?" and answered that he could find only "odd little statements, reports of a case or two, a few useless pages of isolated postmortem records, and these are sandwiched among incomprehensible statistics and farm balance sheets."[20] In spite of his harsh criticism, Weir Mitchell had apparently succeeded in not antagonizing the superintendents: when he finished his address, the delegates elected him an honorary member. Walter Channing later wrote a reply, noting that nearly every point taken up by Weir Mitchell "had in some form previously been discussed by those caring for the insane," and thus implying that they were able to keep their own house in order without any intervention from neurologists.[21]

The main theoretical bridge between the organic and the functional camps in American psychiatry emerged from Adolf Meyer's so-called psychobiological approach. Meyer was open to psychological and social as well as to biological determinants of mental disorders, and—what was most important—his impeccable credentials put him above criticism from the neurologists. In Zurich, he had studied with August Forel, a major figure in neuroanatomy, and was trained as a neurologist and pathologist. When Meyer first came to the United States in 1892, he had worked as a pathologist at Kankakee Hospital in Illinois. During this period, he also had a nominal appointment at the University of Chicago, teaching some courses and influencing, as well as being influenced by, the sociologist W. I. Thomas. Meyer moved to the Worcester Hospital in Massachusetts in 1895 as a pathologist, but became interested in psychiatry—owing, it is believed, to his mother's depression—and accepted a second position in psychiatry at Clark University. His next move was to New York, where he became director of the Psychiatric Institute of the New York State Hospital and also served as the professor of psychiatry at Cornell. This was the period of his major neurological research. In 1909, he accepted a position as professor of psychiatry and director of the newly established Henry Phipps Psychiatric Center at Johns Hopkins Medical School. His career in the United States spanned the period from the 1890s through almost 1950, the year of his death. His influence was enormous, and he was elected president of both the American Neurological Association and the American Psychiatric Association. From about 1930 until his death, he was generally acknowledged to be the "dean" of American psychiatry.

Although Meyer's psychobiological approach gave full recognition to the importance of biological factors in mental disorders, he insisted that these should not be considered apart from the whole person. He emphasized the importance of taking complete life histories, including information about early childhood, the family, school, and community; and of keeping good records so that information would be readily available if a patient was readmitted to the hospital. Meyer was the ultimate eclectic, arguing that psychiatry should not rest exclusively on a "biological conception of man," while asserting that "all mental activity must have its physiological side and its anatomical substratum in the forms of nervous mechanisms, combinations of cells, especially of the cerebral cortex."[22]

Meyer's psychobiological approach, which I shall discuss in chapter 2, could not, in practice, however, reconcile the sharp differences between neurology and psychiatry, because members of the two specialties were not only fighting over ideology but competing for the same patient population. During the 1920s, most neurology textbooks included a section on psychiatric disorders, presumably to prepare neurologists to treat psychiatric disorders. Israel Wechsler, the professor of neurology at Columbia University and author of *The Neuroses* (1929), included a chapter on psychiatric disorders in his successful neurology textbook, first published in 1927.[23] However, as psychiatrists began to question the appropriateness of neurologists treating mental patients, the true basis of the dispute became increasingly overt. Derek Denny-Brown, professor of neurology at Harvard, characterized the reaction of many neurologists to the growing field of psychiatry: "The mounting aggressiveness of psychiatry soon encroached on what had been considered the neurologist's field, and by the 1930s many predicted the extinction of the genus neurologist. Many neurologists and neurologic departments became absorbed by departments of neurosurgery or else adopted the protective coloring in the form of the title 'neuropsychiatrist' and still exist in unhappy dependence on these borderlands."[24] Harvey Cushing, America's leading neurosurgeon, saw further that neurology was being squeezed by psychiatry on one side and neurosurgery on the other: "The neurologist, elbowed out of some of his previous activities by neurosurgeons, shows signs of an increasing interest in the psychoses and a desire to have a larger share of their treatment."[25] Whereas earlier the neurologists, who were often in the intellectual forefront of medicine, criticized the psychiatrists and questioned whether they would ever be able to treat mental patients effectively, it had become clear even to some neurologists that a "thorough grounding in neurology does not fit one for the practice of psychiatry."[26] On its other side, neurology's traditional turf was being invaded by neurosurgery, a field that had been in its infancy, if not in its prenatal period, at the turn of the century. When, in the 1880s, surgeons for the first time operated on tumors of the brain and spinal cord, they depended on neurologists to locate the probable site of the growth in the nervous system (before the 1880s, brain surgery had been attempted only after traumatic injuries). Thus, the first successful removal of a brain tumor was performed in London in 1884 by Rickman Godlee, but it was the neurologist Bennett who diagnosed the condition and persuaded Godlee to operate. Similarly, the first spinal-cord tumor was removed in 1887 by Victor Horsley, but the neurologist William Gowers localized the tumor and invited Horsley to operate. In 1905, Harvey Cushing observed that it hardly seemed possible that "the surgery of the nervous system by itself could furnish material enough to occupy a surgeon's undivided attention and insure him a livelihood."[27] In 1920, when the Society of Neurological Surgeons was organized in the United States, it had only eleven members—all experienced general surgeons.[28] By 1942, however, Percival Bailey, with neurosurgeons in mind, observed that "the arrogance of the physician who would not dirty his hands, but called in the barber for such work, has been succeeded by the arrogance of the man of action."[29]

The reversal of roles that took place between neurology and neurosurgery was not, however, to be explained solely by the "arrogance of the man of action." Neurosurgery was offering an increasing number of ways of treating neurological problems, and neurosurgeons were no longer content to be simply the hands of the neurologists. As pointed out by Paul Bucy, competency in neurosurgery does not imply "merely facile fingers—a certain technical excellence."[30] When operations on the nervous system became a separate specialty, neurosurgeons learned more neuroanatomy and neurology, finding that they could often do their own diagnoses without a neurologist.

For lack of a unique therapeutic method, the domain of neurology was shrinking. Lamenting the state of neurology, Percival Bailey observed that "electrotherapy used to be its standby, especially in Europe, but was so obviously ineffective that it became a laughing stock." Conducting a survey in 1946, Bailey concluded that the able young men were going into neurosurgery, and almost none into neurology.[31] A fear was growing that neurologists were becoming assistants to the neurosurgeons or, as more earthy physicians described them, "the pimps of neurosurgeons." One neurologist lamented that the field had shrunk "to include a few common but untreatable diseases and a large number of interesting but rare and obscure clinical syndromes,"[32] while a cynical observer remarked that the neurologists' business is "to take care of these patients who have nothing the matter with them or for which nothing could be done."[33]

This development had clear economic consequences. The general practitioner was becoming reluctant to send patients to a neurologist with the likely delays and additional expenses, as a patient usually ended up in the hands of a neurosurgeon. Moreover, if the neurosurgeon found the condition untreatable, he would refer the patient back to the neurologist, not to the general practitioner; the latter generally lost a patient if a neurologist had acted as middleman. Thus, as general practitioners tended to bypass them, neurologists were finding it difficult to earn a living practicing clinical neurology alone. Some neurologists obtained training in psychiatry and neurological surgery in order to survive. Percival Bailey noted that the neurologist is obliged "to operate or to acquire an increasing amount of psychiatric work in order to live."[34] The urgent need of neurologists to hold on to the "psychiatry business" was expressed in Henry Alsop Riley's 1932 presidential address to the New York Neurological Society. Commenting on the decline of neurology as a profession, he noted that:

> Many factors have led to the present situation, but now the economic stimulus is added. Perhaps the greatest danger that threatens neurology today is the possibility of the passage of the care of the psychoneuroses into the hands of psychoanalytic psychologists and psychiatrists. If this tendency continues to develop and remains unopposed by members of our specialty, it will not be long before the great mass of patients suffering from these psychogenic disorders will be lost to the neurologist and neuropsychiatrist.[35]

In London in 1941, the neurologist C. P. Symonds attempted to support "the neurological approach to mental disorders" by observing that it was impossible to deny that "the brain is the organ of mind." He added that the distinction between "organic" and "functional" mental illness was a source of confusion, as "disordered function not yet traced to demonstrable lesion may be so traced tomorrow." Grudgingly admitting that psychotherapy can help a patient "make the best of his illness," Symonds added that there is much also to be lost from psychiatry of the wrong kind:

> It was in one sense unfortunate that the Freudian method of approach scored such an immediate success in the interpretation of hysteria. This led to the false assumption that because hysteria was a neurosis, and because the origin of hysteria was "unconscious" motivation, all other neuroses had a similar origin. As a result I dare not guess how many patients with affective neurosis have been subjected to the Analytic Inquisition.[36]

By 1946, Percival Bailey would write that a takeover by the psychoanalysts was "an accomplished fact."[37] The psychoanalyst had taken over the departments of psychiatry in most prestigious medical schools shortly after the end of the Second World War.

Even though their position with respect to neurology had grown considerably stronger, by this time psychiatrists were facing serious competition from nonmedically trained persons who were beginning to practice psychotherapy. In his presidential address to the American Psychiatric Association in 1932, William Russell noted with considerable alarm that "psychologists who are without medical and psychiatric training, social workers and others who have been psychoanalyzed and instructed in psychoanalysis may also be found engaging in the treatment of psychiatric conditions in private practice." Russell warned that there is a mistaken belief among psychologists and other "lay workers" that psychiatric problems "can be diagnosed and adequately treated without the aid of psychiatrists." He found it disturbing that a psychological clinic had recently been established by a department of psychology, and that the National Research Council, with Carnegie Foundation support, had announced that the head of a department of psychology was funded to study "the present status of mental disorders." Russell strongly advised the American Psychiatric Association to condemn this practice as unscientific and dangerous. He also felt compelled to correct the impression that psychiatry and psychology have common roots. Psychiatry is rooted in medicine, he said; while psychology "grew out of philosopy and metaphysics." Psychiatrists in general felt compelled to emphasize the medical—that is, the "somatic"— bases of mental illness to counter the growing competition from "lay practitioners."[38]

Psychiatrists were becoming increasingly sensitive to intrusions into what they now regarded as exclusively their domain. In his 1933 presidential address to the American Psychiatric Association, James V. May elaborated on the theme of Russell's address the previous year. Reviewing the fact that few medical schools considered psychiatry sufficiently important to appoint a professor in the field, May observed that mental illness was treated as a "medical playground" in which physicians ranging from ophthalmologists to gynecologists felt free to speculate about the cause of psychoses. And in the "battle of experts" in murder trials, psychiatrists were rarely used. The major "invasion" into psychiatry had been from neurologists; but more recently, May observed, psychologists had been making increasingly deep incursions into psychiatry. Among the dangerous trends were the large numbers of psychologists attempting to become "lay analysts" and the practice of psychology departments in universities of offering courses in "abnormal psychology"— a subject "which is psychiatry pure and simple, and does not belong within the domain of psychology." May then proceeded to the results of what today would be called "market research," presenting statistics demonstrating that, in the Veterans Administration and in other institutional settings, the overwhelming majority of the neuropsychiatry cases had functional disorders and, therefore, were cases to be treated by psychiatrists, not neurologists. May concluded that the American Psychiatric Association had to establish clear standards—presumably those only psychiatrists could meet—for treating mental illness if they were to assume a position of "supremacy."[39]

The ongoing dispute between the organic and the functional views, particularly between the neurologists and the psychiatrists, had impeded the establishment of clear standards for physicians treating mental patients. Medical students received little training in psychiatry; yet up until the 1930s, any physician could request that his name be listed in the directory of the American Medical Association as a specialist in psychiatry. During the First World War, General Jack Pershing had sent an urgent message from France, requesting that something be done to eliminate the prevalence of mental disorders in replacement troops. In response, Edward Strecker, later head of psychiatry at the University of Pennsylvania, was given the responsibility of organizing six-week "crash courses" in psychiatry at several leading universities. There was, however, no immediate overall improvement in the teaching of psychiatry. Even in the 1930s, when more psychiatry was taught in medical schools, there was little agreement about what constituted sufficient training to treat the mentally ill.

While some medical specialties had already established standards and procedures for obtaining certification or diplomas, the creation of standards in psychiatry was stalled by the conflict between the American Neurological and the American Psychiatric associations. For a while it looked as if each of these groups would go its own way. Then, in 1933, the National Committee for Mental Hygiene called a special meeting to be held at the American Psychiatric Association convention. The neurologists interested in mental disorders were meeting one month later at the Nervous and Mental Diseases Section of the American Medical Association. There was a feeling of urgency at the American Psychiatric Association meeting, lest its members be preempted by the neurologists. William Alanson White followed up on James May's presidential address by trying to propel the group into action. He warned that "somebody else is going to take our place, somebody else less qualified than we are," and added that if the psychiatrists sit idly by, "the next annual meeting may be too late."[40]

The Council on Medical Education called a meeting of psychiatrists and neurologists to work out a compromise (see figure 1.1). The following quotations from leading psychiatrists and neurologists at the meeting—as recorded

Figure 1.1.   Meeting to Organize the American Board of Psychiatry and Neurology, Hotel Commodore, New York, 20 October 1934. Seated (*from left to right*): Clarence O. Cheney, C. Macfie Campbell, Walter Freeman (secretary), H. Douglas Singer (president), Adolf Meyer (vice president), George W. Hall; standing: Franklin G. Ebaugh, Louis Casamajor, J. Allen Jackson, Lloyd H. Zeigler, Lewis J. Pollock, Edwin G. Zabriskie. (Empire Photographers, New York. From W. Freeman, *The Psychiatrists,* New York: Grune & Stratton, 1968. Reprinted by permission.)

by Walter Freeman, one of the representatives of the American Medical Association—indicate how far apart their views were:

DR. ADOLF MEYER: I am perfectly willing to look on my own use of the term neuropsychiatry as a pious wish that there should be as many neurologists as I hope there will be psychiatrists who want to pool their domains. . . . I wish that for every psychiatrist there might be a demand of adequate knowledge of what the nervous system can do, so that we might avoid leaving many thinking that they can take up psychiatry from the spiritual end exclusively.

DR. WILLIAM A. WHITE: We had better hold to our own and define qualifications for more or less specifically psychiatric objectives.

DR. C. CHENEY: There are many persons who wish to be called neuropsychiatrists and that is just the thing it is wished to get away from, and it is being gotten away from by the establishment of qualifying boards.

DR. I. S. WECHSLER: I feel that this splitting off is inimicable to both psychiatry and neurology. Primarily he [neuropsychiatrist] should be thoroughly trained in neurology and psychiatry and then do anything he pleases, just as we thoroughly train them first in medicine.

DR. WALTER FREEMAN: It would be much better to band together, to be known as those who are dealing with the nervous system rather than to split them apart into the organic neurologist and the functional psychiatrist.

DR. H. A. RILEY: As far as I know the feeling of the American Neurological Association, that association would deprecate very much any move which would separate neurology from psychiatry.

DR. EDWIN ZABRISKIE: I think it is simply impossible to separate them.[41]

In the end, however, a compromise solution was found. The American Board of Psychiatry and Neurology was established, *psychiatry* winning out over *neurology* in the title, in spite of the argument that *n* precedes *p* in the alphabet. The overall compromise called for separate examinations and requirements for certification in psychiatry and neurology. A candidate wanting certification in the two fields would have to meet both requirements. This solution was a victory for the psychiatrists: even though any physician was still free to treat mentally ill patients in private practice, this field would eventually be restricted, within medical schools and large hospitals, to board-certified psychiatrists. Psychiatrists had clearly gained politically.

Within psychiatry, psychoanalysts had, during the 1940s, risen to a position of power and influence, with the reputation of being able to treat a wide range of mental illness. Novels, plays, and films had dramatized cures by psychoanalysts who uncovered—often with the help of hypnosis—the deep-seated conflict that was shown to be at the root of a disorder. Hysterical blindness or paralysis of soldiers was depicted as suddenly cured by "depth psychiatry." According to these dramatizations, psychoanalysis would eventually significantly alleviate mental illness—an outcome that psychiatrists knew was an illusion. Moreover, there was a certain irony in that psychoanalysis, while helping psychiatry win its political struggle within medicine, had made the field more vulnerable to competition from the rapidly growing numbers of clinical psychologists and other lay therapists who considered themselves fully capable of practicing various forms of "talking" therapy; indeed, Freud believed that, to practice psychoanalysis, a person does not need to have studied medicine.[42] In order to emphasize their uniqueness, therefore, many psychiatrists remained strongly attracted to somatic treatments—which lay therapists were not licensed to use.

The compromise between the psychiatrists and the neurologists was reached in 1934—only two years before the first lobotomy was performed in the United States. This radical operation was only one of the many somatic treatments for mental illness developed and used on patients in the early decades of the twentieth century. As the history of these somatic treatments is essential for fully understanding why lobotomy was so readily adopted, I shall discuss some of these treatments in the following two chapters.

2

Bizarre Illnesses,
Bizarre Treatment

> Bizarre illnesses may require bizarre treat-
> ment, and in psychiatry they often get it.
> They show so often a stubbornness and re-
> sistiveness to treatment, they expose so clearly
> the ignorance of their pathology and aetiol-
> ogy, that they arouse aggressive reactions in
> the baffled and frustrated therapist.
> —MAURICE PARTRIDGE (1950)

IN the late nineteenth and on into the first four decades of the twentieth
century, there was little hope for severely ill mental patients placed in institutions. Conditions in the institutions for the "incurably" insane had advanced
but little from the days in the late eighteenth century when Pinel in France
and Rush in the United States had first advocated humane treatment for the
insane. Patients were beaten, choked, and spat on by attendants. They were
put in dark, damp, padded cells and often restrained in straitjackets at night
for weeks at a time.

Periodically the public was made aware of the horrendous conditions in
the hospitals, and there were subsequent reforms; but in the absence of any
effective treatment for the patients, conditions soon regressed to their former
state. One of the major periods of reform was triggered early in this century
by Clifford Beers, a former mental patient, who, after his hospitalization,
wrote the remarkable *A Mind That Found Itself*.[1] Published in 1908, the book
not only described the inhumane conditions in most mental asylums but also
contributed many specific suggestions for reform. As Beers had obtained the
support of several psychiatrists and psychologists in advance of publication,
William James consenting to write an introduction, the book could not be
dismissed as just another irresponsible claim of a mental patient.

A Mind That Found Itself received immediate acclaim in the United States and abroad and led to the establishment of the National Committee for Mental Hygiene, a title usually credited to the neuropsychiatrist Adolf Meyer, although the term *mental hygiene* had been used on and off for more than fifty years.* The committee raised money to survey hospital conditions, to hold conferences, and to influence public opinion. Much skepticism and inertia had to be overcome. The *Nation,* for example, applauded Beers's call for improvements in hospital conditions but was pessimistic about the possibility of preventing or curing serious mental illness: "One is bound to face the fact that insanity is in the majority of cases an unpreventable and an incurable disease, and nothing short of Utopia itself can make it very much less so."[2]

It was not easy to change the condition of the insane in the United States: not only were mental hospitals isolated from the rest of medicine, as I noted in the first chapter; but also there was no convincing explanation of the causes of mental illness. Kraepelin's diagnostic labels had little practical significance, as any of the few treatments available were likely to be tried regardless of diagnosis. Physicians commonly resorted to pseudo-diagnostic terminology— "nervous prostration," "nervous breakdown," "cerebral neurosis," and "shell shock"—which implied that they had identified the physical cause of an illness. The American neurologist George Beard, for example, promoted *neurasthenia* in 1880, applying the term to a wide range of different physical complaints now commonly associated with depression. He defined neurasthenia as "a chronic, functional disease of the nervous system, the basis of which is impoverishment of nervous force, waste of nerve-tissue in excess of repair," and wrote, "I feel assured that it [neurasthenia] will in time be substantially confirmed by microscopical and chemical examination."[3] The diagnosis was generally adopted. In Vienna, the *Handbuch der Neurasthenie* was published around the turn of the century; and several neurologists expressed their conviction that neurasthenia was caused by a cerebral tumor.[4] The diagnostic label was used well into the present century, until it was gradually replaced by *psychasthenia,* a term that was more neutral about etiology.

If neurasthenia was caused by "nervous exhaustion," a logical treatment was stimulation of the nervous system. Wilhelm Erb, a leading German neurologist, had pioneered electrotherapy, which soon became widely popular as a treatment for such disorders as neurasthenia and depression (see figure 2.1). In the United States, Beard, in collaboration with A. D. Rockwell, wrote *Medical and Surgical Uses of Electricity,* a monograph that underwent eight editions and was eventually expanded into a 600-page textbook.[5] Morton

---

* In 1843, William Sweetser of the University of Vermont wrote a book entitled *Mental Hygiene.*

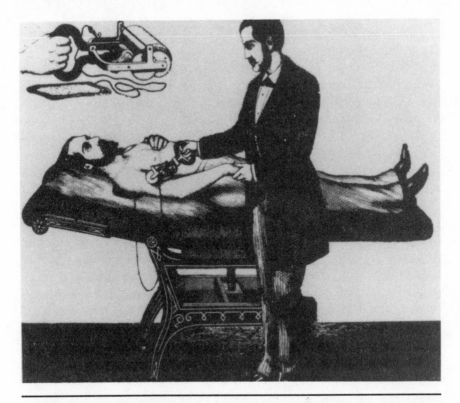

Figure 2.1. Electrical stimulation device invented by Dr. John Butler of New York. It was said to be "especially salubrious in cases of rheumatism, nervous exhaustion, neuralgia and paralysis." (From *Harper's,* 1881.)

Prince, well known for his classical studies of multiple personalities,* wrote an extensive chapter in 1901 on the use of electricity to treat neuroses, and left no doubt about his high regard for electrotherapy:

> Neuroses are, par excellence, the kind of nervous disease in which favorable results are obtained from electrical treatment. More or less doubt may be entertained whether electricity may alter for good or for evil such disease processes as cause organic change in the central nervous system, but any one who has had extensive experience in treating functional nervous affections with electricity must either bear witness to the favorable results observed or else lay himself open to the suspicion that he has not the skill and judgment necessary for its proper employment.[7]

* Prince did much of his work on multiple personalities—a form of hysteria—at the Boston City Hospital. This work was concurrent with Charcot's work on hysteria at La Salpêtrière in Paris.[6]

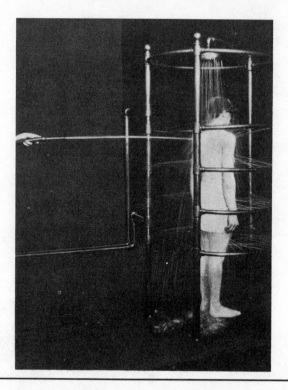

Figure 2.2.   Combined rain douche, horizontal jet, and multiple-circle douche for neurasthenia and other nervous disorders. (From John H. Kellogg, *Rational Hydrotherapy*,
Philadelphia: F. A. Davis, 1902.)

Prince explained in great detail the physiological changes produced by electrotherapy, and described the different results produced by *faradization,* also
called *galvanization,* and static electricity, commonly called *franklinization* in
the United States. He advised applying the cathode (negative terminal) to
excite the nerves and the anode (positive terminal) to produce sedation. Prince
was also aware that electricity, especially the noisy "static machine," could
have a "suggestive power" on patients, and advised exploiting this "power"
fully to augment the physical effect of electricity on bodily processes. As late
as 1929, Richard Hutchings in New York recommended that every physical
therapy department in mental hospitals should have a "low frequency gen-
erator, galvanic-sinusoidal machine, a static machine with at least 16 plates,
and an insulating stool or wicker chair."[8]

Hydrotherapy, which competed with electrotherapy as a treatment for
neurasthenia, had developed into a medical specialty dealing with the various

therapeutic effects of baths, douches, wet packs, steam, spritzers, and hoses (see figure 2.2). The natural baths at spas were also popular among those patients who could afford to combine therapy and social activity. In 1879, one German authority recommended that "the calm waters of the Baltic are preferable for delicate, nervous constitutions, and the North Sea, with its stronger billows, may be recommended for torpid constitutions."9

Throughout this period, many treatments that could just as readily have been considered psychological were often perceived as physical. Thus when, in 1884, S. Weir Mitchell introduced a popular "rest cure" consisting of isolation from family, quiet, diet, and massage, he denied any similarity with religious "mind cures" and insisted that the benefit came from building up the patient's "fat and blood."10 Most neurologists were hostile to any suggestion of "mental therapeutics."

Ironically, George Beard, who had "discovered" neurasthenia and treated it with electrotherapy, was much more inclined to consider psychological variables than were most contemporary neurologists. When he dared to suggest, in a paper presented to the American Neurological Association, that "expectation is itself a curative force" even more than electricity, he had to face an onslaught of criticism. Beard had described the way diseases might appear and disappear under the influence of emotions, and argued that mental qualities, like drugs, could either facilitate or block the action of various therapies. His speculation that it might be possible to facilitate a cure by systematically using a patient's "expectations" was criticized as "unscientific" and "deceptive," and most of the audience regarded the whole approach as unscientific because it involved emotions, whose nature could be neither understood nor isolated. One discussant said that Dr. Beard seems to be close to recommending telling falsehoods to patients. William Hammond remarked that were he to adopt Beard's ideas he would be descending to the level of "humbuggery" and would have to consider throwing his medical diploma away and joining the theologians."11 Even though the neurologists had no treatment to offer seriously ill mental patients, they were then as later extremely reluctant to accept any ideas that might tarnish their image as scientists studying physical realities.

In presenting, as they did, the same list of organic causes of mental disorders over and over, neurologists argued that before long all mental disorders would be found to have a physical basis. They would point to alcohol hallucinations and psychosis and to the other drugs and toxins known to produce psychotic episodes. Support was drawn from the fact that Alois Alzheimer in Munich had found brain abnormalities—"senile plaques," he called them—in certain types of premature dementia (Alzheimer's disease) and had written that undoubtedly there will be more "psychiatric diseases" for which physical causes will be found.*

It was also suspected that faulty metabolism or abnormally high or low amounts of glandular secretions might produce psychiatric disorders. Specific dietary or vitamin deficiencies like those producing pellagra were beginning to be recognized as a possible cause of mental disorder. It was reported that, in some asylums in northern Italy during the early part of the century, as many as 40 percent of the mental patients were actually suffering from pellagra's "three d's"—depression, delirium, and dementia. Even in the 1930s, 19 percent of the admissions to the North Carolina State Mental Hospital in Goldsboro—an institution limited to black patients—had pellagra.[12]

Few patients, however, were helped even when a cause was found. The outstanding exception, and the example most frequently cited by the biological psychiatrist, was general paresis, which was also called "general paralysis of the insane," "general paralysis," and "dementia paralytica" because in its final stages, there often occurred a partial paralysis as well as mental aberrations. Between 1911 and 1919, statistics from New York State mental hospitals indicated that about 20 percent of all male first admissions had paresis. This proportion increased until 1931, when it started to decline. The rate for females was also substantial, even though only one-third of that for men.[13]

Before being recognized as neurosyphilis, general paresis was, in the nineteenth century, commonly thought to have psychological causes—that is, to be a functional disorder—even though there were reports going back to 1850 describing it as neurological owing to the abnormal gait and other physical symptoms of the disease. Nevertheless, speculations that it was caused by "inflammation of the brain" or associated with syphilis were rejected by many. As late as 1902, Max Nonne, a neurologist in Berlin, wrote that he wished "to make it clear that progressive paralysis is not a specific syphilitic disease of the brain."[14]

The functional explanation of general paresis attributed the cause to the mental strain of a life filled with excesses; that is, it was thought to be a disease produced by intemperance and immorality. The evidence that it ran in families and therefore appeared to be inherited did not argue against this interpretation as the tendency toward immorality was also thought to be inherited. The higher incidence in men than women was consistent with this view, men having more opportunities to live an intemperate life. Thus, in his 1897 *Textbook of Mental Diseases,* T. H. Kellogg wrote:

Civilization favors general paresis through the demands which it makes on physical

---

* Alzheimer's paper entitled (in translation) "On a Peculiar Disease of the Cerebral Cortex" was published in Germany in 1907.

and mental powers, competition, reckless and feverish pursuit of wealth and social position, overstudy, overwork, unhygienic modes of life, the massing of people in large cities, the indulgence in tea, coffee, tobacco, stimulants, and social and sexual excesses, and artificial modes of life.[15]

Around the turn of the century, a series of observations gradually revealed the true cause of general paresis, as neurologists and pathologists began to describe the degenerative and inflammatory changes in the brains of patients who died from this disorder. In 1896, after Franz Nissl reported that he could diagnose general paresis on the basis of histopathological evidence, he extended his findings to other mental disorders, saying that "as soon as we agree to see in all mental derangements the clinical expression of definite disease processes in the cortex, we remove the obstacles that make impossible agreement among alienists."[16] However, it was not until Alzheimer's 1904 publication that a full description of the nerve-cell changes in general paresis was available. Several years earlier, Richard von Krafft-Ebing had injected fluid obtained from the sores (lues) of known syphilitics into "general paralytic" patients and concluded, from this highly unethical study by today's standards, that the patients already had syphilis because they had no reaction to the injection.[17]

The next steps occurred in quick succession. First, Fritz Schaudinn identified the spirochete organism from a sample of pus obtained from the sores of syphilitics. Then J. W. Moore sent slides of the brains of patients who had died from either "general paralysis" or syphilis, to Hideyo Noguchi, a Japanese bacteriologist at the Rockefeller Institute for Medical Research. Noguchi was convinced the spirochete involved was the same in both cases. This evidence finally ended the "syphilis-general paralysis question."[18]

In regard to treatment, an important first step was the development of a test for more reliable diagnosis. August von Wassermann, a German bacteriologist, published his blood test (the "Wassermann test") in 1906.[19] In 1910, after 605 failures, Paul Ehrlich, another German bacteriologist, succeeded in discovering "606," his "magic bullet," the arsenic compound Salvarsan (arsphenamine). This drug was later replaced by Neosalvarsan, a more easily administered compound. The "magic bullet" turned out not to have much magic power when given to advanced cases of general paresis, as neither Salvarsan nor Neosalvarsan was effective once the disease had become entrenched in the brain.

The treatment of general paresis that had considerable success was developed by a psychiatrist, Julius Wagner-Jauregg (figure 2.3). He had been working as an assistant in the psychiatric clinic in Vienna in 1883 when he observed that the mental state of a psychiatric patient who had an attack of erysipelas— a skin disease that produces a high fever—improved after the fever subsided.

Figure 2.3.   Julius Wagner von Jauregg (1857–1940). Wagner-Jauregg won the 1927 Nobel Prize in Medicine and Physiology for his introduction of malarial treatment for general paresis. (Courtesy of the National Library of Medicine.)

Wagner-Jauregg recalled earlier reports of mental symptoms of "general paresis" patients being alleviated when they were treated by ointments that produced fever and suppurations (oozing pustules) around the head. Concluding that the fever was therapeutic, he began to explore other ways of inducing fever in mental patients. He tried inoculating them with vaccines of tuberculosis, typhoid fever, and recurrent fever, and even infected patients with erysipelas. He reported some success and speculated that malarial fever might be therapeutic for general paresis; but his early publications attracted little interest and for a while he abandoned this research.[20]

During the First World War, Wagner-Jauregg, then past sixty and chief of the Psychiatric Clinic in Vienna, was presented with a "shellshocked soldier" brought for treatment from the Italian front.* Wagner-Jauregg took

* After the war, in 1920, there were charges that psychiatrists at the Vienna Psychiatric Clinic, where Wagner-Jauregg was director, had used electrical treatment to punish soldiers thought to be malingering. It was alleged that the intensity of the electric current had been increased to such high levels that deaths and suicides occurred during the "treatment." Freud was asked to

the patient's high fever from malaria as a "sign of destiny." Drawing blood from the soldier's vein, he injected it into three paretics. When, after several days, they also began to run a high fever, he drew blood from them and injected six other patients. Six of the nine patients—those in whom the paresis was not as far advanced—seemed to improve substantially. Wagner-Jauregg later reported that, after four years, three of these patients were still working efficiently: apparently the disease had been arrested.[22]

Wagner-Jauregg continued the treatment, using blood from patients with tertian malaria, a relatively mild form that can be controlled with quinine. He reported about 44 percent "cures"; and, in a later collaborative study with a friend who directed a syphilis clinic, he studied patients who had positive Wassermann-test results but as yet no symptoms of the disease. He used Neo-salvarsan combined with malarial-fever treatment on these early cases, and the percentage of success in arresting the disease was above 80 percent.[23] Interest in the method spread rapidly; and in 1927, at the age of seventy, Wagner-Jauregg was awarded the Nobel Prize in Medicine. The fact that he was—as he still is—the only psychiatrist to have been so honored was not lost on neurologists and psychiatrists with an organic view of mental disease.

Wagner-Jauregg's malarial-fever treatment was widely adopted in Europe and the United States and continued to be used into the 1940s. Although there were numerous attempts to apply this treatment to schizophrenics, little success was achieved in spite of unsubstantiated claims that some patients had improved. During the 1930s, several devices for artificially heating patients were introduced after it was clearly established that it was not the malaria, but the fever, that produced the improvement—presumably because the high temperatures and some aspect of the body's reaction to the heat combined to destroy the spirochete. Hot baths, hot air, radiothermy, diathermy, infrared-lightbulb cabinets, and special electric "mummy bags" were all used to induce fever.[24]

At the time, general paresis was central to psychiatry. Smith Ely Jelliffe and William Alanson White's influential *Diseases of the Nervous System: A Textbook of Neurology and Psychiatry* devoted ninety-two pages to the topic of syphilis in its 1923 edition, even though both authors had a psychoanalytic orientation.[25] As late as 1945, Karl Menninger devoted a number of pages of

---

serve on the investigating commission and wrote a "Memorandum on the Electrical Treatment of War Neurotics," stressing the similarity of all traumatic neuroses and insisting that there was no injury to the nervous system in "shellshock" soldiers. Freud also noted that the initial success of electrical stimulation was not lasting, as most soldiers had a psychological breakdown soon after returning to the front. Freud, who had been a classmate and close friend of Wagner-Jauregg, stated that he was "personally convinced that Professor Wagner-Jauregg would never have allowed it [the electrical stimulus] to be intensified to a cruel pitch."[21] The charges were eventually dropped.

Figure 2.4.   Adolf Meyer (1866–1950). (Reproduced with permission from the *Archives of Neurology* 64 [1950]:880. Copyright © 1950 American Medical Association.)

his book *The Human Mind* to general paresis, describing it as a frequently overlooked explanation of why people suddenly become insane.[26] The neurology and psychiatry journals were filled with discussions of how to distinguish the delusions of a paretic from those of other psychiatric patients. Syphilis was called "the great imitator"; and the psychiatrist B. Thom wrote, in 1921, that "there is no psychosis that cannot be caused by syphilis."[27]

While the history of general paresis provided the major support for the organic position on mental illness, not everyone was willing to accept it as an adequate model for all mental disorders. Adolf Meyer (figure 2.4) rejected general paresis as a general model for psychiatry, but was, nevertheless, amazingly open to all kinds of somatic treatment. Meyer's influence, which I have already mentioned, is remarkable in that he had neither an identifiable theory about the cause of mental illness nor a practical therapeutic program. He was highly respected by both neurologists and psychiatrists and unusually open to both psychosocial and biological influences. His psychobiological approach was broad enough to accommodate almost any suggested treatment that came along; and later, in the history of lobotomy, as I shall describe, his refusal to offer any opposition had the effect on several occasions, of providing support. While committed to biological interpretations at a philosophical level, he regarded the adjunctive services of nursing, occupational therapy, and social work (his wife was one of the first psychiatric social workers in the United States) as essential. Meyer emphasized the interconnectedness of psychiatry and the community. His therapeutic program and his ideas about the causes of mental disorders were never stated explicitly, but he did stress using play and work therapy to strengthen the elements of a patient's personality that remained healthy. The keystone of his therapeutic program was socialization.

For psychiatrists not completely committed to either psychoanalysis or an exclusive use of somatic treatment, Meyer's nondoctrinaire, psychobiological approach was the only reasonable position. His influence was apparent in almost all of the successful textbooks of psychiatry during the 1930s and 1940s: for example, Arthur P. Noyes's *Modern Clinical Psychiatry* (1934) and *Practical Clinical Psychiatry* (1925) by Edward Strecker, Franklin Ebaugh, and Jack Ewalt—both of which underwent several successful editions. In discussing the treatment of schizophrenic patients, the latter authors comment:

> We know of no better conception than the psychobiologic interpretation of Meyer. It views the patient critically in the long section of his life history and particularly surveys the series of maladaptations that preceded the final schizophrenic one. It then asks such pertinent and therapeutic stimulating questions as these: What are the resources of the patient? What has he to react with? What is the situation he is called on to meet? Can we modify his resources in order to enable him to better the situation, or can we modify the situation so he may better meet it with his resources? Etc. Finally, the psychobiologic idea does not court exclusively any single therapeutic mistress and it leaves open the door of every reasonable treatment plan.[28]

In spite of this influence, Meyer's "common sense therapy" was not easy to put into practice, as he readily admitted: "To use the patient's assets is a more difficult problem than using something under our control. It means that we have to search for and rouse the relatively normal person-functions so that they may be made to digest what is less normal and needs reassimilation."[29] This goal was not likely to be achieved in the understaffed and underfunded state mental hospitals. Although they espoused Meyer's approach, the leading textbooks of psychiatry through the 1940s mainly provided illustrative case material for each diagnostic category, but offered little that was useful under the heading of "treatments." Suggestions to use nonspecific sedative drugs such as barbituric acid compounds, bromides, chlorals or paraldehydes, rest, baths, and wet packs—the latter providing they were not used as punishment—

were offered without much expectation that they would do more than provide temporary relief.

In this therapeutic and theoretical vacuum, almost any treatment was tried, providing it had the potential for treating large numbers of patients with a minimum of highly trained staff. This situation made somatic treatments particularly attractive. Moreover, the baffled and frustrated therapists, as Partridge implied in the epigraph to this chapter, were often defensive and irritated by their own helplessness—a mood likely to propel them into trying anything.[30] Of the great number of somatic treatments introduced between 1920 and 1950, few were too radical to be rejected out of hand.

Attempts to treat mental disorders by the removal of one or another of a patient's endocrine glands was widely adopted. It had been known for some time that mental states and mood can be influenced by hormonal imbalances, and that depression and irritability can be caused by under or over activity of the thyroid or the adrenal gland. "Mental instability" was often attributed to the endocrine changes experienced by women during the menstrual cycle or following pregnancy.[31] Thyroidectomies, ovariectomies, male castration, and removal of all or parts of other glands of mental patients numbered in the tens of thousands around the world during the 1920s. In 1935, Walter Freeman considered the subject of sufficient importance to do an extensive autopsy study attempting to relate the condition of the endocrine glands to the personality of patients who had died at St. Elizabeth's Hospital. It was generally acknowledged that extreme hormonal imbalances could cause irritability, mental and physical sluggishness, or emotional instability; but Freeman found little else of interest: "As far as determining whether an individual shall be a proud, sensitive, suspicious, paranoid individual or a timid, shut-in, dreamy schizoid person; a boisterous, jolly, hail-fellow-well-met cycloid, or a moody, pedantic, egocentric epileptoid individual, the endocrine glands would seem to have little to say in the matter."[32]

Sleep therapy or "prolonged narcosis" as it was commonly called was widely used in Europe during the 1920s and 1930s and also had its proponents in the United States. In Europe, the German psychiatrist Jakob Klaesi was usually credited with introducing sleep therapy in the early 1920s, because he published the first substantial series of cases; but the use of drugs to induce prolonged periods of unconsciousness to treat the psychoses can be traced back to the 1870s.[33] The rationale for this treatment lay in the belief that sleep would restore a stressed and exhausted nervous system back to a more normal state. In the United States, where it was used in the 1930s in the departments of psychiatry at the universities of Wisconsin and Pennsylvania medical schools (as well as elsewhere), its roots could be traced back to Weir Mitchell's "rest cure," popular before the turn of the century.[34]

By means of barbiturates or opium derivatives, mental patients were kept in comatose sleep for periods usually between one and two weeks, but in some instances as long as a month. The patients were generally allowed to awaken sufficiently for brief periods during the day for nutrition, for bowel and bladder relief, and for routine nursing care. Cures or substantial improvement of manic-depressive and schizophrenic patients were not infrequently reported to be in the range of 70 percent to 80 percent.

Although some psychiatrists in the 1930s emphasized that sleep therapy made patients amenable to psychotherapy, most were convinced that the effectiveness of the treatment should be attributed to hypothetical physiological changes such as "improved cellular oxidation," "healing of brain inflammation," establishment of a normal balance between either "cortical and subcortical brain activity," or between "the brain's inhibitory and excitatory forces." As late as 1961, on a visit to the Soviet Union I spoke to psychiatrists who stimulated the brain electrically as a means of maintaining a prolonged sleep state in mental patients. Electronarcosis therapy, which originated in the Soviet Union, was also at the time being used by psychiatrists in the German Democratic Republic.*

The possibility of treating mental illness by "psychic stimulation" induced by stimulating respiration and metabolism was also tried in the 1920s. Arthur Solomon Loevenhart, a professor at the University of Wisconsin, started experimenting with different ways of producing what he called "cerebral stimulation." In early research, he tried injecting small doses of sodium cyanide and reported some improvement in schizophrenics. In 1929, he and his colleagues had patients breathe a gas mixture containing 30 percent carbon dioxide (instead of .03 percent normally found in atmospheric air). The high carbon dioxide levels excited the nervous system:

> We have succeeded in what we term cerebral stimulation in cases of dementia praecox, manic depressive insanity and involutional melancholia. Every case treated has shown a positive response. . . . The most favorable and striking sometimes occurred in those patients who had been mute and inaccessible for long periods of time . . . this inaccessibility disappeared and the catatonia passed off. A number of patients have carried on conversations.[36]

Loevenhart's report stimulated wide interest in "carbon dioxide therapy." Patients were given extended series of inhalation sessions, sometimes as many

---

* Electronarcosis was used as early as 1936 in the Soviet Union. During the five-year plan between 1946 and 1950, there was a strong commitment to somatic therapies and physiological explanations of mental illness. In addition to electroshock, insulin coma, sleep therapy, and lobotomy, mental patients were treated by X rays, by injection of various pharmacological agents into the cerebrospinal fluid, and by shock produced by administering incompatible blood, horse serum, or other foreign proteins (with adrenalin held in readiness to combat anaphylactic shock).[35]

as 150. By the latter part of the 1940s, the rationale for this therapeutic procedure included psychoanalytic concepts: improvement was said to result from
an activation of unconscious fears that could then be assimilated or integrated
into conscious thought. After Loevenhart's death, colleagues carried on the
work by attempting to produce more lasting cerebral stimulation with drugs,
particularly sodium amytal—later called the "truth drug" because it stimulates
talking.[37] Psychoanalysts also found sodium amytal a useful way to unlock
the unconscious, as noted by Smith Ely Jelliffe: "The amytal did away with
censorship, did away with the hypervigilant control of repression. . . . so the
drug-analytic method, which has been more or less extensively utilized, is a
very valid extension of our capacities for understanding, or for getting at,
repressed unconscious material."[38]

Another hypothesis, which was seriously considered in one form or another
over many years, was that schizophrenic patients might be suffering from a
"mental inertia" caused by sluggish brain metabolism. In 1931, *Time* magazine
reported that Walter Freeman, the "brilliant young chairman of the American
Medical Association's Section on Nervous and Mental Disease," had argued
that mental disturbances were caused by oxygen insufficiency.[39] Freeman had
suggested that increased oxygen might benefit schizophrenics, and described
"encouraging" results after placing catatonic patients in a hyperbolic chamber
at the Navy Laboratory. Interest in cerebral metabolism and mental disorders
continued to be explored, and many articles were published on measures of
oxygen utilization in the brains of schizophrenics and normals.*

In 1935, Robert Carroll, medical director of the Highland Hospital in Ashe-
ville, North Carolina, treated schizophrenics by injecting, into their cerebrospinal fluid, blood drawn from horses—an idea originating with a Swedish
psychiatrist's early report that the number of leucocytes (white blood cells)
in schizophrenics was low when their symptoms were most severe and high
when they improved. This early observation was supported by Adolf Meyer
who, after observing that patients improved after being given injections that
produced leucocytosis (an increase in leucocytes), wrote that the idea of introducing a "fresh regenerative impulse" was worth pursuing.[40] Carroll's
technique for producing leucocytosis was to inject, by lumbar punctures, 25
cubic centimeters of "inactivated horse serum" into a patient's cerebrospinal

---

* In 1961, when I was in Georgia in the Soviet Union, I talked to psychiatrists who had
established a clinic for mental patients at 11,000 feet above sea level on Mt. Elbrus, the highest
mountain in Europe. These psychiatrists reported that the adjustment to oxygen deprivation,
caused by the rarefied air, produced in catatonic schizophrenics metabolic changes that led to
improvement when they were returned to sea level. More recent studies, using CAT-scanning
equipment for visualizing the brain and blood vessels, have claimed to have found a lower rate
of blood flow in the frontal regions of the brains of schizophrenics compared with normal subjects, but it is not clear whether this should be considered a "cause" or an "effect" of
schizophrenia.

fluid—an equal amount of fluid having been first drawn out. This procedure was repeated two to five times to produce an "aseptic meningitis" with some fever—but, according to Carroll, no other disturbances, except for mild backaches, headaches, and some vomiting for a few days. Following a series of such injections, Carroll reported that schizophrenic patients became lucid, some for only a short period; but in others, the improvement was said to have been permanent.[41]

Inducing hypothermia—cooling the body—was not widely used, but it *cold* was one of the many different somatic therapies explored as treatment for schizophrenia even in such prestigious medical institutions as the Massachusetts General and the McLean hospitals. In 1941, John Talbott and Kenneth Tillotson described experiments on "the effects of cold on mental disorders" which involved placing sedated patients in a Therm-O-Rite Blanket—a "mummy bag" through which a refrigerant was circulated. Body temperature was drastically reduced and one patient died from cardiovascular failure during the procedure, but the investigators reported promising results:

> The observation that mute, aggressive, combative and uncooperative patients suffering from schizophrenia may enjoy periods of mental clarity is significant. It suggests that during the first years of schizophrenia irreparable morphologic damage to the central nervous system does not occur. If the clinical course can be altered during those important years, then there may be a modicum of hope in the treatment of an essentially incurable malady.[42]

One of the most radical somatic therapies was based on Henry Cotton's "focal infection theory" of mental disorders, which attracted considerable attention during the 1920s. According to this theory, toxins produced by bacteria at infection sites in different parts of the body are transported to the brain where they often produce mental disturbances. Henry A. Cotton, medical director of the New Jersey State Hospital at Trenton (figure 2.5), began to treat psychiatric patients by performing surgery to remove the source of infection.* Starting in 1919, he continued with great enthusiasm for several years and, in spite of considerable controversy, was able to convince others to adopt his radical surgical procedures for treating mental patients.

Cotton introduced his theory on a progressive note, asserting that one of the most harmful ideas that had crept into psychiatry was the belief that the psychoses are inherited. Not only does this belief tend to make therapists pessimistic, Cotton argued, but also, the support for this theory is not well

---

* Cotton did most of his work at this institution, which had a long and notable history. It was established in 1848—as the New Jersey State Lunatic Asylum—by Dorothea Dix as part of her campaign to reform treatment of the mentally ill. She had personally selected its site—a beautifully landscaped area that included a 900-acre farm—and chose to spend her last years there as a guest of the institution she founded.

Figure 2.5.   Henry A. Cotton (1869–1933). (Photograph of painting courtesy of Trenton
Psychiatric Hospital.)

founded. He pointed out that "insanity in the family" does not prove it has
been inherited: heredity may have some influence but "should not occupy
the exalted position it has previously held."[43]

Turning to his own theory, Cotton asserted that it was essential to locate
the foci of infections that are the source of the toxins affecting the brain.
Establishing his credentials as a pathologist by mentioning that he had previously worked with Alzheimer, Cotton described his methods for locating
infection. Psychotic patients, he declared, without exception "all have infected
teeth." The physician must be familiar with "modern dental pathology," as
infections are often hidden in previously filled teeth and in those that may
appear healthy:

All crowns and fixed bridge work have been condemned by the best men in the dental profession and we voice the same opinion. So in order to rid a patient of focal infection a very thorough job must be done. . . . I would emphasize the fact that a thorough elimination of focal infection can be obtained only by extraction. . . .

The removal of all infected teeth is imperative. . . . The success or failure of these methods will depend largely on how thorough the dental work is done. Taking out a few infected teeth and leaving questionable teeth, especially those which are capped and devitalized, and claimed by the dentist to be causing no trouble, often results in failure.

Cotton then proceeded to the subject of tonsils:

Chronic infection of the tonsils is as important as infected teeth and the mouth cannot be considered free from infection unless infected tonsils are removed. . . . That the children of the present generation are having their infected tonsils enucleated, will, we believe have a definite influence on the elimination of systemic and mental disorders later in life.[44]

The sinuses also need to be carefully examined, he said, but are infected only in a few cases.

Moving lower in the body, Cotton discussed infections in the gastrointestinal system, especially the stomach and colon. A relatively high incidence of infection was found in the stomachs of mental patients, compared with the very low incidence said to have been found in nonpsychotic "control" patients. Twenty percent of the patients in the "functional" psychoses group, especially female patients, were reported to have serious lesions in the colon. In females, the cervix also was found to be infected in "about 80 percent of the cases."[45]

The treatment followed logically from the presumed cause: the source of infection had to be removed. Cotton reported especially good results in females after "enucleating the cervix"; and in one remarkable case, "the patient gained 70 pounds after the cervix had been enucleated and she was restored to a normal mental condition." In some instances, the infection was said to have spread to the fallopian tubes and ovaries, which also had to be removed before improvement was seen. Cotton reported that the uterus had to be removed less frequently, as "only 38 hysterectomies out of 758 operations were necessary." In males, infected seminal vesicles were treated "by excision," but these were found to be infected only "occasionally."[46]

If none of the preceding treatments produce improvements, a thorough examination of the lower intestines was made:

Exploratory laparotomy is performed, and, if necessary the colon is resected. In our early operations we resected only the right side of the colon, anastomosing the ileum to the transverse colon. . . . In some cases brilliant results were obtained and in others none. Examination, in the unsuccessful cases, proved that the left side was infected, so that it was wiser to remove it. In the last year we have found it necessary to resect the whole colon to the lower sigmoid region. . . . our statistics would approximately be as follows: In 250 operations, 62 have been recovered, or 25 percent; the death rate has been 30 percent; improved cases, 15 percent; and 30 percent unimproved.

The 30 percent mortality rate following colon resection was eventually reduced. "It is obvious," Cotton noted, "that the success of our program depends on having competent consultants in different surgical procedures."[47]

He reported that over fourteen hundred patients had already been treated by his methods; and of these patients, only forty-two were still in the hospital. With respect to the "functional group," which included the "manic-depressive insanity, dementia praecox, paranoid condition and the psychoneuroses," only 37 percent recovered during the ten years prior to the introduction of his treatment of "focal infections." In the four years afterward, Cotton reported, the recovery rate "based on the same criterion, has averaged 80 percent."[48]

Anticipating criticism during the discussion period scheduled to follow his lecture at the American Psychiatric Association meeting in Quebec, Cotton mentioned the support of Adolf Meyer, who had written the foreword to Cotton's book, *The Defective, Delinquent and Insane: The Relation of Focal Infection to Their Causation, Treatment, and Prevention* (1921), based on his Louis Clark Vanuxem Foundation lectures at Princeton. After quoting from Meyer's foreword—"he [Cotton] appears to have brought out palpable results not attained by any previous or contemporary attack on the grave problem of mental disorder"[49]—Cotton had only to add: "Such an endorsement coming from Dr. Meyer has been very encouraging and stimulating, and we think such an endorsement is well justified by the results obtained at the Trenton State Hospital in the last four years." Cotton finished his lecture with the observation that his "method of detoxication" had saved the state over $300,000.[50]

The discussion following his presentation was lengthy, reflecting both the interest in the subject and its controversial nature. The comments were almost identical to those made fifteen years later when the result of prefrontal lobotomies were first presented. One discussant spoke about the great interest in the focal infection theory, quoting from an editorial in the London *Lancet*. Other discussants, however, expressed concern about whether serious surgery was being undertaken too casually:

> Now to my mind a colostomy or a cholectomy is a somewhat serious operation. Dr. Cotton speaks of them in a way that almost leads me to think the operation is as simple and devoid of danger as the extraction of a tooth. . . . Neither would I

regard the amputation or enucleation, to use Dr. Cotton's term, of the uterine cervix as an operation to be performed in support of a theory unless there were indisputable grounds for believing such a procedure necessary.

One participant questioned the validity of Cotton's statistics. Abraham Brill, the prominent psychoanalyst, was critical and used arguments almost identical to those he would raise eighteen years later when Walter Freeman described prefrontal lobotomy:

> There is no doubt that Dr. Cotton's work has caused a great deal of stimulation to the profession, and he deserves much credit, but when I heard of one case of dementia praecox that had her lower intestine cut and died from a general infection, I said to myself, "Well, after all, it was a case of dementia praecox and the family is relieved of a great burden." But when I saw a next case of cyclothymea whose depressive moods formerly lasted no longer than three weeks or a month, and who, after all his upper teeth had been extracted, merged into a deep depression, which has lasted over a year, I did not feel so kindly to this treatment. . . . We all know that the manic-depressive attack is practically self-limiting, all that one does is to help the patient in his struggle by preventing him from committing suicide or some other serious act, and by keeping up his vitality; that is why a great many of these patients attribute their recoveries to Christian Science, to some particular pill or to some funny ceremonial.[51]

In spite of the strong criticism of Cotton's theory and treatment, more than one-half of the discussants were supportive. A Canadian physician observed that too many people regard mental illness as hopeless and added that we should be "glad that we have the optimist in medicine." He expressed the hope that Dr. Cotton's work would encourage younger medical men in his country, because certainly this will influence "the discharge list of the hospitals in the future." Another discussant attacked the critics for standing in the way of "therapeutic innovation" and discouraging the work of pioneers:

> The medical profession in my opinion is largely responsible for the presence in the community of quacks of all kind who gain a foothold amongst us, chiropractors and other types of irregular practitioners; we are to blame for their presence for the reason that we, many of us, stand in a negative attitude about these things.

This comment was echoed by another:

> If there is anything in it we want to help him [applause from the audience]. We do not want to put ourselves in a position of opposition to anything that promises benefit or good to our patients. . . . We need to do more of the work Dr. Cotton is doing. . . . We need men, brains, and money to do it.

One physician in the audience reported that at his institution, where they had used Dr. Cotton's methods, results, especially from the gynecological procedures, were "in harmony with the opinion that Dr. Cotton expressed." The tide had turned, and the earlier critics were attacked not only for standing in the way of progress but, as a later discussant implied, for not behaving like gentlemen: "I merely wanted to say that when a man comes before this house and makes a statement I think we feel that it should be accepted as a fact until we are in a position to refute it. Dr. Cotton has had four years' experience in this work." And, finally, in reference to the influence of the popular press, a discussant spoke of the need for more research so that physicians would know how to respond: "When a friend of one of your patients comes to you and says, 'I have read in the *Review of Reviews* and other magazines about the work being done at Trenton and would like to know why you cannot do the same.' "[52]

In 1923, a carefully controlled study indicated that the patients who underwent the surgery recommended by Cotton did not improve any faster than an unoperated comparison group.[53] Interest in the focal infection theory declined, as psychiatrists became aware of the unreliability of the statistical data reported by state hospitals in the 1920s. There was evidence, for example, of instances where the same patient—readmitted and discharged several times—appeared in the data as four or five "cures."

The criticism of Cotton's theory and the validity of his evidence apparently did not hurt his reputation. In 1933, he wrote an article, "The Physical Causes of Mental Disorders," for H. L. Mencken's *American Mercury* magazine and was described therein as one of the leading psychiatrists in the United States.[54] Mencken delighted in embarrassing psychiatrists and was pleased when he thought some physical explanation exposed their theories as ridiculous.* The *American Mercury* article revealed that Cotton had not changed any of his arguments or claims during the preceding ten years. He reviewed all of his arguments and summarized his almost fifteen years of experience:

> We have estimated that about 80% of the so-called functional type of mental disorders are due not only to infected teeth and tonsils, but also to congenital malformations of the colon, or large intestines. . . .
> . . . Relics of medical superstition and barbarism are being supplanted by up-to-date conceptions as to the true relation of structure and function. The inhuman neglect that has resulted from the old discredited philosophical dualism is being overcome by the idea of unified mind and body.[56]

However barbaric Cotton's methods may seem to us today, he was not a

---

* Mencken wrote that "the discovery of thousands of cases of so-called mental disease were actually victims of the small, but extremely enterprising spirochete pallida. The news threw a bomb-shell into psychiatry."[55]

peripheral figure, nor did he lack training or recognition. He had an impressive professional vita, having studied not only with Alzheimer but also with Emil Kraepelin in Munich. As a young man, Cotton was greatly influenced by Adolf Meyer, who later became a good friend. Cotton was a member and officer of important professional organizations in the United States and was honored by being elected foreign member of the Medico-Psychological Association in Great Britain. When he died of a heart attack, at the age of fifty-seven (shortly after the *American Mercury* article appeared), obituaries were written by distinguished psychiatrists, among them Adolf Meyer.[57]

The use of surgery to treat mental illness was not restricted to the removal of endocrine glands and the operations stimulated by the focal infection theory: there were several early attempts to intervene directly into the brain. Before Wagner-Jauregg developed his fever treatment for paresis, Claye Shaw, working at the Bamstead Asylum in England in 1899, drilled holes in the skull of paretics and penetrated the dura matter over the motor cortex in the frontal lobes. Shaw hypothesized, in the *British Medical Journal* (September 1891), that patients suffering from "progressive paralysis" have, in this region of the brain, an elevated intracranial pressure caused by an inflammatory process.[58] Several other physicians attempted to treat "general paralysis" by trepanning the skull—among them Wagner at the New York State Lunatic Asylum in Utica and a William Fuller of Grand Rapids, Michigan.[59] Although paretic patients were claimed to have improved after these early surgical interventions into their frontal lobes, René Semelaigne, in Paris, reviewed the results and concluded that there was little convincing evidence that any of these operations had been effective.[60]

While this early brain surgery was primarily on patients suffering from paresis rather than psychiatric disorders, the distinction was, as has been noted, not at all clear at the time. Brain operations were, moreover, also performed on schizophrenics. Indeed, psychosurgery (though not yet so named) may be said to have begun as early as 1890 when Gottlieb Burckhardt, the director of the asylum in Prefarigier, Switzerland, performed six brain operations on mental patients.* These operations had aroused considerable controversy, especially after a patient died following surgery, and he was forced to discontinue. Nonetheless, he insisted that two patients had improved after the operation, and anticipated that others would pursue "the path of cortical extirpation with even better and more satisfying results."[62] Among these others

* The trepanization of skulls which was widely practiced at a much earlier period did not, at least intentionally, involve operations directly on the brain. This practice was carried on in different forms, as is evident from this passage (translated from the Latin) by Rogerius Frugardi ("Roger of Salerno"), an eminent twelfth-century surgeon: "For mania and melancholy, the skin at the top of the head should be incised in a cruciate fashion and the skull perforated to allow matter to escape."[61]

was the Estonian neurosurgeon Lodivicus Puusepp, who had tried to treat mental patients early in 1900 by cutting the nerve tracts connecting the frontal and parietal lobes. Although he regarded this attempt as unsuccessful, he did not abandon the hope that some kind of brain surgery might be more beneficial. In 1910, he made four or five holes in the skull over the frontal lobes of paretics and inserted under the dura either a cyanide-mercury mixture or Neosalvarsan. Puusepp reported that while the patients were not cured, his treatment produced a "notable improvement."[63] While little was learned about the frontal lobes from these early operations, they had left the impression that it was possible to perform surgery in this area in humans without dire consequences.

As will be seen, interest in brain surgery to treat mental illness continued up into the 1930s, when Moniz started to perform psychosurgery. Thus, even from this incomplete account it is evident that many psychiatrists and neurologists were willing to try almost any somatic treatment claimed to ameliorate mental illnesses. For many physicians, risking any therapeutic possibility was preferable to confessing helplessness.

3

"Anything That Holds Out Hope Should Be Tried"

We were a strange lot to move [in 1927] into
the Psychiatric Institute. . . . We worked exclusively with the brain. In our leisure time,
we talked about the brain, its normal struc-
ture, its gross and microscopic pathology and
the changes produced by death. We knew
little of the symptoms and classification of
mental disease. Of treatment, we did not have
to know much. In the early 1920s, there were
only two known treatments in psychiatry:
The malarial and other fever therapies introduced by Wagner-Jauregg (in 1918) to treat
general paresis (neurosyphilis); and prolonged
sleep treatment, introduced by Klaesi (1922)
to treat cases of schizophrenia.
—LADISLAS VON MEDUNA

THE great breakthrough in the treatment of mental illness seemed to have
arrived in the 1930s with the introduction of four radical somatic therapies:
three modes of shock treatment and one surgical procedure. The latter, the
main focus of this book, was first reported by Moniz in 1936. Insulin-coma
and metrazol-convulsion therapy were first reported in 1933 and 1935, respectively; while, in 1938, electroconvulsive shock was initiated. All the theories attempting to explain how these treatments worked proved to be wrong
and, as they were stated, so vague as to be impossible to take seriously. Yet,
as soon as reported, each of these therapies was put into practice, and hailed

Epigraph is from "Historical Article: Autobiography of L. J. Meduna," *Convulsive Therapy* 1
(1985): 43–57.

Figure 3.1.   Manfred Joshua Sakel (1900–1957), discoverer of insulin-coma (shock) treatment for schizophrenia. (Wide World Photos.)

enthusiastically and energetically by the medical profession and the public. All were used extensively throughout the 1940s and early 1950s. Today, only electroshock continues to be used to any large degree.

Insulin-coma therapy was first reported by Manfred Joshua Sakel, a Viennese physician (figure 3.1). A strange, withdrawn, and sometimes difficult man, Sakel later in life became obsessed with establishing his priority for discovering not only insulin-coma treatment—which was never contested—but all of the shock therapies. Born in 1900 in the Jewish community of Nadivorna, Poland, then part of the Austro-Hungarian empire, he belonged to a family with a long rabbinical tradition and was simultaneously acutely sensitive to anti-Semitism and proud of his Jewish ancestry. Sakel claimed to be a direct lineal descendant of Moses Maimonides, the twelfth-century rabbi, physician, and philosopher. After first studying at Brno in Czechoslovakia, Sakel finished his medical education at the University of Vienna and started working at the Lichterfelde Sanatorium in Berlin.[1] The sanatorium specialized in treating drug addiction and was frequently used by theatrical people, artists, and physicians seeking a cure. One of these patients, a famous actress, was a diabetic as well as a drug addict. In accidentally giving her an overdose of insulin, Sakel induced a mild coma (insulin lowers the level of glucose in the blood; an excess of insulin deprives the brain of energy and thus induces a coma). Because the actress's craving for morphine subsided after she regained consciousness, Sakel was soon giving insulin to all the drug addicts. His first publication on the subject, in 1930, describing insulin therapy as a cure for drug addiction, might have served as a warning against any future claims of his. He reported fifteen patients cured of morphine addiction without a single failure, but he neither provided sufficient information about what he considered a cure to be, nor specified how long the patients were followed up after treatment.[2]

During this early work, Sakel induced a deep coma in a patient, again accidentally, by injecting a high dose of insulin. The mental state of the patient, who was psychotic as well as a drug addict, seemed to improve after recovery from the coma. Encouraged by this observation, Sakel initiated animal experiments, which he said he did in his own kitchen, and established to his satisfaction that comas could be safely induced by insulin if they were terminated by an infusion of glucose. He began inducing comas in schizophrenic patients and continued this experimental therapy program after moving to the Neuropsychiatric University Clinic in Vienna. He first reported the success of his method in a short paper published in 1933. In less than four months— between November 1934 and February 1935—he published thirteen reports in the *Wiener Medizinische Wochenschrift,* expounding his methods, results, and theories with insulin coma.[3] Sakel claimed an improvement of over 88 percent—a phenomenal record in that most psychiatrists regarded schizophrenia as incurable.*

Joseph Wortis, a young biologically oriented psychiatrist and research fellow at Bellevue Hospital, was in Vienna during this period. He had been given a grant to undertake a "learning analysis" with Sigmund Freud, and thus had an opportunity to visit the clinic where Sakel was working. Wortis wrote to his colleagues at Bellevue that insulin coma was one of the most remarkable therapies he had ever seen.[4] Similarly impressed was Bernard Glueck, another psychiatrist and the founder of Stoney Lodge, a private sanatorium in Ossining,

---

* One of the patients Sakel treated was Vaslav Nijinsky, the famous Russian ballet dancer. While Sakel claimed "some success," Nijinsky never improved sufficiently to resume a normal life.

New York. He visited Sakel around the same time and, while assisting him for three weeks, observed twenty psychotic patients undergoing insulin therapy. In an article published in the *Journal of the American Medical Association* in 1936, Glueck wrote that if the results he witnessed proved to be permanent, it would be "one of the greatest achievements of medicine."[5] By this time, favorable reports were beginning to come out of Switzerland, the Netherlands, Poland, and Germany. Insulin therapy was also claimed to have been used successfully at the Moscow Psychiatric Clinic.

When in 1936, New York State's Commissioner of Mental Health, Dr. Frederick W. Parsons, heard of Sakel's success, he found the funds to bring Sakel to the United States. At the Harlem Valley State Hospital in Wingdale, New York, Sakel gave a course in insulin therapy for twenty-five psychiatrists from different New York state hospitals.[6] Before long, the treatment was being used at many institutions throughout the country.

Wortis's 1938 translation of Sakel's monograph from German into English also helped to promote insulin-coma treatment in the United States.[7] The cost of the translation was underwritten by grants from the National Committee for Mental Hygiene and the Fraternal Order of Masons. The introduction, written by Nolan Lewis of the New York State Psychiatric Institute, stated that the method "has already revealed enough to justify continuing its use."[8] In a preface to the monograph, Foster Kennedy, professor of neurology at Cornell and Bellevue Hospital, favorably compared Sakel's approach to Freud's:

> In Vienna at least one man had revolted from the obsession that only psychological remedies could ever benefit psychological ills; had refused to turn his back on modern weapons to fight mental ailments, to use only the same weapons that the Greeks had used—though the Greeks used them better. The scholasticism of our time is being blown away by a new wind and whatever may be the verdict of the next decade in reference to Manfred Sakel's contribution to the treatment of so-called "schizophrenia" we shall not again be content to minister to a mind diseased merely by philosophy and words.[9]

By this time, the metrazol-shock treatment had been introduced by Joseph Ladislas von Meduna. Meduna's background may partly explain Sakel's later antagonism. He was born in 1896 in Hungary to a family—originally of Marrano (Jews who had converted to Christianity) ancestry—that had over the centuries become titled and conservative. Meduna's formative years were spent in Catholic boarding schools. The First World War interrupted his medical education, and he spent part of the war fighting in Italy. After the war and before returning to medical school, he joined the anticommunist forces led by Admiral Miklos Horthy.[10]*

Meduna eventually obtained a clinical-research appointment at the Inter-academia Brain Research Institute, directed by Karl Schaffer. Examining brains of former epileptic and schizophrenic patients, Meduna convinced himself that he was able to detect subtle differences in the nerve cells of these two patient populations. These "observations," which have not stood the test of time, served to confirm his suspicion of a mutual antagonism between epilepsy and schizophrenia.

It was only a small logical step from belief in an antagonism between epilepsy and schizophrenia to the idea of inducing convulsions as a treatment for schizophrenia. In animal experiments, Meduna tried several chemical agents for inducing convulsions and finally decided to use camphor, an extract obtained from a particular laurel bush.† In early experiments, he injected camphor into schizophrenic patients and later switched to metrazol, a synthetic preparation (called cardiazol in Europe). His first patient was a catatonic schizophrenic who had hardly moved in four years and whose bodily needs (including tube feeding) had to be taken care of by the hospital staff. After the fifth convulsion, the patient got out of bed, dressed himself, and began to show an interest in everything.

After treating five patients in rapid succession, Meduna concluded that he had discovered a new treatment. He wrote a paper describing his method and results and showed it to Karl Schaffer, who had by this time moved the group to the department of psychiatry at the University of Budapest. In his unpublished autobiography, Meduna described Schaffer's reaction: "He called me a swindler, a humbug, a cheat. . . . How dare I claim I had cured schizophrenia, an endogeneous, hereditary disease? He knew what was in my mind—to publish, get newspaper publicity, and make money! 'If you dare publish this paper, I disown you.' "[12]

Meduna published the first report, in January 1935, in a paper cautiously entitled "Attempts to Influence the Cause of Schizophrenia by Biological Means"; but he claimed "recovery" in ten patients, "good results" in three, and no change in thirteen. Within a year, interest in the therapy had spread

* In fighting the communist regime of Bela Kun and the Rumanian occupation army, Horthy's forces became increasingly reactionary and fascist. In taking reprisals against the labor unions and Jews—Bela Kun was of Jewish origin—Admiral Horthy was said to have replaced a "red terror" with a "white terror." By this time, however, Meduna had resumed his career at the Royal University of Science in Budapest.

† In the 1780s, camphor had been used on a "melancholic" patient—accidentally inducing a convulsion and providing temporary relief.[11] The extract had also been used centuries earlier in China and Japan as a stimulant and is still so used today.

widely to many countries, partly stimulated by Meduna's extensive traveling throughout Europe. In 1938, the American Psychiatric Association published a supplement to its journal which included the proceedings of a symposium on shock therapies held by the Swiss Psychiatric Association in May 1937. Articles by Meduna, as well as by Sakel and many other physicians, were published at the time, clearly indicating the great interest in the new somatic therapies. In March 1939, Meduna came to the United States, ostensibly to give a lecture in Chicago at the American Psychiatric Association meeting, but he had already made up his mind to try to emigrate. With the support of a former Hungarian colleague, he was able to remain in the United States and was given an appointment at Loyola University.[13]

In a few years, metrazol was being used extensively in the United States; and by 1940, almost all major mental institutions included it in their therapeutic armamentarium. According to one count, between 1935 and 1941, more than one thousand articles were written on metrazol convulsion treatment.[14] When electroshock, which was a more convenient way to induce convulsions, began to replace metrazol, Meduna switched his interest to the carbon dioxide breathing treatment, which I described in the preceding chapter.[15]

There never was a convincing explanation of how metrazol convulsions worked. The original belief that schizophrenia and epilepsy do not occur in the same person turned out not to be valid. Moreover, as metrazol convulsions proved more effective with depressed patients, as recognized in 1937 by P. Verstraeten,[16] the presumed anatgonism between epilepsy and schizophrenia became irrelevant. The idea that there might be, however, antagonism between a convulsion and schizophrenia had been widely accepted in the 1930s; and there were even some attempts to treat schizophrenics by injecting blood drawn from epileptics immediately after a convulsion.[17]*

The third major convulsive treatment, electroshock, was developed by Ugo Cerletti in collaboration with his younger colleague Lucio Bini. Born in Conegliano, Italy, in 1877, Cerletti was broadly trained, studying medicine in Rome and Turin and afterward doing research in neuropsychiatry in Paris, Heidelberg, and Munich. He had many interests, including archeology and pre-history; and during the First World War, he was credited with inventing a delayed-action fuse used by the Italian army. Like Meduna, Cerletti also worked with epileptic patients, and became convinced that the body produces a "vitalizing substance" in response to the stress of a convulsion. Later Cerletti called this hypothetical substance "acro-amines" (a substance produced by

---

* The many convulsive treatment methods introduced during this period included a nitrogen breathing technique.[18] The Italian psychiatrist Amarro Fiamberti used the drug acetylcholine to induce a "vascular-induced convulsion"; and more recently, Indoklon, an inhalant, has been used to induce convulsions.

extreme struggle) and tried to produce it by shocking animals with a device built by Bini. While they could produce convulsions without any difficulty, many of the animals died. It was Bini who realized that the animals were dying not from the convulsion but from the electric current passing through the heart. When the position of the electrodes was changed from the mouth and anus to the sides of the head, none of the animals died. Later at the slaughterhouse in Rome, Cerletti learned that pigs were not killed with electricity, as he had assumed, but only stunned by it and killed afterward by other means. He and Bini then demonstrated on animals that electroshock convulsions could be administered safely, as the amount of current needed to produce a convulsion was considerably below the lethal intensity.

In April 1938, Cerletti and Bini were presented with an opportunity to try the electroconvulsion technique on a human. The police commissioner of Rome had sent over a man, presumed to be a schizophrenic, who had been wandering around the train station in a confused state. When the current was applied to his head, his body jolted and stiffened, but he did not lose consciousness. Clearly, the voltage was too low, and they decided to try again the next day. Overhearing the two doctors, the patient said, "Not another one! It's deadly!" As those were the only comprehensible words they had heard the patient utter, they ignored his protest and decided to try again immediately with a higher current:

> We observed the same instantaneous, brief, generalized spasm, and soon after, the onset of the classic epileptic convulsion. We were all breathless during the tonic phase of the attack, and really overwhelmed during the apnea as we watched the cadaverous cyanosis of the patient's face; the apnea of the spontaneous epileptic convulsion is always impressive, but at that moment it seemed to all of us painfully interminable. Finally, with the first stertorous breathing and the first clonic spasm, the blood flowed better not only in the patient's vessels but also in our own. Thereupon we observed with the most intensely gratifying sensation the characteristic gradual awakening of the patient "by steps." He rose to sitting position and looked at us, calm and smiling, as though to inquire what we wanted of him. We asked: "What happened to you?" He answered: "I don't know. Maybe I was asleep." Thus occurred the first electrically produced convulsion in man, which I at once named electroshock.[19]

Like metrazol, electroshock treatment, though first used with schizophrenic patients, was later found to be most effective as a treatment for depression. Cerletti did not believe that the convulsion produced the improvement, but thought it was due to the hypothetical acro-amines formed in the body in reaction to the stress. He wanted, in fact, to eliminate the convulsions, which he found objectionable. He tried to isolate the acro-amines from shocked animals in order to inject it into patients. A similar theory had prompted some French psychiatrists to inject blood from shocked patients into schizophrenics.

By 1941, insulin and metrazol shock and electroshock were being used extensively across the United States. Electroshock, introduced last, had by 1940 been used by only 10 percent of mental institutions; but by the next year, this proportion had jumped to 43 percent, and its use was still rapidly increasing. Electroshock was considered safer, easier to administer, less expensive, and "pleasanter" than metrazol. Many hospitals had used metrazol shock in combination with insulin coma, as Sakel had recommended in some of his publications; by 1941, they were rapidly switching to insulin plus electroshock or to electroshock alone.

In 1942, at the American Psychiatric Association meeting in Boston, Lawrence Kolb and Victor Vogel reported the results of their survey of the use of insulin, metrazol, and electroshock treatments in the United States.[20] Replies from 305 hospitals indicated that over 75,000 patients had received at least one of the shock treatments between 1935 and 1941. By 1938—three years after its introduction to the United States in 1935 and when its use was at its peak—54 percent of mental institutions were using insulin treatment. The numbers had declined only slightly by the end of 1941. Metrazol treatment, first used in 1936, was in 1939, its peak year, being used in 65 percent of institutions, but this percentage declined rapidly after electroshock was introduced.

The popular press played a major role in promoting the shock therapies. Enthusiastic articles describing insulin and metrazol therapies appeared in such widely circulated magazines as *Time, Newsweek,* and *Reader's Digest,* and also in *New Republic, Scientific American, Hygeia, Science Newsletter,* and *Science Digest.* In early 1937, *Time* reported that Sakel "has cured hundreds of cases of schizophrenia at his Vienna clinic by means of insulin injections," and added that "dozens have been cured in private and public mental hospitals in Switzerland, the Netherlands, Germany, Poland, Russia, England." *Time* noted in 1939 that Sakel "frankly declared he does not know how and why his cure works" but that it "indubitably works." The magazine also reported that "young Dr. Joseph Wortis" of Manhattan had had about a dozen cures in the past few months. An earlier article in *Newsweek* had been a little more cautious, noting that the statistics coming from Europe were hard to evaluate as there was no data on the rate of improvement of comparable patients not given insulin. *Newsweek* also observed that United States physicians "have reported less convincing results." A 1939 *Reader's Digest* article, however, had few reservations in its report of this new "miracle" treatment. The article described a "hopeless" patient—a Viennese physician—who after twelve days of insulin treatment had only occasional irrational moments; after thirty-three comas, "the patient's mind cleared completely and he was allowed to resume his practice." Furthermore, it was reported that Sakel found that 70 percent of the patients who had been ill less than six months made full recoveries. Recovery rates were reported to be the same from "Korea to Iceland and from Edinburgh to Galveston."[21]

The popular press took little note that the convulsions induced by metrazol and electroshock were often violent and not uncommonly produced fractures, mostly of the femur, arm, scapula, spine, jaw, and acetabulum (the socket in the hip bone). More fractures were produced by metrazol than by electroshock. Partly for this reason, but primarily because it was easier to administer, electroshock rapidly replaced metrazol convulsions.

The incidence of fractures and dislocations produced by metrazol was reported to be 39 out of 1,000 applications; and by electroshock, only 9 per 1,000 fractures, and the mortality rate was only 4 out of 7,207 cases. Starting in the 1940s, the drug curare was gradually introduced to block neuromuscular connections and eliminated the violent spasms produced by electroshock, eventually reducing to zero the incidence of fractures; in spite of this improvement, the treatment has remained controversial. Major complications in insulin treatment occurred when a comatose patient could not be restored to consciousness by infusing glucose in time to avert brain damage. Such damage occurred in 8.5 cases out of 1,000. The death rate with insulin averaged about 6 per 1,000, and was two-and-one-half times greater in public than in private hospitals.

Of these three shock treatments, the history of insulin coma most closely parallels that of prefrontal lobotomy. Like Walter Freeman with lobotomy, Sakel pursued the application of insulin as a therapy for mental illness with enormous energy and single-minded purpose: the wide adoption of insulin coma and lobotomy was largely the result of the respective efforts of these two men. Reports of success with both treatments were not only exaggerated but were accepted uncritically by an enthusiastic popular press and by the public, which responded with more demands for each new "cure." The proponents of both insulin coma and prefrontal lobotomy—along with metrazol and electroconvulsive shock, of course—claimed to alleviate a serious mental problem that lacked alternative treatment. Finally, neither therapy was subjected to rigorous or effective criticism within psychiatry.

After Sakel had trained several psychiatrists to use his insulin treatment, he was invited to present a paper at a joint meeting of the New York Neurological Society and the New York Academy of Medicine on 12 January 1937. He described his experience with over three hundred schizophrenic patients treated with insulin, and presented statistics based on his first one hundred patients, all of whom had been ill for less than six months. Sakel claimed a "full remission" in 70 percent of these cases, and explained that by "full remission" he meant not only that patients were free of all symptoms and fully capable of returning to work but also had "full insight into their illness" and normal emotional reactions. He also claimed that 18 percent of the remaining patients had "good remissions," which meant, he said, that they were able to work even though they still had some symptoms. Still other patients improved, he reported, but not to the point where they could resume work. The results with chronic patients, while much less dramatic, were still significantly better than the rate of "spontaneous remissions," which Sakel simply asserted was between 5 percent and 25 percent.[22] Since figures for "spontaneous remissions" were never justified, there was no way of knowing whether this was an accurate estimate for the type of patient Sakel was treating. He never presented a control group of his own, nor did he ever mention how long patients were followed after treatment; thus, it was impossible to know whether his "cures" were long-lasting.

Following this presentation, the psychiatrists recently trained by Sakel also described their results. There were reports from Joseph Wortis at Bellevue, from Bernard Glueck in Ossining, and from members of the psychiatric staff at the Worcester State Hospital in Massachusetts. All described good results, although several observed that their success rates were not nearly as high as Sakel reported. The consensus of the psychiatrists reporting results may be epitomized by such remarks as: "All patients undergoing hypoglycemic therapy benefited to a greater or lesser extent"; "Insulin produces amelioration in the majority of cases"; and "There is not a shadow of a doubt in my mind that there has been definite improvement in a number of cases." Several psychiatrists were much more enthusiastic:

> Orientation and interest in current events and in personal appearance improve regularly; patients who in some instances have had to have meals in their rooms for years are able to join others in the dining room . . . the emergence of affability and friendliness and the capacity to cultivate transference relationships. . . . It is difficult to believe that one is dealing with the same persons.[23]

The psychoanalyst Smith Ely Jelliffe was not shy about offering an interpretation of insulin-coma treatment even though he had not yet had any experience with the procedure:

> One sees various clinical pictures of regression to primitive, infantile behavior of a type approaching the fundamentally normal aspect of very early intra-uterine life. . . . [The coma] brings the subject practically into an intra-uterine bath of primary narcissistic omnipotence. By the hypoglycemic threat one can envisage a type of phylogenetic dissection by a metabolic tool from the frontal forebrain back to the medullary respiratory and vagal nuclei and a decerebration experiment of a subtle

form by a pharmacodynamic type of instrumentation. This onslaught by the death impulse, with its consequent alteration of autistic, negativistic, hostile behavior has been noted for centuries.

Here is something impressive made doubly so by the increased weight of the Freudian libido hypothesis as it pertains to the psychodynamics of schizophrenic behavior . . . one sees complete release from the more superficial layers of the tyranny of the superego. The world becomes a loving, not a threatening, world, and early erogenous zones are permitted a certain freedom of functioning . . . [this] form of psychotherapy can be, it seems to me, of great value in synthesizing a fragmented ego.

Jelliffe concluded by suggesting that a more lasting effect might be accomplished by switching, at the end of the treatment, to a slower-acting and longer-lasting form of insulin, such as protamine insulin, to keep the patient in just the right balance "at the edge of the death threat."[24]

Jelliffe—who would make similar comments when lobotomy was introduced in the United States—was seventy-one at the time and a flamboyant and influential figure in psychiatry. He had been Karl Menninger's teacher and a close friend of William Alanson White, with whom he had written a successful textbook. Jelliffe had edited the *New York Medical Journal* and then virtually took over the *Journal of Nervous and Mental Disease,* the *Nervous and Mental Disease Monographs,* and the *Psychoanalytic Review.* A Rabelaisian speaker with a prodigious memory, he mixed his own brand of psychoanalysis with the classics and Greek mythology. His comments at meetings were always bold, sometimes insightful, and not infrequently outlandish nonsense. He had a thundering voice and clearly enjoyed being in the limelight; and indeed, his lectures were performances. He had served as an expert witness for the defense in the celebrated "murder case of the century"—the shooting, in 1906, of the architect Stanford White; and afterward, the newspapers widely covered the lawsuit he instigated to collect his $10,000 fee from the defendant, whom he had helped get acquitted on an insanity plea. Along with Bernard Glueck and William White, Jelliffe was again an expert witness in the highly publicized Leopold and Loeb case in Chicago in 1922.

Also attending the New York Academy of Medicine meeting was Adolf Meyer, who observed that it was not necessary to describe insulin coma in Jelliffe's "highly interpretative" statement of bringing the patient to "the edge of the death threat." Nevertheless, Meyer complimented Sakel, referring to him as an "ingenious originator" and a "conscientious worker." Meyer did recommend caution, noting that after the popular accounts of the treatment, he had received "a flood of letters appealing for immediate application of this panacea." Giving the usual caveats, Meyer stated that careful studies in qualified centers were needed and added that they will be made possible "by virtue of the fine cooperation of Dr. Sakel."[25] Shortly afterward, Meyer introduced the insulin treatment to the Henry Phipps Clinic at the Johns Hopkins School of Medicine. His remarks about insulin coma are almost identical to those he made the same year when Walter Freeman reported on the first prefrontal lobotomies in the United States (page 144).

Although Abraham Brill, presiding chairman of the session, had often been critical of somatic therapies, he said at the meeting that "the results reported by my Swiss friends are of a character that make it really an obligation to try out."[26]* The year before, he had written to William Alanson White that "schizophrenia is so hopeless that anything that holds out hope should be tried";[27] but in 1936 White was still skeptical. Winfred Overholser, who succeeded him as director of St. Elizabeth's Hospital, also feared that insulin therapy was becoming overly "popularized and prematurely hailed as a panacea."[28] Most hospitals, however, were quick to adopt insulin-coma therapy.

The rapid adoption of this treatment is especially phenomenal in light of the amount of care it required. I still vividly recall watching it being administered at the Winter Veterans Administration Hospital in Topeka, Kansas, in 1950. The treatment started early in the morning so that patients could be monitored for any delayed reactions that might occur during the day. The ward was semidarkened, and about thirty patients were lying on beds to which they were bound with folded sheets so as not to roll out. Alongside each bed was a table with sweetened orange juice, dextrose bottles, and syringes filled with different amounts of insulin. Patients just starting the six- to ten-week treatment series received low doses of insulin, while patients further along were given high doses. The injections were given in different parts of the body, usually alternating between the cheeks of the buttocks on successive days. Treatment was given five days a week, with two days' rest, and might be continued for as long as ten weeks, with fifty comas for patients who had not yet responded sufficiently.

Shortly after an injection, the patients became quiet as the insulin began to lower blood sugar and deprive the brain of energy. Some of the patients started to perspire and salivate, drooling down their chins. By the end of the first hour, the patients who had higher insulin doses had lapsed into a coma; many were tossing, rolling, and moaning, their muscles starting to twitch; and some had tremors and spasms. Here and there an arm would shoot up uncontrollably. Some of the patients started to grasp the air, reflexively. I noticed other "primitive movements," including rapid licking of the lips. Patients furthest along in the treatment series, with the highest insulin doses,

* Abraham Brill had many personal contacts in Switzerland, as he had, at one time, been at the Psychiatry Clinic in Zurich.

might be having violent convulsions.* With all these people—tossing, moaning, twitching, shouting, grasping—I felt as though I were in the midst of Hell as drawn by Gustave Doré for Dante's *Divine Comedy*.

Many trained nurses were needed to monitor this whole process. Pulse and respiration had to be taken at regular intervals. Physicians had to exercise considerable judgment about when a coma should be terminated. Sakel never gave standardized instructions, insisting that it was necessary to respond to the individual patient. Nevertheless, he maintained that the length of a coma was the critical determinant of the success of the treatment. Yet a patient left in a coma too long could suffer brain damage or death. Usually after several hours, the treatment was terminated by glucose: a conscious patient could drink a glucose-sweetened solution, but a comatose patient would receive glucose directly into the stomach from a nasal tube that had been inserted in advance. In an emergency, glucose might have to be injected intravenously. In England, where wartime sugar rationing was more severe, potato soup with molasses was given to patients who could take it by mouth. Sometimes there was a delayed reaction, and a patient would go into a second coma after the termination stage, so patients had to be carefully watched for much of the day.

As for Sakel's theory, it was so vaguely stated and so impossible to relate to any known—let alone measurable—biological process that it was all but incomprehensible. He had speculated that the disease process injured the phylogenetically younger brain pathways so as to allow the older, more primitive pathways to dominate; and argued that, to restore a more normal balance, "one has to either limit the supply of energy, or else to diminish combustion in the cell by muffling it." He apparently believed that insulin produces a selective blockade of the most active cells—those underlying the psychosis—and, at the same time, strengthens the latent "normal pathways." Sakel compared the role of the convulsions to the artillery in a military campaign, and that of the hypoglycemia to the infantry: "the artillery never conquers and occupies hostile territory. It can only open the way for the infantry." Apparently recognizing that many people would be unwilling to accept such a vague theory, Sakel blocked criticism by insisting that "the mistakes in theory should not be counted against the treatment itself, which seems to be accomplishing more than the theory itself."

I have a high regard for strict scientific procedure and would be glad if we could follow the accustomed path in solving this special problem: it would have been preferable to have been able to trace the cause of the disease first, and then to follow

---

* Sakel distinguished between the "wet shock," which was really a comatose state accompanied by profuse perspiration and salivation, and "dry shock," which was the convulsion.

the path by looking for a suitable treatment. But since it has so happened that we by chance hit upon the wrong end of the right path, shall we undertake to leave it before better alternatives present themselves? For even if the hypoglycemic treatment of psychoses accomplishes only a part of what it promises, it nevertheless has a value beyond its therapeutic claims, for it should perhaps now enable us to work backwards from it to the nature and cause of schizophrenia itself.[29]

Even without a reasonable explanation, the results of experimental trials in the United States did seem to indicate that schizophrenic patients improved during the treatment. The New York State Department of Mental Hygiene conducted a large-scale study of the effectiveness of insulin treatment. As the procedure was not standardized, a group of psychiatrists participating in the study were sent to the Harlem Valley State Hospital for training by Sakel. They were then instructed in the use of a standardized form for evaluating patients before and after treatment. The completed forms were sent to a central agency where they were analyzed by statisticians; and in 1938, Benjamin Malzberg published the results based on over one thousand patients treated with insulin. The main problem with the study was the absence of a concurrent control group of schizophrenic patients who did not get insulin. The insulin group had to be compared with patients admitted to the same hospitals during the previous two years. Malzberg's report indicated that, following the Sakel treatment, the "recovery rate" increased from 4 percent to 13 percent and the percentage of "improved" increased from 11 percent to 27 percent. Over all, about 65 percent of the patients treated with insulin exhibited some improvement, compared with only 22 percent of the control group. Although the results were not close to matching Sakel's success, Malzberg concluded that the evidence "provided encouragement."[30] A year later, when Malzberg did a follow-up study of the patients, the difference had shrunk considerably, although the insulin group remained slightly better.[31] This generally favorable report was bolstered by the data presented in 1939 by Edward A. Strecker, a prominent psychiatrist at the University of Pennsylvania. According to *Time* magazine, Strecker, who worked at one of the "best equipped shock clinics in the United States," reported that 30 percent to 40 percent of insulin-treated patients returned to normal living.[32]

While interest in insulin therapy diminished slightly in the 1940s in the United States, the therapy continued to be used extensively. In 1946, Earl Bond and Jay Shurley, who had taken over Strecker's shock clinic at the University of Pennsylvania, reported that the four hundred schizophrenic patients given insulin treatment had a significantly better outcome than untreated patients. Bond and Shurley acknowledged that insulin treatment is a drastic procedure, sometimes with serious complications including death (about twelve in their series), but concluded that an insulin unit may be a worthwhile investment "for the next few years"—a comment that seemed to suggest that enthusiasm had started to wane.[33] Nevertheless, insulin treatment continued to be touted as the best treatment available for schizophrenia, particularly if initiated early in the disease. There were even claims that the remissions achieved with insulin were more lasting and of "better quality" than those that occurred "spontaneously." By the mid-1950s, however, the use of insulin had declined significantly in the United States—mostly because of the new antischizophrenic drugs—but was continued more extensively in Great Britain. In 1960, Willy Mayer-Gross and his colleagues in England wrote in their textbook that insulin treatment "is still recognized as one of the most effective methods of treating early schizophrenia."[34] Before the decade ended, however, insulin treatment was not even mentioned in most textbooks.[35]

Sakel was in many ways a tragic figure. He was untrained in the methodology of science and never learned to present his ideas in a scientifically acceptable manner. Although a keen observer, he read little of the medical literature. When he first came to the United States, he had been in great demand as a speaker and was given an honorary degree by Colgate University in 1936.[36] Around this same time he received an offer of a university professorship in the Midwest, but preferred to live in New York, where he managed to attract enough private funds to establish the Manfred Sakel Foundation. This foundation was supported by the Gimbel Foundation but never did any scientific work, its sole purpose apparently to train people in the "Sakel method." Sakel remained a bachelor, lived simply, gave money generously to charities, was interested in the work of Quakers, and hoped to get funds to build his own hospital. In 1946, he suffered an almost fatal heart attack and withdrew into private practice, maintaining little contact with professional organizations.

He did attend the First World Congress of Psychiatry in Paris in 1950, which was dominated by discussions of biological therapies. In the publication of the proceedings, over five hundred pages were devoted to shock treatments and lobotomy.[37] The discussion of insulin-coma treatment at this World Congress of Psychiatry was extensive, and Sakel could easily have considered the meeting the crowning event of his career. Unfortunately he had become bitter, and his rambling talk combined scientifically naïve statements, self-serving remarks about past failures to recognize genius, and personal recrimination:

> Those psychiatrists anxious for as quick a participation as possible in the new approach . . . chose the simplest technique. . . . The medical and historical fact is that there is just one shock treatment, introduced by me in 1933. . . . Convulsions produced by whatever means is not a distinct "shock" or a "new treatment." . . .

The severe shock therapy was officially announced to the general public in 1933, and anybody could read it, including Meduna. Cerletti considered it of such importance that he came to Vienna in 1933 to see about it for himself.

... Meduna himself confirmed this situation explicitly in his monograph ... but unfortunately in his monograph he proclaims the inducement of convulsions without insulin as a "new treatment."

... I was strongly adverse to airing in public, scientific controversies which happen to be inseparable from personal matters. But now I feel it necessary to speak publicly.[38]

Sakel managed to alienate almost everyone by his inappropriate behavior.*

After the Paris congress, Sakel was even more isolated from the rest of psychiatry. Interest in insulin therapy was clearly waning and would soon be replaced completely by the new antischizophrenic drugs. Having no students or followers, he never benefited from friendly criticism of his theories or his style of writing. His final review of insulin treatment, published in 1954, was so long and incoherent that it reflects poorly not only on him but also on the quality of editorial review in psychiatry that would permit this article to be published in such a form. Sakel dismissed the recent evidence of the ineffectiveness of insulin treatment by attributing the failures to improper training in his technique and to "inexperienced physicians." He observed that surgeons are "forced to acquire their mastery in a rigid and complete apprenticeship," and implied that this should also be the case for those who would use insulin coma. Opposition to his methods, Sakel wrote, was due to the "law of inertia"—a tendency to resist all new ideas; and, noting that history was full of examples—Galileo in astronomy, Robert Koch's difficulty in establishing the basis of infectious disease—he said of his own work that "the same is true in psychiatry."[41]

Although his reputation had declined in the United States, a symposium commemorating the thirtieth anniversary of his first insulin experiments was held at a European meeting of psychiatrists in Vienna in September 1957. Sakel was honored and referred to as "the first man in history to show that the worst form of insanity could be treated by physiological methods."[42] Three months later, on 2 December 1957, he died from a second heart attack. In the United States, only Joseph Wortis wrote an obituary in a professional journal.[43] The *New York Times,* however, ran a prominent obituary, which recalled that Sakel had treated Nijinsky and that he had been referred to as

---

* Ironically, although Sakel had accused Cerletti and particularly Meduna of usurping his method, he avoided mentioning that Professor Hans Steck and others had started using insulin hypoglycemia to treat psychoses before 1929.[39] Meduna, I should note in fairness, did not hesitate to criticize Sakel, noting that insulin therapy had no sound theoretical foundation; and since his theory does not "explain the nature of schizophrenia, it cannot throw light on the manner by which schizophrenia is cured."[40]

the "Pasteur of psychiatry."[44] A year later, two books Sakel had written were published posthumously; one on schizophrenia, the other on epilepsy.[45] The books were not taken seriously, and reviewers severely criticized them for the dogmatic way Sakel asserted opinion as facts without providing any convincing evidence. Both books were also criticized for their author's failure to consider any alternative theory or conflicting data and for their lack of organization.[46] Neither book had any impact on the treatment of either epilepsy or schizophrenia and both were virtually ignored.

In 1949, Lothar B. Kalinowsky and Paul H. Hoch wrote in their *Shock Treatments and Other Somatic Procedures in Psychiatry* that they believed the new somatic treatments, including lobotomy, to be qualitatively different from any apparently similar past treatment: "Therapeutic procedures applied in former centuries, which emphasized fright and intimidation, are often erroneously called the predecessors of the present day shock methods. Actually they were strictly psychological procedures and have no place in a survey of organic treatment in psychiatry."[47] Those past therapeutic interventions, however, were not qualitatively as different as Kalinowsky and Hoch suggested. When, in 1812, the "father of American Psychiatry," Benjamin Rush, strapped mental patients into his mechanical "gyration device"—a whirling chair—he was not trying to frighten or torture them. He also had in mind a physiological theory, and wrote that the gyration "opposes the impetus of the blood toward the brain, it lessens muscular action everywhere, and reduces the force and frequency of the pulse." Rush even thought that the vomiting and nausea had a beneficial effect on the circulation of the brain.[48]

To capture the mood during the 1930s, let me cite a 1938 review of progress in psychiatry in the prestigious *New England Journal of Medicine*. This review opened with a reminder of the magnitude of the problem—"one bed for mental disease for each bed for all other diseases in America. Of those cases accumulating in state hospitals roughly 75 percent have been diagnosed dementia praecox, later called for reasons not too obvious, schizophrenia." Of the state of the field, the author observed that "organicists and psychogenecists vied with each other in elaborating minutiae brought forth by their endeavors, but no one really claimed to understand the cause, nature, or cure of this dread malady." Recent developments, however, provided hope: "By far the most important development is the so-called shock treatments for dementia praecox and other psychoses." All of these physiological developments, it was concluded, seem "to bring the field of psychiatry a little closer to that of general medicine."[49]

4

A Portuguese Explorer:
Egas Moniz

I was always dominated by the desire to accomplish something new in the scientific
world. Persistence, which depends more on
will-power than intelligence, can overcome
difficulties which seem at first unconquerable.
—EGAS MONIZ (1949)

ON 12 November 1935, in a Lisbon hospital, a neurosurgeon drilled two holes
into the skull of a mental patient and through them injected absolute alcohol
directly into the frontal lobes of her brain. In the following five weeks, six
similar operations, each somewhat different from the one before, were performed on six other patients. In the eighth, the procedure was radically
changed. Instead of destroying nerve cells with alcohol, the surgeon inserted
into the brain a specially constructed instrument and rotated it to cut or crush
the nerve fibers in its path. Owing to the procedure's similarity to coring an
apple, it was called the *core operation*. Egas Moniz, the Portuguese neurologist
who devised the operation, named it *prefrontal leucotomy* from the Greek *leuco*,
meaning "white matter" (referring to nerve fibers), and *tome*, meaning "knife."
The instrument (described in chapter 6) was called a *leucotome*.

In a matter of months, Moniz published the results of the first twenty
operations; and within the year—1936—his operation was being performed
on mental patients in more than half a dozen other countries. Before long,
psychosurgery—a term also introduced by Moniz—was being tried in almost
every country of the world. Although there was strong criticism of psychosurgery from the beginning, the very rapidity of its adoption demonstrates
the keen interest throughout the medical profession in somatic treatments
for mental illness. In the rest of this book, I plan to explain the broad acceptance
of this drastic treatment, based as it was on the flimsiest of theories and on
completely inadequate evidence. Of the many contributing factors, not least

Figure 4.1.   Egas Moniz around forty years of age. (Courtesy of the National Library of Medicine.)

are the backgrounds and personalities of Egas Moniz, Walter Freeman, and other eminent physicians.

Born in Avanca, Portugal, on 27 November 1874, the eldest son of an aristocratic family, Moniz was instilled from his youth with stories of his family's history and his nation's past glory, and with the dream of carrying on that tradition. Indeed, he did reach great prominence not only as a statesman and man of letters but also as the "father of cerebral angiography" and as Portugal's only recipient of the Nobel Prize, for the "discovery of prefrontal leucotomy" (see figures 4.1 and 4.2). Legends about him, some of them fostered by himself, abound. In tracing his story, I found it not always easy to separate myth from fact.

Avanca is a coastal village about 40 kilometers from the city of Oporto in the north of Portugal—a region steeped in history. Prince Henry the Navigator was born in Oporto, and Ferdinand Magellan came from this region. Later, Wellington defeated Napoleon's troops on the fields just south of Oporto. Avanca has changed little from the time of Moniz's birth, remaining a small agricultural and fishing village with a population of less than five thousand; when I visited there in 1982, it was still possible to see the evaporation flats

Figure 4.2.　Portuguese stamps honoring Egas Moniz. (Photograph by the author.)

and "salt cones" readied for the fishermen. The village's railroad station and many of its buildings are faced with attractive blue and white tiles, several glorifying historical events. One of these commemorates Moniz and his accomplishments. It is a colorful, picturesque town but, now as then, offering few opportunities to most of its inhabitants.

Moniz was baptized António Caetano de Abreu Freire, but the honorific name Egas Moniz was added at his christening. His father, a prominent landowner and a member of the aristocracy, traced the family back to the original Egas Moniz, a twelfth-century patriot who had fought against the Moors.* During most of his youth, Moniz attended Jesuit boarding schools or lived in Pardillho, a nearby town, with his uncle, Abadelde, a priest, who was the greatest influence in his early life.† As the eldest son, Moniz was the principal focus of his uncle's efforts to preserve the family name and tradition. Abedelde taught him to read before he started school and filled him full of tales about the family's history and Portugal's past glories, often reading epic poems aloud to him and instilling in him a love of literature and poetry. His uncle carefully screened his schools, checking on the curriculum, the teachers, and the students. When the boy started school, he talked so much about Egas

---

　* It is characteristic of some of the myths surrounding Moniz's life that the original Egas Moniz is described only as "a Portuguese patriot who roused the country to drive the Moors from the mountains of the Iberian peninsula." Actually, the original Egas Moniz remained loyal to the House of Castille, representing Spanish domination, even though he was nominally in the service of Henrique, a nobleman who had assumed the title of king and led a rebellious movement for Portugal's independence.

　† Like all his family, Moniz was a Roman Catholic, but he was not particularly religious and did not fulfill his uncle's hope that he would follow in his footsteps and enter the priesthood. Upstairs in Moniz's house in Avanca is a small chapel, which was used mainly by his wife.

Moniz that his classmates were soon calling him "Moniz." By the time he became a student at Coimbra University, he was signing his name "António Egas Moniz" and, before graduation, had shortened it to "Egas Moniz."

Moniz's family, whose income from an agricultural equipment business was not always dependable, was land rich, employed many servants, and owned two large houses, the Casa do Marinheiro ("House of the Sailor") in Avanca and a second one in Torreirra. Of the four children, a brother died in infancy, leaving Moniz's younger brother, Miguel, and his sister, Luciana. When Moniz was sixteen, the year before he entered Coimbra University, his father died. For some time, the family's income had been steadily declining owing to the floundering Portuguese economy, and the family had been forced to sell Casa do Marinheiro. Abadelde purchased the property to keep it in the family and also assumed the cost of Moniz's education. What happened to Moniz's father then is not clear. In his memoirs, Moniz wrote only that his father went to Mozambique, a Portuguese colony, on some business venture; and soon after, word came that he had died from a "severe malaria attack" in April 1890. Within a few years, all the remaining members of Moniz's family died. First, Luciana succumbed to tuberculosis. Then Miguel abandoned college and followed in his father's footsteps, taking off for Mozambique on a speculative venture. Again, Moniz did not describe any of the circumstances in his memoirs, except to write that within a year he received word that Miguel had been killed in a "hunting accident." Shortly afterward, his mother became terminally ill, and her death was soon followed by that of his uncle Abadelde, leaving Moniz, in his early twenties and an undergraduate at Coimbra, the family's only survivor and heir.

While he excelled academically at the university, Moniz also was active in campus politics (the university had a tradition of mobilizing political action throughout the country) and soon was known for his pamphlets, which he signed "Egas Moniz." Although his family had traditionally been monarchist, Moniz favored a republican form of government.* Within the wide spectrum of political opinions at Coimbra, his views were moderate, usually the same as those of the Centrist political party. He was a natural leader, and his classmates elected him president of the Tuna Academica—the most prestigious of the many student societies in the university's cultural and political life.

In his senior year, Moniz decided to study medicine at Coimbra, and he continued to excel. Then at the age of twenty-four, the year before graduation, he was suddenly stricken with gout, which left his joints swollen and painful and remained a serious problem throughout his life. After graduation from

---

* For several decades, the monarchy in Portugal had been on the verge of collapse and, in 1891, was overthrown by a revolution that arose in Oporto. With the help of the military, however, the monarchy regained control.

medical school in 1899, he was appointed lecturer, teaching basic courses in anatomy, histology, and general pathology.

In 1901, while still a lecturer, Moniz married Elvira de Macedo Dias, whom he had first met when he was living with his uncle and she was visiting neighboring relatives. Her wealthy Portuguese family had made its fortune in Brazil, and her dowry and later inheritance relieved Moniz of all financial concerns and made it possible for him to entertain in an increasingly lavish style. Moreover, in a few years he was receiving a good, steady income from a successful series of books on sexual behavior and physiology. *A Vida Sexual,* first published in 1901, was initially considered shocking in conservative Portugal because it included information on the bizarre sexual practices of Krafft-Ebing's patients and some of Freud's revolutionary ideas on infant sexuality. This first book was followed by others on different aspects of sex and finally by a highly successful, abridged, one-volume version of all these books, which eventually went through nineteen editions.[1]

Moniz obtained his training in neurology in France, where he visited regularly throughout his life. There he attended clinics, first in Bordeaux with Pitres and later in Paris at La Salpêtrière with prominent French neurologists, including Pierre Marie, Joseph Babinski,* and Jules Dejerine, all former students of Charcot. The clinics in Paris were among the best in the world— primarily because of the reputation Charcot had established—and drew as students a steady flow of neurologists from many countries. Moniz, however, worked most closely with Jean Sicard, at the Necker Hospital in Paris. Sicard later encouraged him to begin his research on cerebral angiography. Moniz always regarded his neurology training as completely French and later, when he had contributions of his own to report, he went first to France.

Moniz remained active in politics and, in 1900, at the age of twenty-six, was elected to the Portuguese parliament as a deputy from Estarreja, the district that includes Avanca. He was able to hold this public office while continuing as a lecturer at Coimbra. He was re-elected to the parliament many times during the fifteen years that followed his graduation from medical school.

During this early period, he accomplished little of distinction in neurology. Indeed, when he was given the newly established professorship of neurology at the University of Lisbon in 1911, some colleagues at the university resented the appointment, considering it political rather than earned, and believing that António Flores had contributed more to neurology and deserved the professorship. The position gave Moniz a larger income and made it possible for him eventually to expand and remodel the family home, Casa do Mar-

---

* Moniz spent considerable time with Henri Babinski, the neurologist's brother. Henri was an expert on wines and cognac and the author of a highly successful cookbook, *Gastronomie Pratique* (published under the pseudonym Ali-Bab).

Figure 4.3.   Casa do Marinheiro, Moniz's home in Avanca, now the Egas Moniz Foundation and Museum; here are displayed all the awards he collected during his life. (Courtesy
Ática, S.A.R.L., Editores Livreiros, Lisbon, Portugal.)

inheiro, in Avanca (figure 4.3).* Up to this time, his publications in neurology
consisted of a few unfocused articles (mostly case reports) on "Jacksonian"
epilepsy, encephalitis, and a thalamic pain syndrome.

   In 1910, the Republicans won majorities in both Lisbon and Oporto, leading
to an uprising that caused King Manuel to flee to England. During this period
of chaos and street fighting among different political factions, Moniz maintained his moderate pro-Republican position, disagreeing with the more radical
policies of the Unionist party. During the First World War, he was asked by
Sidónio Pais, the head of the government and a former colleague at Coimbra
University, to serve as the Portuguese ambassador to Spain—a position that
provided Moniz with an opportunity to meet the Spanish neuroanatomist and
Nobel laureate, Santiago Ramón y Cajal. At this time also, Moniz managed
to set aside several hours a day to write a book on the neurology of war
injuries. This book, which described traumatic injuries to the nervous system
and their treatment, was not an original contribution and had little influence.[2]

   When the war ended in 1918, Moniz was appointed minister for external
affairs and represented Portugal at the Peace Conference at Paris, with the

* For this work Moniz employed the well-known Italian architect Ernesto Korrodi, who had
previously restored the Portuguese National Castle in Sintra.

difficult task of trying to secure colonial concessions and indemnities for his country in spite of its small contribution to the defeat of Germany and the other Central Powers. The negotiations were often heated; and at one point, the French premier Georges Clemenceau, angry because Moniz's Centrist party had not actively opposed Germany, refused to talk to him. After the French field marshal Joseph Joffre arranged for the two statesmen to meet however, they discovered interests in common—including medicine, which Clemenceau had studied in his youth. Moniz, always alert to the advantages of potentially useful social contacts, maintained a correspondence with Clemenceau and once sent him a case of fine port wine from his home district.

It has frequently been asserted that during the heated negotiations over indemnities at the Paris Peace Conference, Moniz and a member of his own delegation, Afonso Costa, the leader of the radical anticlerical Democratic party, fought a duel.* Actually, it was only a "Latin duel"—much blustering and challenge, but nothing more.†

Throughout this period of political and economic chaos, Moniz was able to maintain a prominent place in politics in spite of frequent changes in the government. Because he devoted so much time to politics, he was unable to make any significant contribution to neurology. In fact, many people believed that Moniz aspired to become prime minister. However, the partisan disputes within the delegation he headed in Paris, the power politics displayed at the conference table, the economic unrest which triggered national discontent and a series of assassinations and changes of government, all combined to dampen his interest in politics. In 1919, he described his frustrations with political life in *Um Ano de Politica,* a copy of which he presented to his wife with the inscription: "My Dear Elvira—with the hope that it would not remind her of the pain that politics has given to her."³‡

For another six or seven years, Moniz struggled to combine politics with his university activities. He was serving as dean of the medical school when radical students raised a red flag over the campus, but he refused to intervene even when the minister of health ordered him to stop the demonstration. When army troops entered the campus and some students were injured, the medical staff insisted on treating them. In the subsequent confrontation, Moniz was arrested along with some students and members of the medical staff. The arrest was merely perfunctory, and he was soon released. This was his third

* The two men had disagreed before, when Moniz, as minister of external affairs, had re-established diplomatic relations with the Vatican after they had been severed by a more radical government.

† The exaggeration probably resulted from a careless translation of the word *duelos,* which the Portuguese use to refer to a very strong disagreement or argument.

‡ The inscription, dated 19 August 1919, is in a copy of the book in Moniz's library in his former home—now a museum in Avanca.

arrest because of political activity—the previous two having occurred during student demonstrations at Coimbra.*

In Portugal, a succession of weak governments followed the murder in 1921 of António Machado Santos, the founder of the republic, until a second military revolt gained control in 1926. This government appointed as finance minister António de Oliveira Salazar, economics professor at Coimbra University, who eventually became prime minister and consolidated his control over the country. The Salazar dictatorship, which remained in power for the rest of Moniz's life, was the final blow to any political aspirations that he still retained.

Moniz had not, during these years of political ferment, been idle intellectually. By 1925, he had finished a small book on clinical neurology, to celebrate the anniversary of the school of surgery in Lisbon,[4] and a biography of Padre Faria (a Goan monk who had studied hypnotism)[5]† and had edited a two-volume collection of the works of the Portuguese physician-novelist Júlio Dinis.[6] Outside of neurology, Moniz would always write about Portuguese heroes, scientists, or artists. In 1929, he published a biography of John XXI, the thirteenth-century Portuguese physician and scholar who became Pope.[7] Later, in the early 1940s, he wrote a monograph about the Portuguese sculptor Mauríco de Almeida[8] and what was supposed to be an introduction to a friend's book on the card game Boston; but his contribution—a two-hundred-page history of playing cards, with almost that many pictures of cards used throughout the ages—dwarfed the book itself.[9]

By 1926, when Moniz was fifty-two, his private practice in neurology was flourishing and, with his other sources of income, he was able to acquire in Lisbon a huge, palatial house equal in size to most embassy buildings.‡ He enjoyed entertaining in a grand style, planning the meals, selecting the wines from his well-stocked cellar, and even designing the uniforms of his servants. Many men at his age and stage of life would have been content to rest on their laurels, to practice medicine as it suited, to dabble in the arts, to travel. But not Moniz: childless, embittered by politics, and dissatisfied with his accomplishments, he turned to neurology for both vindication and fulfillment.

Not a basic scientist but a clinician, Moniz needed an important practical problem to work on and, quite naturally, sought it in Paris. His friend and former mentor Jean Sicard (1872–1929) had been experimenting with opaque substances that could be injected into patients to make organs visible under X rays and, together with his former student Jacques Forestier, had in the early 1920s introduced myelography, a technique for making the spinal cord

* One of the previous arrests was with a large group of Coimbra University students and staff members demonstrating against João Franco (1885–1929), a regent, who had illegally assumed dictatorial powers and was secretly supporting the floundering monarchy.
† Moniz had been long fascinated with hypnotism and used it therapeutically.
‡ Today it is the Nunciatura Apostolica, the residence of the Pope's representatives in Portugal.

visible. Since, after injection of an opaque substance, abnormalities in the spinal cord could often be seen on the X-ray pictures, this technique appeared to have a promising future as a diagnostic tool. In Germany[10] and the United States,[11] a similar technique was being used on peripheral blood vessels of the arms and legs.

The only techniques available at the time for seeing anything within the brain, without major surgery, were ventriculography and pneumoencephalography, which were developed by Walter E. Dandy in 1918 and 1919, respectively. Both techniques make visible the ventricles (fluid canals) within the brain and spinal cord and differ only in where and how opaque substances are injected. Usually cerebrospinal fluid is withdrawn with a hypodermic needle and syringe and replaced by air or another gas that is opaque under X rays. In ventriculography, this is done through a burr hole drilled in the skull; while in pneumoencephalography, the needle is inserted between the vertebrae of the spinal cord into the spinal canal. Both of these techniques make it possible to see whether the ventricles are distorted by something, such as a space-occupying tumor within the brain. Myelography for visualizing spinal tumors had also been tried with air injections, but Sicard and Forestier had obtained better pictures when they injected lipiodol, an oily, radio-opaque substance.[12]

Moniz made up his mind to try to develop a technique for visualizing the brain by injecting an opaque substance into the blood vessels in the neck. If the blood vessels in the brain could be made visible by X ray, they would probably be distorted, as the ventricles were, by tumors and other abnormalities within the cranium. Moreover, distortions in the blood vessels might make it possible to localize the injury with more precision than could be done with ventriculography. It might also be possible to see non-space-occupying injuries, such as vascular accidents (strokes, embolisms), which would not necessarily distort the ventricles.

Moniz knew that injecting blood vessels with opaque substances had already been done successfully, but no one had yet dared to use this technique on the brain as a diagnostic tool. There had, however, been early attempts to treat patients with general paresis (see pages 41–45) by injecting substances into the brain through arteries in the neck. In 1919, the German neurologist Von Knauer, who had injected Neosalvarsan into the carotid artery of paretics, reported that the procedure was safe.[13]

Building on this earlier work, Moniz began his experiments in 1926. He was assisted by the young surgeon Pedro Almeida Lima (1903–), a recent graduate of the University of Lisbon Medical School. Lima was from an old Portuguese family and the son of an eminent physicist who was president of the university. Lima was to be Moniz's assistant on all of the cerebral angiography work; and later, when prefrontal leucotomy was started, he also performed this surgery. (Moniz's gout, which frequently flared up, made him dependent on other people for anything requiring manual dexterity. At times even a handshake could be excruciating.) Moniz and Lima began the research on rabbits, then switched to dogs. The first task was to find a substance both sufficiently opaque to X rays and safe to inject into the blood vessels supplying the brain. Moniz tried large quantities of lithium bromide and strontium bromide intravenously and by mouth, but these failed to outline the brain sufficiently. Since the veins carry blood away from the brain, his logic in some of these initial explorations is obscure. He may have misunderstood earlier work in which the gall bladder was outlined by opaque substances, administered by mouth or intravenously, but whose success depended on the capacity of the gall bladder to concentrate the circulating substance. The brain has no such capacity.[14]

Moniz then switched to injecting substances into the cerebral arteries. He considered lipiodol, which Sicard had used successfully, but rejected it lest its oily base produce a brain embolism. By the end of 1926, after trying different substances in various concentrations, Moniz obtained the first cerebral angiogram in a dog with a solution of strontium and lithium salts. He then began to inject these substances into the arteries in the necks of cadavers. This also was not an unprecedented step: E. Haschek and O. T. Lindenthal had injected opaque substances into the blood vessels of cadavers in Vienna as early as 1895, only one year after Roentgen had first described X rays, and produced useful radiographs.[15]

Working under less than ideal conditions, Moniz injected the cadavers in the pathology department of the Institute of Anatomy and then transported the detached heads in his chauffeured limousine through the city streets to his laboratory in the Santa Marta Clinic, always fearing, as he later remarked, what would happen if a car accident revealed what he was carrying. Many of the X-ray pictures from the cadavers proved promising, even enabling him to see anatomical details, which had not previously been described, of the internal carotid artery within the human brain.

Moniz had rapidly progressed from rabbits to dogs to cadavers. In early 1927, he began injecting patients with a 70 percent solution of strontium bromide. It was difficult to inject the strontium solution into the artery, and often all or most of it leaked out into the surrounding tissues. In other cases, too much was injected, and patients developed neurological problems.* On the sixth attempt, Moniz and Lima thought that the right amount had been injected, but the patient died a few hours later. Moniz later wrote that he was

* There were several cases of Horner's syndrome, characterized by a "sinking" of the eyeball and by ptosis (drooping eyelid).

tormented by the accident; but after several weeks of additional experimentation on animals, he switched to a 25 percent solution of sodium iodide and injected it into another patient. He and Lima decided to forget the "cosmetics" and to be certain of the injection by making an incision in the neck to expose the carotid artery, rather than trying to inject through the skin. They also pinched the carotid artery below the site of the injection to prevent any of the solution from going down the artery, away from the brain. In a couple of cases, they apparently pinched the artery of a patient too long, and the occlusion deprived the brain of oxygen long enough to produce a "temporary hemiplegia" (paralysis of half the body). A few patients had epileptic convulsions (probably caused by the strong salt solution irritating the brain), and some patients experienced severe pain during the procedure.

After further modifications of the technique, Moniz's persistence was rewarded. On 28 June 1927, his other assistant, Eduardo Coelho, ran out of the darkroom, shouting with excitement and holding the first clear picture of the cerebral arteries of a patient who was suspected of having a pituitary tumor. The X ray showed that the blood vessels just above the pituitary gland had been displaced in a way that clearly indicated a growing mass in the region. Wasting no time, Moniz took the first train to Paris, after instructing Lima to obtain additional cerebral angiographs from another patient—one thought to have a temporal lobe tumor—and to send them to him immediately if anything looked interesting.

The usual account of these events—whose origin has been lost, but which was also told by Moniz in his own memoirs—has it that Moniz, happening to be in Paris in early July, heard Sicard casually remark how important it would be to have a good technique for diagnosing brain tumors, and thus decided to show his cerebral angiogram picture. This version of the story makes Moniz appear unconcerned about credit and priority. The record is clear, however, that he rushed to Paris with his first successful picture so that he could immediately announce his discovery and establish his priority.

At a meeting of the Neurological Society in Paris, Moniz described his technique and showed his only successful picture to date, obtained just ten days earlier.[16] As recorded by Moniz, when he finished reading his paper, Joseph Babinski stood up and said that if carotid injections proved to be safe, we would all be grateful for this discovery that "would enable us to localize intracranial tumors, the site of which is often difficult to determine." Sicard observed at the meeting that the ability to localize a "morbid focus" would make it possible to "intervene surgically in the right place," and added enthusiastically that they should all "congratulate Egas Moniz."[7]

Babinski invited Moniz to give a more extensive presentation the following week at a session of the prestigious French Academy of Medicine. By this

Figure 4.4. Moniz, at the age of fifty-three, lecturing on cerebral arteriography at the Faculty of Medicine meeting in Lisbon (1927). By this time, Moniz, who was as vain as he was ambitious, was concealing his baldness with a toupee. From his mid-twenties he suffered increasing pain from the gout-ridden joints in his hands and legs, but throughout the pain he never stopped working. (Courtesy Ática, S.A.R.L., Editores Livreiros, Lisbon, Portugal.)

time, Moniz had received a second picture, which Lima had sent from Lisbon by the night train. Although the patient had died before the angiogram was taken, Lima had made the injection anyway and obtained a good X-ray picture from the cadaver. The picture suggested a temporal lobe tumor, and for the first time, a diagnosis from an arterial angiogram was confirmed at autopsy.

Moniz—well aware that several other groups were planning to do similar research, as the essential techniques already existed—continued to work with amazing speed. Although the first useful picture from a patient had not been obtained until the end of June 1927, he had published 8 articles on the subject before the year ended, and 17 in 1928—a year in which he also made an extensive journey to Brazil to lecture on cerebral arteriography (see figure 4.4). By 1934, he had written 2 books and 112 articles on the subject.

Moniz earned great praise and eventually many honors for his work in cerebral angiography—recognition that, despite his gentle and unassuming demeanor, he deliberately sought. According to the Medical Nobel Archives, early in 1928, less than six months after the first useful angiogram had been obtained, the Nobel Committee received two letters nominating Moniz for the prize in medicine. Both letters, dated January 1928, were sent by colleagues of his—Azevedo Neves and Pedro Antonio Rapose—at the University of Lisbon. The nominating letters were so brief—one consisting of only two sentences—that it is difficult to avoid the conclusion that they were written to fulfill an obligation, rather than out of any conviction.* Since at that time it was widely known that one death and several serious complications had been caused by cerebral arteriography, the Nobel Committee concluded that the value of the procedure and its safety had not been adequately proven to justify serious consideration. The committee also took note that other people had made significant contributions to this development.

Not only did Moniz plan carefully to assure getting maximum credit for his accomplishments, but he also went to extremes to avoid sharing any of the credit—even though his work was clearly an extension of that of others. Moreover, his technique had several serious problems, making it appropriate for others to try to improve it. The substances Moniz initially injected were dangerous and often painful. Also, it was suspected, as editorials in the British medical journal *Lancet* argued, that cerebral arteriography at this period could detect only gross brain damage, usually evident by other means.[18] Nevertheless, Moniz was very angered by German and also Japanese investigators, who he was convinced were trying to steal the credit for his discovery. He was particularly incensed when, in 1932, W. Lohr and W. Jacobi in Magdeburg, Germany, described a technique for combining pneumoencephalography and arteriography so that the arteries and the ventricles could be seen in the same picture.[19] Although they did cite Moniz's work, Lohr and Jacobi mentioned only a 1931 paper, not his earlier publications. They also criticized his use of a 25 percent sodium iodide solution, pointing out that, even by his own admission, it sometimes produced complications, and claiming that Thorotrast, another contrast medium, produced better and safer results.

Having only recently switched to using Thorotrast himself, Moniz felt that Lohr and Jacobi were trying to get priority for the entire discovery by improving the technique. As he later described in his memoirs, Moniz wrote to Professor Max Nonne in Hamburg to intervene: "The notable master of Hamburg [Professor Nonne] took under consideration my claims and named [Georges] Schaltenbrand, another professor of renown in the specialty, to help judge the case. Both parties adduced their arguments." Moniz described the results of the investigation:

Egas Moniz has priority in fact and publication of cerebral angiography with respect

---

* It is more than passing interest in this connection that twenty years later, in 1948, Moniz wrote to Walter Freeman asking him if he would be willing to nominate him for the Nobel Prize for his work on leucotomy (see page 225).

to arteriography with Thorotrast and this was confirmed by Professor Lohr and Jacobi.

Even though the issue had been resolved, the two authors together, and later Lohr independent of Jacobi—a Jew from whom nothing more was subsequently heard—continued to omit my name and not refer to my publications.[20]

Although his memoirs were written after the Nazi atrocities had been revealed to the world, Moniz took no account of the conditions in Germany in the 1930s that might have made it impossible for the "Jew" Jacobi to do further work.*

Moniz interpreted these events only from his own perspective, colored by his need for recognition. Since he was not responsible for developing Thorotrast—others having used it first as a contrast substance for X rays—he was overreacting. When Lohr was later in Lisbon—"under the pretext of being a tourist," as Moniz put it—he refused to talk to him and sent Almeida Lima instead.[21] Ironically, it was not long before Thorotrast, a radioactive substance, was suspected of being unsafe. Within a decade, after many cases in which Thorotrast granulomas (nodules of inflamed tissue) were found in the brains of patients, its use was discontinued.

Moniz had also been angered by the Japanese investigators Makoto Saito, H. Yanagizawa, and Kazunori Kamikawa in Nagoya. They had cited his earliest publications, but they pointed out that the bromide and iodide solutions he was using caused pain, other complications, and even death. These researchers developed an oily suspension they called L'ombre. Moniz had initially rejected oily solutions, but Saito and his colleagues reported in 1930 that this substance could be used safely. They also introduced other revisions in the technique that would extend its usefulness to parts of the body besides the brain. The Japanese investigators clearly acknowledged Moniz's priority, but, having made advances in the technique, they titled their 1930 article "A New Method of Blood Vessel Visualization."[22] Since the article was published in the United States, where it would receive considerable attention, Moniz feared that the Japanese investigators were trying to present their own research in a way that might make his priority in the field less clear:

Even before Lohr and Jacobi, in 1931, the Japanese Saito, Kamikawa and Yanagizawa tried with audacity not less than that of Lohr and Jacobi, to harvest praise by taking over the method. They introduced a new material, a lipiodol emulsion, that we never succeeded to reduce to an injectable liquid despite the specified formula. . . .

---

* Ironically, Jacobi was not Jewish. He was the son of a priest and the husband of a woman who was half Jewish. Either to protect his wife and children or to advance his career, Jacobi joined the Nazi party and the S.S., just months after Hitler came to power. He became director (Führer) of the Society of German Neurologists, but after his wife was found to have falsified her documents he was suspended and censored. Jacobi committed suicide in 1938.

always the same tendency of minimizing that which went before. They attempted not only to extend the work of cerebral angiography, but to present themselves as the discoverers. . . . At this point, I began to recognize the value of my work as in different countries some experimenters without scruples, and always with the same voracity, wanted to appropriate it.[23]

Surely an egocentric perception of these events, especially from a man whose work was clearly dependent on the earlier contribution of others. No less obsessed with priority than the "father of insulin therapy" (see chapter 3), Manfred Sakel, Moniz knew better how to use his social network and was much too skilled in diplomacy to make a public accusation. He maneuvered effectively behind the scenes.

Moniz's maneuvering to secure maximum credit for his work should not, however, obscure the many contributions he eventually made to angiography. If the introduction of cerebral arteriography was not truly innovative, his talent for selecting important problems that were ready to be solved, his persistence, his keen observational powers, and his ability to persuade others to join him in exploiting and extending the basic techniques all contributed to accomplishments of the greatest significance.

Moniz and a large group of colleagues came to be called the "Lisbon School of Angiography." His own papers described the normal variations of the vascular system in the brain, the way angiograms distinguish different types of brain tumor, the diagnosis of arterial thrombosis, the possibility of visualizing different parts of the brain by injecting various arteries, the rate of cerebral blood flow, and much more. This work was reported in many articles and five book-sized monographs, in the preface of one of which Joseph Babinski wrote that Moniz had set forth on this enterprise as courageously as his compatriots of the fifteenth century, the explorers Bartholomeu Diaz and Vasco da Gama.[24] The first two of the monographs were published in Paris; the third, in Turin; the fourth, in Berlin; and the last, in Barcelona in 1941.[25] Moniz saw to it that his work was widely known.

Several of his colleagues, notably Reynaldo dos Santos and Lamas e Caldas, made important independent contributions. The basic technique was applied to diagnostic problems of the heart, the lungs, the kidneys, and the lymphatic system. By 1932, Moniz was able to obtain the first X-ray phlebography (picture of veins) showing the veins in the brain; he stated that, in the future, cerebral angiography should include pictures of the veins as well as the arteries. A "radiocarousel" was developed, which was capable of taking six serial angiograms at one-second intervals and was useful for calculating the rate of blood flow. Surprisingly, Moniz restricted all his pictures to a lateral view of the brain; but researchers in Europe, Japan, and the United States varied the views and made other significant contributions to this rapidly growing field. By the late 1940s—after the Second World War—many patients were able to walk out of the hospital a few hours after an angiogram.

Moniz was considered again, in 1933, for the Nobel Prize for his work in cerebral angiography. And once again he received only two nominations, both from colleagues on the medical faculty of the University of Lisbon.* This time one of the nominators had made a sincere effort to write a supporting letter for Moniz that could justify the prize. Their deliberations indicate that the Nobel Committee members thought more seriously about his nomination than they had in 1928. While still acknowledging that many other people besides Moniz had made important contributions to the development of the methodology, the committee recognized his original applications of the technique. The Nobel Committee found Moniz's measurements of cerebral blood flow potentially valuable, but implied that the physiology was not sophisticated and that the results needed further verification. In the end, the committee members decided that they could not consider giving Moniz the prize unless it was shared with Walter Dandy who, well over a decade earlier, had developed ventriculography and pneumoencephalography. As no one had nominated Dandy, the committee, according to its rules, could not consider him and therefore dropped Moniz from further consideration.[26]

Not only had others made important contributions to this field that predated his own work, but during the 1930s, investigators in Sweden had substantially improved both ventriculography and cerebral angiography, making it seem more unlikely that Moniz would be singled out for the Nobel Prize.[27] He had, nonetheless, achieved international recognition in neurology, but the prize he had sought still eluded him.

It was in this mood, at the age of sixty-one, that Moniz attended the Second International Congress of Neurology held in London in August 1935. He was by then a celebrated name among the neurologists and neurosurgeons who came to the congress from all over the world—and he was an impressive figure, standing, meticulously and elegantly attired, next to his exhibit, appearing as much the diplomat as the physician-scientist. One whole wall of the exhibition hall—an unusually large space for one person—had been allocated to him for exhibiting his work on cerebral angiography. Walter Freeman's more modest exhibit on ventriculography was nearby. It was at this congress that the two met for the first time.

There was in these years intense interest in the frontal lobes of the brain (which I shall discuss further in chapter 5), and an all-day symposium on the subject had been scheduled. Moniz and Freeman both attended. Most of the

* The nominations were from J. Salazar de Souza and Lopo de Carvalho.

presentations were by clinical neurologists, reporting changes in patients after damage to their frontal lobes. Participating in the symposium were some of Europe's leading neurologists, including Kurt Goldstein, Henri Claude, Delmas-Marsalet, and Clóvis Vincent. The American neurologist Richard Brickner and the neurosurgeon Wilder Penfield described patients studied for several years following extensive destruction of their frontal lobes. There was also an experimental report by Carlyle Jacobsen and John Fulton from Yale on the effect of removing a large part of the frontal lobes of chimpanzees.

A year earlier, in September 1934, Freeman had written to John Fulton requesting information about the behavioral changes seen in monkeys and chimpanzees following destruction of much of their frontal lobes. Carlyle Jacobsen replied describing the deficits observed in problem-solving ability and added that there were also personality changes resembling those described for "general paresis" patients. When Fulton invited Freeman to come to Yale to observe him doing surgery, Freeman accepted, replying: "I care much less about operating than I do witnessing the tests that you will apply to animals. I hope therefore that Dr. Jacobsen will be manifest."[28]

During the 1935 symposium, according to almost all accounts, Moniz was deeply impressed by an anecdote described by Jacobsen. Although Jacobsen's report mainly concerned the impaired learning capacity of two chimpanzees after destruction of a large part of their frontal lobes, he also described the emotional changes in one animal after the operation. This animal, an emotional female chimpanzee, had become increasingly upset during the testing prior to the surgery. She began to have temper tantrums and refused to go into the test chamber. Following the surgery, however, the chimpanzee seemed to approach the test almost cheerfully. Fulton, who was chairing this session, later reported Moniz's reaction after the presentation:

> Dr. Moniz arose and asked if frontal lobe removal prevents the development of experimental neuroses in animals and eliminates frustrational behavior, why would it not be feasible to relieve anxiety states in man by surgical means?
> At the time we were a little startled by the suggestion, for I thought that Dr. Moniz envisaged a bilateral lobectomy [removal of a large part of the frontal lobes on both sides of the brain], which though possible would be a very formidable undertaking in a human being.[29]

Afterward Moniz refused to acknowledge that this symposium had encouraged him to undertake prefrontal-lobe operations on mental patients, but insisted that he had been thinking for several years about performing such operations. Indeed, many years later when Freeman, embarking on a history of psychosurgery, asked Moniz about the importance of the London Frontal

Lobe Symposium, the latter "pleaded forgetfulness for the details, but admitted that he got to work immediately after returning home."[30]

And that is what Moniz did. With a minimum of preparation, with no animal experiments to test the safety of the procedure, he initiated the operations described at the opening of this chapter, less than three months after his return from the London Neurological Congress. I shall now turn to the theory Moniz put forward to justify his undertaking and then, in chapter 6, will discuss in detail the individual operations and their results.

5

The Emperor's New Clothes: Moniz's Theoretical Justification for Psychosurgery

It seems possible that with additional experience and a minute study of the pathologic changes seen in the brain, the knife may be the means of restoring to reason many cases now considered incurable.

—EMORY LAMPHEAR (1895)

MONIZ'S CLAIM that he had spent several years thinking about psychosurgery before the first operation was not supported by anything he ever wrote on the subject. On the contrary, his theoretical justification for performing prefrontal leucotomy was such a vague and loosely reasoned argument, it should have persuaded no one that the risk was worth taking. Prefrontal leucotomy was readily adopted, not because of the strength of Moniz's argument, but—in the absence of any alternative effective somatic treatment—because many psychiatrists and neurologists were willing to accept claims of success uncritically, especially when made by a neurologist as renowned as Moniz.

Before describing Moniz's justification for resorting to such drastic treatment as brain surgery to treat mental illness, let me first consider Moniz's general views on the causes and treatment of psychiatric disorders, his style of solving problems, and the importance for him of theoretically justifying these operations.

Although Moniz was a neurologist, he was not inexperienced with patients suffering from psychiatric disorders, and by this time—he was sixty-one—his ideas about the causes and treatment of mental illness were firmly set. By the mid-1920s, Moniz, like most neurologists, saw many patients with mental and emotional problems. In fact, probably the majority of the patients he saw in his successful private practice had psychiatric problems. As a thoroughly committed "organicist," he had little respect for psychological explanations and believed that Freudian concepts were more appropriate for literature and the arts than for the practice of medicine. Typically, he would refer to functional psychiatry as unscientific and accused it of getting lost "in the dark forest of logic and psychopathology and now and then even making incursions into the metaphysical."[1] Convinced that progress could be made only by an organic orientation, he thought that the results of psychiatric therapy were not very encouraging, "with the exception of malarial treatment in general paralysis." Moniz observed that "the diseases in the field of neurology called functional have disappeared successively. The same will happen in psychiatry and true progress in this science will only be made with an organic orientation."[2]

When he treated patients with mental disorders, Moniz used somatic treatments almost exclusively and, as I have said, used hypnotherapy occasionally to alleviate symptoms by posthypnotic suggestion, not as a tool to uncover intrapsychic conflict. Primarily, however, he used drugs, hormonal treatment, and electrotherapy—the latter, as I described in chapter 2, having gone through a period of great popularity especially as a treatment for neurasthenia. In his own variation of this therapy, Moniz treated patients with what he called an "electric shower."* He had gotten the idea for this treatment while at La Salpêtrière, where the neurologist Gilles de la Tourette (who identified and named the Tourette syndrome) had used a "vibrating" or "shaking" helmet of similar shape for treating neurasthenic patients and those suffering from pain and vertigo (see figure 5.1). Tourette's teacher, Charcot, having observed that a journey on a shaking train could alleviate the tremors of patients with Parkinson's disease, had developed a vibrating chair to treat them. Moniz modified Tourette's device, eliminating the mechanical shaking and substituting a noisy electrical hum. The metallic "funnel" or "hat" was simply positioned over, but not touching, the patient's head. Any benefit achieved with this device has to be attributed to a patient's suggestibility—a factor Morton Prince had recognized as important in electrotherapy thirty-five years earlier (see page 26). Several of Moniz's colleagues at the University of Lisbon

---

* His letterhead promoted his "Institute of Electrotherapy for the Treatment of Nervous and Mental Disorders." However, the "Institute" was mainly himself and occasionally António Fernandes and Cancela de Abreu, both of whom sometimes worked with him.

Figure 5.1. The "shaking helmet" used by Gilles de la Tourette at La Salpêtrière in Paris. A similar helmet was used by Moniz for his "electric shower" treatment. (From P. Guilly, "Gilles de la Tourette," in F. C. Rose and W. F. Bynum, eds., *Historical Aspects of the Neurosciences.* Copyright © 1982, Raven Press, New York. Used by permission.)

thought his methods ludicrous and behind his back called him the "Electrician of Alecrim [Rosemary] Street," referring to the address of his private office.

Moniz was much more empirical than theoretical in his approach to problem solving. The logic of his arguments was typically weak, if not fallacious. Indeed, as the epigraph to chapter 4 implies (see page 62), he placed little faith in theory; when he selected young people to work with him, he avoided those too theoretically inclined, preferring assistants who could substitute for his gout-ridden hands and were willing to persevere until some needed technique was made to work. His lectures were not easy to follow, not because they were profound, but because they were more intuitive than logical. Students, however, respected and generally liked him as he was gentle and paternalistic toward them, but his lectures were not popular. Moniz had, however, acquired the French style of presenting clinical cases with a literary flair that could be effective. He had the ability to pick important problems but did not attack them with great originality or creativity. His problem solving was characterized by trial and error, persistence, and a willingness to take risks, and his reputation and position made it easier for him than for most people to take these risks.

Moniz was determined to establish that he had been thinking about psychosurgery for several years. He was too experienced a diplomat not to be acutely aware of the possibility that he might be accused of having acted impulsively, or even irresponsibly, in performing brain operations in haste. In his first report on leucotomy, Moniz forestalled possible criticism by acknowledging at the outset that brain surgery on psychiatric patients might appear "excessively daring," but that "the results which have already been attained will vindicate our audacity." On almost every appropriate occasion, he made a special effort to remind his audience that he "did not perform surgical intervention in the mentally ill at random," and insisted that he was "guided by well thought out theoretical considerations."[3]

Even as late as 1948, Moniz felt compelled to speak about the origin of leucotomy, at the First International Conference on Psychosurgery in Lisbon. By this time, psychosurgery was widely accepted: witness the twenty-seven countries represented at the conference. Moniz began his lecture, "How I Came to Perform Prefrontal Leucotomy," by saying: "It was owing to no sudden inspiration that I performed the surgical operation which I called prefrontal leucotomy," and reiterated that the plan developed over a number of years while he was "engaged in solitary meditation." [4]

In addition to emphasizing his "solitary meditation," Moniz often mentioned that he had discussed his ideas about prefrontal leucotomy with Almeida Lima. In his memoirs, Moniz later wrote: "Only to Almeida Lima did I trust the secret of my aspirations and speculations and we kept the matter secret."[5] On different occasions, Moniz said that the professor of psychiatry Sobral Cid and the neurologist Alexander Cancela de Abreu also participated in some of these "secret" conversations. Although he never explained why these conversations had to be secret, Moniz implied that he did not want to give jealous colleagues an opportunity to attack him.

Although his claim that he had planned the operations for several years was generally accepted, none of the participants in these "secret conversations" wrote anything about them. Almeida Lima neither mentioned such conversations in his biographical sketches of Moniz, nor could he recall these events when I specifically asked him in June 1982. Lima was eighty years old at the time, and his memory of the events may have faded, but I find it hard to believe that a secret conversation about a plan to perform prefrontal leucotomy—a serious surgical procedure and, moreover, one that led to Portugal's only Nobel Prize—was not only never recorded but completely forgotten by the man who worked most closely with Moniz.

What was this theory that Moniz said had germinated over several years? Actually, not much of a theory at all. Typically Moniz would begin his justification of prefrontal leucotomy by reviewing the early "psychiatric disorders" that had subsequently been shown to be caused by brain pathology. As I have noted, he always used general paresis as a prototypical example, but also ran through the usual litany of known biological causes of mental disturbance, such as alcohol psychosis and senile dementia. The implication was that it would be only a matter of time before all of the so-called psychiatric disorders would be considered neurological problems to be explained by brain pathology. He never considered whether these examples of mental disorder caused by brain pathology were adequate models for schizophrenia, depression, and mania; he simply used them to demonstrate the superiority of the organic over the functional approach in psychiatry.

Moniz's argument for prefrontal leucotomy was based solely on a series of general, loosely connected, and essentially untestable statements. First, he asserted that the frontal lobes are the seat of man's "psychic activity," and that thoughts and ideas are somehow stored in the nerve-fiber connections between brain cells. Moniz then stated that all serious mental disorders are the result of "fixed" thoughts that interfere with normal mental life. Here he was apparently generalizing the concept of an *idée fixe,* which Pierre Janet considered to be the cause of hysteria, and applying it to all mental disturbances. Moniz argued that "fixed thoughts" are maintained by nerve pathways in the frontal lobes which have become pathologically "fixed" or "stabilized." Effective therapy, according to Moniz, required the destruction of these abnormally "stabilized" pathways in the frontal lobes. (See figure 5.2 for a labeled diagram of the brain.)

In designating the frontal lobes as the region of the brain responsible for intellectual activity, Moniz presented only the barest outline of the huge literature on this part of the brain. For over a century, the frontal lobes had been the main battleground of a long and hotly disputed controversy over whether thought processes and even sensory and motor functions are localized in specific regions of the brain or governed by holistic brain processes. A review of this history indicates clearly that much more was known about the frontal lobes than Moniz ever used in his theorizing.

Early proponents of the holistic view, like Pierre Flourens in France, concluded from experiments—performed between 1820 and 1850, often on frogs and pigeons rather than mammals—that one region of the brain could substitute for another, and provided examples of animals recovering lost function after damage to specific brain regions.[6] While most protagonists of the holistic viewpoint acknowledged the importance of the cerebral cortex, especially for intelligence, their experiments led them to insist that intellectual activity could not be localized in any one region of the cortex.

By 1870, however, Eduard Hitzig and his collaborator Gustav Fritsch had become convinced that motor functions are highly localized within the frontal part of the cerebral cortex. They had studied the bodily movements evoked in dogs in response to electrical stimulation of discrete regions along a strip located at the posterior part of the frontal lobes—a region now called the "motor area" of the cerebral cortex. There appeared to be precise point-to-point connections between specific brain regions in the motor area and muscles in different parts of the body. The debate began to heat up, as protagonists on each side presented conflicting evidence based on experiments using different techniques, different species, and different behavioral observations.

The advocates of the holistic view suffered a major setback as a result of a dramatic debate between Friedrich Goltz of Strassburg and David Ferrier of the West Riding Asylum in Yorkshire. The occasion was the Third International Medical Congress held in London during the first week of August 1881. It was a highly publicized meeting, and the Prince of Wales delivered the welcoming address. Over three thousand physicians, from many parts of the world, attended the congress; and among them were Charcot, William Keen (Freeman's grandfather—see pages 122–23), John Hughlings Jackson, and most of the other major international figures in neurology and neurosurgery.[7]

Goltz began his talk, "Discussion of Localization of the Vital Functions in the Cerebral Cortex," with characteristic directness and self-assurance.[8] "A fruit," he stated, "can look extremely tempting and nevertheless be wormy at the core. It is not difficult to detect the wormy core in all the hypotheses of cerebral localization."[9] He continued on the offensive by criticizing the electrical stimulation technique of Hitzig and Ferrier, asserting that it was not a technique that enabled one to draw reliable conclusions about localization. Electrical current can produce muscle twitches from many places, he said, including the spinal cord. The only method that is valid for determining localization, Goltz insisted, is the removal of areas claimed to be essential for certain capacities: such removal disproves the claim that there is a high degree of specific localization within the cerebral cortex. Goltz had brought along a dog that he claimed had had all of its so-called sensory and motor cortex removed, and he planned to demonstrate what the animal could do that afternoon in the laboratory at King's College. All those interested were invited.

Ferrier then went to the rostrum and declared that, while not disputing Goltz's observations, he disagreed with his conclusion. Ferrier pointed out that the degree of localization is not the same in all species. The amount of localization in the cerebral cortex of dogs, let alone in the brains of frogs and pigeons, is much less than in monkeys, apes, and especially humans. In lower animals, many of the sensory and motor functions are controlled by parts of the brain below the cortex. Besides, Ferrier declared, Goltz's technique for destroying the cortex—a water-jet system—was very imprecise. Although Ferrier acknowledged that in most of his earlier experiments he had used exclusively the electrical stimulation technique, he pointed out that more recently he had supplemented the earlier results with experiments involving destruction of specific brain areas. He reported that shortly before the meeting he had started using a new "hemostatic scalpel" which made it possible to produce highly localized damage to the cerebral cortex. That afternoon, he, too, would have a demonstration, and he invited the audience to see two monkeys that he had prepared for the occasion.

A crowd of eager spectators attended the afternoon demonstrations. Goltz began by showing his dog to the audience, pointing to its deformed head and inserting his fingers into holes in the skull. He explained that in five successive operations he had removed the greater part of the cerebral cortex on both sides. He then put the dog through its paces, demonstrating that it could run and jump, use its eyes to avoid objects, could hear (it cringed when a whip was snapped), and smell (it turned its head when Goltz had a physician in the audience blow cigar smoke at it).

Figure 5.2.   Brain Structures Significant in Psychosurgery.

A.   A lateral view of the human brain showing the four lobes (frontal, parietal, temporal, and occipital) of the cerebral cortex. The lobes represent convenient anatomical designations but not homogeneous functional units. The more posterior portion of the frontal lobe contains the motor cortex involved in voluntary movement and Broca's area, one of the brain regions important for speech. The region in front of Broca's area is the prefrontal area. The amygdala and the hippocampus (see B) are located below the cortical surface within the temporal lobes.

B.   A midsaggital (dividing the brain into left and right halves) view. The approximate position of the ventromedial (orbital) region of the prefrontal area can be seen. The ventromedial region and the anterior part of the cingulate cortex (also shown) eventually became the principal targets of lobotomy. The cingulum, the amygdala, and the hippocampus are major structures comprising the limbic system—the so-called emotional, or visceral, part of the brain. The limbic system is believed to function, together with the ventromedial area of the frontal lobes and parts of the thalamus and the hypothalamus, to regulate emotional responsivity.

C.   The ventricles are a canal system within the brain. This system continues into the spinal cord, carrying cerebrospinal fluid, which provides an exchange of biochemicals with the brain and cushions brain and spinal cord against injury. The lateral ventricles (one in each hemisphere) extend into the frontal lobes and were sometimes damaged during early lobotomies. The ventricles are often distorted by tumors and other causes of brain damage and can be seen under X ray if radio-opaque material has been substituted for some of the cerebrospinal fluid.

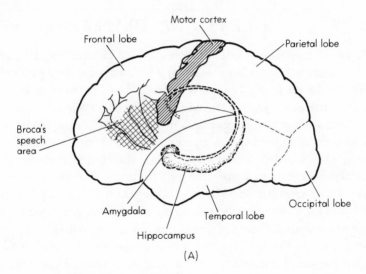

Motor cortex

Frontal lobe

Parietal lobe

Broca's
speech
area

Amygdala

Temporal lobe

Occipital lobe

Hippocampus

(A)

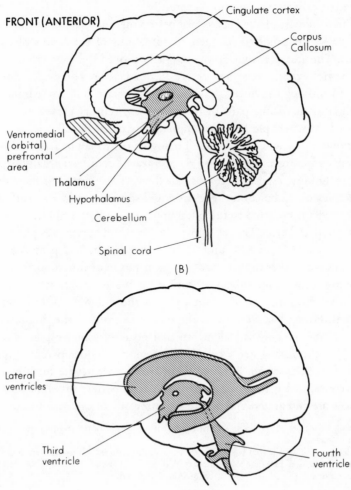

FRONT (ANTERIOR)

Cingulate cortex

Corpus
Callosum

Ventromedial
(orbital)
prefrontal
area

Thalamus

Hypothalamus

Cerebellum

Spinal cord

(B)

Lateral
ventricles

Third
ventricle

Fourth
ventricle

(C)

It all seemed convincing until Ferrier began his demonstration. He signaled for an attendant to bring a large monkey, which was obviously paralyzed on one side, to the platform. The monkey dragged one leg, and its arm hung helplessly. Charcot was reported to have uttered loudly, "Why, it's a patient."[10] Ferrier demonstrated that the impairment was specific. The animal was alert and healthy in all other respects, rapidly taking food with its good hand. A second monkey was then brought out. This animal seemed unimpaired until the attendant, on Ferrier's signal, shot off a cap pistol. The second animal did not even flinch, while the first monkey, filled with terror, tried to escape but toppled onto the floor with one foot and one arm flailing grotesquely. Ferrier explained that a precise lesion in the "motor cortex" on one side of the brain had been made in the first monkey; while in the second animal, a discrete lesion had been made bilaterally in the part of the temporal lobes that Ferrier had started calling the "auditory center."

In the third part of the debate, just before the final session of the congress, the pathologists reported on their examinations of the brains of the three animals. The area destroyed in the monkeys proved to be relatively restricted to the regions that Ferrier had described in advance. The destruction of the cortex in the dog, however, was much less precise; and most important, it seemed clear that some parts of the motor and sensory cortex had not been destroyed. Most people left the congress convinced that Ferrier and the localizationists had won the day, and many began to appreciate that such behavioral symptoms as Ferrier had demonstrated could eventually be used to locate pathology, especially tumors, in the cerebral cortex of man; and that if infections could be controlled,* it would soon be feasible to perform corrective surgery restricted to the offending area.† Charles Sherrington would later write that, before Ferrier's time, the "cerebral cortex was pictured as an uncharted sea of featureless uniformity," but that he had provided a "solid basis of proved experimental fact" for localization of functions.[12]

The arguments and evidence bearing on the localization of intellectual functions were much more complex than those about sensory and motor functions. Hitzig had found that the region in front of the "motor area"— the prefrontal area—was "silent" in that no observable responses could be evoked by stimulating nerve cells in this region and no apparent sensory loss followed its destruction. As this "silent area" must have a function, he and others considered it to be an "association area" where different sensory experiences are integrated and where planning for future actions takes place.

---

* Joseph Lister's recent work in combating infections was honored at this same International Medical Congress.

† Indeed, in the decade that followed, several operations on brain tumors were performed for the first time. Walter Freeman's grandfather William Keen wrote a remarkable description of these operations in the context of justifying animal vivisection.[11]

Hitzig inferred that the capacity for abstract thought, as distinguished from intelligence based on memory, was localized in the more anterior part of the frontal lobes: "I believe that intelligence, or more correctly the treasury of ideas, is to be sought for in all parts of the brain, but I hold that abstract thought must of necessity require particular organs and these I seek provisionally in the frontal brain."[3] There is, indeed, a long history of associating the most frontal parts of the brain with intelligence: for example, the exaggerated, bulging forehead of the "immortals" admired by the Chinese for their wisdom.

Pierre Paul Broca, whose name is identified with the frontal-lobe speech area, had reached a similar conclusion as early as 1861:

> The majesty of the human is owing to the superior faculties which do not exist or are very rudimentary in all other animals; judgment, comparison, reflection, invention, and above all the faculty of abstraction, exist in man only. The whole of these higher faculties constitute the intellect, or, properly called, understanding, and it is this part of the cerebral functions that we place in the anterior lobes of the brain.[4]

The association of intellectual activity with the frontal lobes became more generally accepted following publication of reports by Ferrier and by Leonardo Bianchi in Italy. After Ferrier destroyed this prefrontal region in monkeys and apes, he found that the animals might appear normal at first; but less superficial observation revealed clear deficits:

> I could perceive a very decided alteration in the animal's character and behavior. The animals operated on were selected on account of their intelligent character. Instead of, as before [the frontal lobe operation], being actively interested in their surroundings, and curiously prying into all things that came within their field of observation, they remained apathetic or dull, or dozed off to sleep, responding only to the sensations or impressions of the moment, or varying their listlessness with restlessness and purposeless wanderings to and fro.[15]

Later Bianchi, at the University of Naples, described the deficits he observed following frontal-lobe damage in experimental animals and patients: "Removal of the frontal lobes does not so much interfere with the perceptions taken singly, as it does disaggregate the personality, and incapacitate for serializing and synthesizing groups of representations."[16] Bianchi, like many other observers—some of whom I shall refer to later in this chapter—was convinced that animals and humans were impaired following damage to the prefrontal area of the brain, even though it was difficult to characterize the essential nature of such deficits.

During the early part of this century, a series of European studies of patients with cerebral tumors were published.[17] These reports were in general agreement that, following bilateral injury to the frontal lobes, there were often dramatic changes in personality and intellect that were not seen after injury to other brain areas. Among the personality changes commonly listed were juvenile or "puerile" behavior; lowered moral standards and tactlessness; inappropriate affect (exaggerated euphoria was called *Witzelsucht* in the German literature); restlessness; unrestrained talkativeness (loquacity); apathy; a decrease in motivation, initiative, and will (*Mangel an Antrieb* in German); and irritability. Not uncommonly, the same patient exhibited wide swings in mood from time to time and without apparent reason. Thus, bilateral frontal-lobe injury could produce exaggerated activity and lethargy (even stupor) in the same patient at different times. Similarly, a patient might change rapidly from being very irritable and short-tempered to being good-humored and even euphoric.

Many of these emotional changes after damage to the frontal lobes had been observed in the remarkable case of Phineas Gage, a patient described in 1848 by a small-town physician in the United States. Gage, a twenty-five-year-old construction foreman employed by the Rutland & Burlington Railroad in Vermont, was preparing a hole for blasting powder, when an accidental explosion drove a 3½-foot-long iron tamping rod completely through his left cheek and the frontal lobes of his brain. He was taken to a hotel in an oxcart and was attended by the local physician, John M. Harlow. To the surprise of everyone, Gage recovered, surviving for fifteen years. After he died, Harlow reported the remarkable changes in Gage's personality:

> The equilibrium of balance, so to speak, between his intellectual faculties and animal propensities, seems to have been destroyed. He is fitful, irreverent, indulging at times in the grossest profanity (which was not previously his custom), manifesting but little deference for his fellows, impatient of restraint or advice when it conflicts with his desires, at times pertinaciously obstinate, yet capricious and vacillating, devising many plans of future operation, which are no sooner arranged than they are abandoned in turn for others appearing more feasible. A child in his intellectual capacity and manifestations, he has the animal passions of a strong man. Previous to his injury, though untrained in the schools, he possessed a well-balanced mind, and was looked upon by those who knew him as a shrewd, smart business man, very energetic and persistent in executing all his plans of operation. In this regard his mind was radically changed, so decidedly that his friends and acquaintances said he was "no longer Gage."[8]

Patients with frontal-lobe damage revealed intellectual impairment as well as emotional changes. Some of the patients were too easily distracted from any task, and a poor attention span caused them to perform inadequately. Also, the behavior of these injured patients often seemed overly determined by what was present at the moment rather than by long-range goals; and they had difficulty keeping in mind more than one thing at a time. Bianchi had described this as an impairment in the capacity "to synthesize separate experiences."[19] Performance on memory tasks and on the type of "informational" questions included in standardized intelligence tests was usually not impaired, whereas the ability to abstract principles or to solve problems—especially when a solution required the integration of unfamiliar elements—was often very poor.

With the advances in neurosurgery after the First World War, many more patients survived brain operations and were, therefore, available to be studied. Monographs summarizing the changes in hundreds of brain-damaged patients, operated on to remove tumors or repair scar tissue after traumatic injury, appeared in the 1920s and 1930s, and all essentially confirmed the major conclusions of the earlier reports. In addition, there were reports of a few patients who had undergone frontal-lobe operations and were subsequently observed and tested extensively over several years. Studies of this type by the neurosurgeon Wilder Penfield, the psychiatrist Spafford Ackerley, and particularly by the neurologist Richard Brickner had great impact on Moniz, although, as I shall describe, he never seemed to appreciate how seriously the patients were impaired.

As early as 1923, Erich Feuchtwanger in Germany had published his classic study of First World War soldiers who had gunshot wounds in the brain.[20] What characterized the soldiers with frontal-lobe damage were changes in mood and drive as well as deficits in judgment and in the capacity for planning ahead, rather than any loss of memory or most of what was assessed by "intelligence tests." Several early investigators, especially Karl Kleist of Freiberg reported evidence of some specificity within the prefrontal area. They found that damage to the medial portions of the frontal lobe primarily produced emotional changes and less intellectual impairment, while destruction of the lateral aspects of the frontal-lobe areas seemed to produce the reverse pattern.[21]

There was no doubt that frontal-lobe damage could produce considerable impairment. The American neurosurgeon Percival Bailey perhaps revealed more of his social bias than he intended when he commented on this fact in his 1933 monograph:

> I hesitate before amputating a frontal lobe. This procedure is always followed by more or less great alteration in character and defects in judgment. In a washerwoman these results may be of little concern, but when the patient is a professional business man, who must make decisions affecting many people, these results may be disastrous.[22]

In his review of the literature on the frontal lobes, however, Moniz simply extracted what was useful to his argument and ignored the rest. He observed that the frontal lobes are important to higher "psychic functions," and called attention to the fact that the prefrontal region is most highly developed in primates, especially in man. Nevertheless, Moniz wrote that he did not believe in strict localization of higher mental activity within the brain. In a rambling, virtually pointless comment, he observed that all parts of the body—brain, autonomic nervous system, peripheral nerves, and glands—contribute to intellectual activity and personality:

> It is not conceivable that we can discover a center of intelligence, of memory, of personality, of conscience, and of will. Psychic activity is a function of the nervous system as a whole . . . Memory, the intelligence, the will, the affectivity, the conscience, etc. are phenomena of cerebral activity, in which the peripheral nerves, the cranial nerves, the spinal cord, the medulla oblongata, the isthmus of the encephalon, the central nuclei, the cerebellum and neurovegetative systems are involved. The whole organism, and particularly the endocrine glands, the products of exogenous toxic disintegration and the autointoxications, are very important factors which influence the psychic life.[23]

Thus, Moniz came down with one foot firmly planted in each camp. Taking a holistic position when asserting that the whole body contributes to "psychic activity," he also agreed with the "localizationists" in maintaining that the prefrontal area plays the critical role in such activity. There was no evidence from Moniz's early publications or in his selection of target sites for prefrontal leucotomy that he considered the evidence presented by Kleist and other investigators that the lateral and medial area of the frontal lobes may serve different functions—a distinction eventually recognized as crucial.

In his justification for prefrontal leucotomy, Moniz was particularly influenced by the patient described by the American neurologist Richard Brickner.* From the beginning, Moniz stressed the importance of Brickner's patient, Joe A, and cited this case more frequently than any other. Although briefly acknowledging that Joe A may have had some difficulty in "synthesis" and some loss of "social sense," Moniz clearly minimized these deficits by repeating Brickner's statement that "the psychic functions of this patient were altered more quantitatively than qualitatively."[25] Taken out of context, Brickner's statement could be very misleading: all he meant to imply was that no intellectual function was completely absent after removal of Joe A's frontal lobes. There could be no doubt that Brickner's patient was seriously impaired both qualitatively and quantitatively, as is clear from the following summary of the case.[26]

Before his surgery, Joe A had been a successful stockbroker with his own seat on the New York Stock Exchange; and his life, up to the age of thirty-

---

* Almeida Lima also stressed the important influence of Brickner's patient on Moniz.[24]

eight, had been relatively normal. Then he started to have chronic headaches, which became progressively severe; eventually, he was found to have a large and growing tumor, a meningioma, in both frontal lobes. The neurosurgeon Walter Dandy removed the tumor in two successive operations, leaving Joe A, at the age of thirty-nine, with extensive damage to the prefrontal area on both sides of his brain.

For several months after the operation, Joe A was extremely restless, but his behavior eventually stabilized, and many of his intellectual functions seemed to return to normal. His orientation in time and place and his memory seemed unimpaired, and he understood the nature of his illness and the efforts made by his family and Dr. Brickner on his behalf. When tested, Joe A demonstrated that he understood proverbs, and he retained his skill in playing checkers. At certain times, he could be charming, displaying impeccable manners; and on one notable occasion, a group of visiting neurologists detected no abnormalities even after a fairly long conversation with him.

While preserving many of his former abilities, especially well-rehearsed old habits, Joe A was totally unable to function even close to his former level. He was extremely distractible and had great difficulty focusing his attention. He spoke frequently about going back to work, but made no effort to do so and remained dependent on his family. Brickner's report made it apparent that Joe A had difficulty planning for even daily activities, let alone pursuing more distant goals.

Moreover, Joe A's emotional reactions were often inappropriate. At times, he harshly criticized people in their presence. He could become irritable and aggressive when frustrated. For no discernible reason, his mood might switch from good humor to an obvious attempt to hurt people's feelings by imitating, in a childish manner, everything they did or said. He was also likely to make childish boasts about his physical prowess. Hearing about a forthcoming swimming race, he said, "I think I'll get my bathing suit and enter the meet and win a few events, and I can do it too. I'm some swimmer—I'm one of the best swimmers around here." Or when he overheard a conversation about a few boys playing a baseball game, he blurted, "I think I'm going to take up professional ball playing. I can sock that ball better than anyone else." He frequently bragged about sexual adventures; while in reality, his interest in sex, previously normal, had almost completely ended.

Joe's performance on psychological tests indicated that he had deteriorated intellectually. His IQ score varied between 80 and 90, depending on his willingness to cooperate, but he could achieve even this low score only after considerable urging and prompting. The examining psychologist observed that he performed well only with familiar material. When he had to form new associations or make new judgments, serious defects were unmasked.

Thus, Moniz's statement that Brickner's patient functioned "appreciably better than the majority of the insane"[27] completely ignored the serious nature of Joe A's impairment. Walter Freeman later commented on Brickner's case and all the other clinical evidence presented at the London Conference:

> There is little doubt but that the audience was impressed by the seriously harmful effects of injury to the frontal lobes and came away from the symposium reinforced in their idea that here was the seat of the personality and that any damage to the frontal lobes would inevitably be followed by grave repercussions upon the whole personality.[28]

In later papers, Moniz continued to refer to Brickner's case in the same manner, but also cited other clinical cases such as the patients described by the neuropsychiatrist Spafford Ackerly and by Wilder Penfield at the London congress. Ackerly, who was on the staff at the University of Louisville, had presented the case of a thirty-year-old Hungarian-American woman most of whose prefrontal area was destroyed after she had undergone surgery to remove a tumor. Although Ackerly was impressed with what the woman could sometimes do, considering the extent of her brain damage, her intellectual life was drastically restricted. Capable of concentrating on one thing at a time as long as it was in front of her, the woman usually failed to consider anything more remote.[29] Fifteen years later, another investigator tested this same woman and found her to be still much impaired.[30]

At the London congress, Wilder Penfield described several patients who had undergone frontal-lobe surgery because of a brain tumor. One of the patients had been his own sister. Distressing as it was for Penfield to describe her case, he justified his presentation by saying that "if she were alive I am sure she would approve of such an analysis in the hope that it might help others."

> One day about fifteen months after the operation she had planned to get a simple supper for one guest [Penfield] and four members of her own family. She looked forward to it with pleasure and had the whole day for preparation. This was a thing she could have done with ease ten years before. When the appointed hour arrived she was in the kitchen, the food was all there, one or two things were on the stove, but the salad was not ready, the meat had not been started, and she was distressed and confused by her long continued effort alone. It seemed evident that she would never be able to get everything ready at once. With help the task of preparation was quickly completed and the occasion went off successfully with the patient talking and laughing in an altogether normal way.[31]

Although her insight, capacity for introspection, and ability to follow instructions appeared to be preserved, Penfield concluded that his sister's ability to plan action, among other disabilities, was definitely impaired following the surgery.

In 1975, after considering all the evidence of serious impairment following damage to the frontal lobes available to Moniz, Antonio Damasio, a neurologist formerly of Lisbon, concluded: "It took cold blood to go ahead."[32] It was not a matter of "cold blood," however: Moniz simply never came to grips with the issue. While he spoke vaguely about the frontal lobes being the anatomical location of psychic activity and the part of the brain that most distinguishes humans from animals, a moment later he would argue—in spite of evidence to the contrary—that damage to this brain area does not produce serious impairment.

As for the chimpanzee studies of Jacobsen and Fulton, Moniz rarely referred to them or did so only in a cursory way. He seems, indeed, to have made a conscious effort not to refer to the chimpanzee that stopped having "temper tantrums" following destruction of its frontal lobes, even though the description of this event had prompted him to ask whether it would "be possible to relieve anxiety states in man by surgical means." Almost everyone who has written about the origin of psychosurgery has concluded that Jacobsen's report at the London conference was, at the very least, the catalyst that impelled Moniz into action.* Moniz may well have avoided the subject because he appreciated that any reference to it would make his decision to perform prefrontal leucotomy appear to be sudden, rather than the result of many years of thought.

Aside from the issue of its influence on Moniz, the Fulton-Jacobsen experiment has often been grossly distorted. It is almost always stated that this experiment proved that following extensive damage to the frontal lobes, "experimental neuroses" cannot be produced. Even the Nobel Prize Committee's summary of Moniz's discovery of prefrontal leucotomy stated that research had demonstrated that it was not possible to demonstrate "experimental neuroses" in animals with damaged frontal lobes. Actually, Jacobsen and Fulton were studying not "experimental neuroses" but the importance of the frontal lobes for a chimpanzee's capacity to solve problems and to learn what they called "complex adaptive responses." Figure 5.3 shows one of the test problems Jacobsen used to evaluate the capacity of animals to learn. The major conclusion of the study was that, following bilateral damage to the frontal lobes, chimpanzees can no longer solve problems they could do easily before the surgery.

During his investigation, Jacobsen made the observation that has become

---

* In a recent book, Barahona Fernandes, a younger colleague of Moniz, argues that the clinical cases involving surgical removal of tumors—particularly Brickner's patient—had a much greater influence on Moniz than did the Fulton-Jacobsen study. Fernandes, however, presents no new evidence that I have not considered in the present analysis.[33]

Figure 5.3.   The delayed-response test used by Carlyle Jacobsen and John Fulton to assess the problem-solving ability of chimpanzees. After a chimp had observed the experimenter place a food treat under one of the cups, an opaque screen was lowered in front of them. Following an interval ranging from a few seconds to minutes, the screen was raised and the animal was allowed to choose one of the cups. A normal chimpanzee can usually choose the correct cup after more than a five-minute delay. Following bilateral destruction of the frontal lobes, the chimpanzees in this test were not able to do better than chance even when the delay was only five seconds. (Drawn by R. Spencer Phippen.)

the source of the distortion. One of the subjects, a female chimpanzee named Becky, was an emotional animal when she first arrived at the Yale Laboratory, before any experiments had begun. Later, during the testing, she often had "temper tantrums" when she lost out on a food treat because she failed to make the correct response. Becky was atypical in never learning to solve problems even before the surgery. As the experiment progressed, Becky stopped even trying and had to be dragged into the testing chamber, where she often rolled on the floor, while defecating and urinating. Following bilateral destruction of the frontal lobes, she, of course, still could not solve problems but stopped having "temper tantrums" and eagerly entered the test chamber. Jacobsen described Becky's changed behavior by stating that she appeared to have joined a "happiness cult."[34]

While these changes in Becky's emotional behavior made for an interesting anecdote, it did not constitute a serious study of "experimental neuroses"— a term used by Pavlov and others to refer to the systematic production of a neurosis by giving animals unsolvable problems. The major point, however, is not semantic: Becky was the only animal to behave that way after surgery.

Lucy, the other chimpanzee in the study, showed just the opposite results. Before the operation, Lucy exhibited only a minimal amount of emotion when she occasionally made an error. After the surgery, however, she had frequent temper tantrums, because she started making many errors and was often deprived of her food treats. When Lucy made a series of errors in a row, she "set up a piercing vocalization, drumming violently with hands and feet on the floor of the cage, banging on the wire and on the doors of the apparatus. The experimenter could bring about this response when she made an error, by showing her the food she had missed."[35] Thus, the conclusion that it is impossible to produce experimental neuroses in animals deprived of their frontal lobes was not at all justified by the results of this study.

Although Moniz may have been willing to risk performing prefrontal leucotomy because of his selective view of Brickner's case and the Jacobsen-Fulton chimpanzee experiment, neither the case nor the experiment could justify anticipating that such an operation would benefit psychiatric patients. In every article in which Moniz wrote anything about his rationale for prefrontal leucotomy, he referred to the research of the great Spanish neuroanatomist Santiago Ramón y Cajal, whom Moniz had met while Portuguese ambassador to Spain and greatly admired. Ramón y Cajal, who had been awarded the Nobel Prize in 1905, had described the neural pathways—the "wiring diagram"—of the brain and spinal cord in greater detail than had anyone before him. He also studied brains of young animals, observing that nerve fibers grow progressively more elaborate branches and establish more connections with other nerve cells during development. Ramón y Cajal also investigated the relative capacity of injured nerve fibers to regenerate in young and adult animals.

Moniz never described any specific aspect of Ramón y Cajal's research that he thought applicable to prefrontal leucotomy, but continually implied that it was relevant to his belief that mental illness results from abnormally stabilized thoughts maintained by "fixed" neural pathways. Contrasting normal persons with the mentally ill, Moniz wrote:

> The normal psychic life depends on the functioning of celluloconnective systems; but the groups formed by the cells, the axons and the protoplasmic processes are not fixed physiologically; they change, they become complicated or simplified under the action of extraneural and intraneural stimuli. . . .
>
> The mentally ill have delusions, there are no variations of thought . . . groups of cells take on a special fixity in these cases. We term them established groups. They constitute the path of their psychic life; nobody can divert the course of thought of these patients.[36]

In this passage, while not quoting directly from Ramón y Cajal, Moniz appears to have had in mind the former's statement: "Once development was ended, the founts of growth and regeneration of the axons and dendrites dried up irrevocably. In adult centers, the nerve paths are something *fixed,* and immutable; everything may die, nothing may be regenerated."[37] Here, the Spanish neuroanatomist was referring to his belief that, in adult mammals, injured nerve cells lose their capacity to regenerate axons and dendrites. Ironically, this is one of the few instances where later research proved him to be in error.* In any case, Ramón y Cajal was writing about the capacity of damaged neurons to regenerate physically, not about the potential for functional changes in intact neurons in the adult. He certainly knew that the capacity to learn and to change behavior patterns is not completely lost in adulthood.

There was neither a logical nor an empirical connection between Ramón y Cajal's descriptions of the physical growth of neurons and Moniz's speculation about "fixed ideas" underlying mental disorders. Ramón y Cajal's contributions were enormous, but he did not study how ideas might be stored in the brain. It was Moniz's prerogative to speculate that "fixed ideas" are maintained by abnormally stabilized connections between nerve cells, but his use of the Spanish neuroanatomist's name in this context was misleading and irrelevant. Moniz apparently cited names and ideas as a kind of intellectual "window dressing" without establishing any logical relationship to prefrontal leucotomy. While he did not refer to Pavlov in his earliest publications on leucotomy, Moniz included an increasing number of references to the Russian physiologist in later articles—again, without describing what he thought the linkage was between Pavlov's research and prefrontal leucotomy. Even in his scientific memoirs, *Confidências de um Investigador Ciêntífico,* published in 1949, Moniz described only Pavlov's most basic observations on conditioned reflexes, never making clear what the relevance was to his own work:

> The creation of new connections or synapses between neurons in a chain had never been accomplished before. This is not accomplished immediately and without difficulties. It is necessary that certain conditions occur. First, it is essential that the experimental animal be free of any source of excitement other than from the stimulus being studied. Because of this, Pavlov's laboratory was built in such a way as to be truly a "tower of silence" isolated from any noise. Second, it is necessary that the stimulus, for example, a sound, light, etc. be presented a little before the absolute (unconditioned) stimulus, the meat. Third, the animal must have its brain intact, particularly the cerebral cortex.[39]

The research of Pavlov and Ramón y Cajal had thus virtually no bearing on any of Moniz's ideas about prefrontal leucotomy. Clearly, the detailed

* Experiments have demonstrated that mammalian nerve cells do retain some capacity to regenerate even in the adult.[38]

maps of the brain circuitry that the latter described so elegantly were not needed for Moniz's conclusion that "psychic life is the result of the integrative action of nerve cells and their complex connections."[40] Nor for that matter was Pavlov's systematic analysis of conditioned reflexes required for Moniz to conclude that the brain is the organ responsible for the establishment of new associations. (Ironically, one of the Soviet Union's arguments supporting its decision to prohibit psychosurgery in 1951 was that psychosurgery was in conflict with Pavlovian theory.) Certainly, neither the research of Pavlov nor of Ramón y Cajal could justify the destruction of any particular group of nerve fibers in the prefrontal area of the brain. By referring to these two world-famous authorities, both Nobel laureates, Moniz created the illusion that there was more scientific support for his intuitive hunches than actually existed.

Indeed, most of the experimental and clinical evidence on the effects of destruction of the frontal lobes should have deterred Moniz from trying prefrontal leucotomy. His speculation that the neural substrate for "fixed ideas" is in the prefrontal area not only had virtually no support but was contradicted by the clinical evidence that memories are intact after frontal-lobe damage and by the animal studies of the American psychologist Karl Lashley, who was not able to localize memories in any specific brain region.[41] Even if it were assumed that specific thoughts are stored in discrete sets of neural fibers within the frontal lobes, there would be no way of knowing in advance which set of nerve fibers was storing any particular idea.

Although Moniz's rationale for prefrontal leucotomy was so vague as to constitute no theory at all, his explanation was repeated so often that it—like the emperor's new clothes, in Hans Christian Andersen's famous story— acquired a veneer of truth and was accepted (or at least repeated) by many other people. One person who refused to accept Moniz's theory was his University of Lisbon colleague Sobral Cid, the professor of psychiatry who, in a speech to the Medico-Psychological Society in Paris in 1937, did not mince his words: "As for the hypothesis of functional fixation by which the author [Moniz] explains the good results of his method, which cuts the conduction path of the morbid ideas, it rests on pure cerebral mythology."[42] Later, in a letter to Freeman in 1946, Moniz commented at length on Sobral Cid's remark: "My friend Sobral Cid classifies even the idea of functional fixity of certain cell-connective complexes—so evident in the reflexes described by Pavlov in the dog—as arising from a pure cerebral mythology."[43] Moniz was oblivious to the huge gap between Pavlov's observable events, such as saliva flow and leg withdrawal reflexes, and his own speculations about the neural mechanisms underlying "fixed ideas."

In general, however, the weaknesses in Moniz's theoretical justification for performing prefrontal leucotomy were not discussed (what criticisms there were I shall describe in the next chapter). Probably because of his earlier important contributions to cerebral angiography and his distinguished career as a politician and diplomat, Moniz was generally treated uncritically and described in only the most complimentary terms, with but an occasional hint of criticism. For example, during the introductory remarks at the opening session of the First International Congress of Psychosurgery in Lisbon in 1948, António Flores, a highly respected neurologist at the University of Lisbon, complimented his colleague Egas Moniz, but was compelled by honesty to say that his accomplishments were a tribute to "the power of intuition" and that his work was "based on a debatable and perhaps too neurological a theory."[44] One of the few realistic evaluations of Moniz's theory was made by two Indian physicians who hinted at their opinion, but backed away from elaborating on it: "It is difficult to subscribe entirely to Moniz' view regarding the formation of new association fibers after this surgical procedure: this view of the physiology of the frontal fibers is too naive."[45]

Moniz's admirers generally claim that his creative genius enabled him to see beyond all of the possible risks and opposing arguments.[46] Careful examination of all his publications reveals, however, that his rationale for leucotomy consisted in its entirety of a highly selective review of the effects of damage to the frontal lobes, of the belief that mental disorders are caused by "fixed ideas" dependent on abnormally "stabilized" nerve-cell connections, and of superficial and mostly irrelevant comments on the research of Ramón y Cajal and Pavlov.* There was nothing compelling about any of his arguments that should have persuaded a prudent man to attempt psychosurgery. It was not genius that enabled Moniz to see beyond the risks; rather, it was his willingness to take these risks.

* In 1944, Moniz went into more detail about his theory in his preface to a 1944 monograph of Almeida Amaral.[47] Although Moniz introduced, in his discussion, a few more of Pavlov's terms, such as "irradiation of conditioned responses," he added little of substance.

6

"Seven Recoveries, Seven Improvements, and Six Unchanged"

This [Moniz's monograph] is a detailed study
of twenty patients in which Moniz performed
partial destruction of the white matter of both
hemispheres. . . . Its importance can scarcely
be overestimated . . . Moniz counted 7 recoveries, 7 improvements and 6 unchanged.
From a reading of the detailed protocols, this
seems very conservative.
—WALTER FREEMAN (1936)

WHATEVER the theory, whatever the risk, Moniz had two prerequisites before
he could try prefrontal operations: willing hands to take the place of his own
gout-distorted ones, and patients upon whom to operate. For the one, he
needed Almeida Lima; for the other, Sobral Cid. There was no doubt that
Lima would cooperate: he was much younger than Moniz—there was almost
thirty years' difference between them—and the older man was his professor.
In their symbiotic relationship, Lima had for many years benefited from Moniz's support and influence. It was Moniz who had arranged for Lima's surgical
training under Hugh Cairns in England. On the other side, Lima was always
available to assist Moniz and was willing to be obscured by Moniz's shadow
in order to receive his patronage. Although the younger man had done all
the surgery for the cerebral angiography studies, Moniz had published over
a dozen papers on the subject before Lima's name appeared on any of the
articles. The latter, however, expected to be rewarded eventually by being
named Moniz's successor as professor of neurology—a position Lima could
never hope to get without influential backing. For Lima there was no question

about participating in the project; as he put it to me, "Moniz was my chief and he needed my hands."*

Sobral Cid, however, was in conflict. He was against the idea from the beginning;[1] and as the professor of psychiatry, he could have been firm and refused to give Moniz access to patients. The two men had been friends, however, from their days as classmates at Coimbra University; and Sobral Cid found it easier to go along than to offer opposition, as he thought the experiment had little chance of success and would soon be abandoned.

In deciding on the specific brain area to destroy and on the method to use, Moniz performed no preliminary animal experiments. As he later observed, "in the domain of mental illness, tests with animals are not possible."[2] He was also convinced, from his interpretation of the literature, that frontal-lobe operations on humans could be performed with impunity and, therefore, that animal experiments were neither informative nor a necessary precaution. After briefly considering alternatives, he decided that injecting absolute alcohol would be a safe and effective method of destroying nerve tissue. In this he was following the lead of his friend Sicard, who had been using alcohol on patients suffering from facial neuralgia;† Moniz also used the same type of hypodermic syringe to inject a small amount of alcohol—$\frac{2}{10}$ cubic centimeter—into each side of the brain.

In selecting the target, Moniz was guided mainly by where nerve fibers—apart from their cell bodies—are most concentrated in the prefrontal area. Deciding that it was safer to destroy nerve fibers, because the entire nerve cell would not necessarily be destroyed, he also believed that the abnormal thoughts of mental patients reside in the nerve-fiber pathways between cells. Since his major concern was to avoid damaging large blood vessels, he selected the centrum ovale, a frontal-lobe region traversed by many nerve fibers but few large blood vessels. This region is also a safe distance from the frontal-lobe areas important for speech and voluntary movement. Having selected the target, Moniz decided that "all indications were that both sides of the prefrontal lobes must be attacked"‡—in reference presumably to the evidence from clinical reports and animal experiments that damage to one frontal lobe seemed to produce little change in behavior.

The preparation for the operation could not have been more casual. Almeida

---

* Personal communication in Almeida Lima's home in Lisbon, 18 June 1982.

† Absolute alcohol acts as a sclerotic agent destroying nerve tissue by dehydration. Besides Sicard, who had injected alcohol into the Gasserian ganglion, other surgeons had before this time injected it into the rachidian (spinal) canal to relieve back pain. Moniz had briefly considered injecting osmic acid or using electrocoagulation—today, as it turns out, the most frequently used method (especially radio frequency waves) in psychosurgery.

‡ All quotes and references to Moniz's surgery throughout this chapter are, unless otherwise noted, from his monograph *Tentatives opératoires dans le traitement de certaines psychoses* (Paris: Masson, 1936). Translated from the French.

Lima's sister, who was working as a laboratory technician, was asked to bring up a brain from the morgue. With Lima looking over his shoulder, Moniz inserted a writing pen through the cortex several times until the two men were satisfied that they knew the approximate angle and depth that would place a hypodermic needle in the middle of the centrum ovale.* They were now ready for the first operation.

The patient, a sixty-three-year-old woman, was transferred from the Manicome Bombarda Asylum to the Neurology Service of the Santa Marta Hospital on 11 November 1935. The operation was scheduled for the next day. The woman had been diagnosed as an involutional melancholic with anxiety and well-established paranoid ideas.† She had been in the Bombarda Asylum about three years, but had also been hospitalized there twenty years earlier for one and a half years. Between her first and second admissions to the asylum, she had had a long history of psychopathology. She believed she was being persecuted by her neighbors and the police and accused her pharmacist and physician of trying to poison her. She also had auditory hallucinations, strange bodily sensations, and episodes of severe anxiety, crying, restlessness, and insomnia. Before her second hospitalization, she had secretly practiced prostitution in her apartment until the other female boarders forced her to leave.

The patient's hair was cut the night before the surgery. The next morning, her scalp was cleaned with alcohol; Novocaine, for local anesthesia, and adrenaline, to reduce bleeding, were infused. The holes were drilled (trepanned), the syringe was inserted, and then alcohol was injected into both sides of the brain. The operation, which took only thirty minutes, was performed by Almeida Lima with the assistance of Ruy de Lacerda.

Of what he was thinking at the time, Moniz wrote little except to remark that, in his anticipation of the importance of success, he could not worry about failure:

> On the eve of my first attempt, my justified anxiety and all fears at the moment, were swept aside by the hope of obtaining favourable results.
>
> If we could suppress certain psychological complexes by destroying the cell-connecting groups . . . this would be a great step forward—making a fundamental contribution to our knowledge of the organic basis of psychic functions.[3]

He was not totally unconcerned, however, about possible failure, lest it be

---

* I was told this story independently by several of Moniz's younger colleagues during my visit to the Centro de Estudos Egas Moniz in Lisbon, in June 1982.

† Involutional melancholia is prolonged depression first occurring in the middle or late years. The symptoms include worry, agitation, anxiety, paranoia and other delusions and often somatic complaints.

used against him by professional rivals in Lisbon (and he made an effort to keep the operation a secret). Besides Lima, Ruy de Lacerda, and Moniz, only the chief nurse was present at this first operation.

About five hours after the surgery, the patient was questioned briefly. She was able to respond to some of the simple questions Moniz put to her:

> "Where is your house?"
> "Calcada of Desterio."
> "How many fingers?"
> "Five," she responded with slight hesitation.
> "How old are you?"
> Long hesitation. She was not precise.
> "What hospital is this?"
> She did not respond.
> "Do you prefer milk or bouillon?"
> "I prefer milk."

The patient had a low fever for a few days, but as soon as it subsided, she was transferred back to the asylum. Moniz reported that she still cried, "but not with the previous intensity." There were no comments from Professor Sobral Cid about the patient's condition; but on 8 January 1936—about two months after the operation—Barahona Fernandes, a young psychiatrist, gave his evaluation, later recorded in Moniz's monograph:

> The patient behaved normally. She is very calm, anxiety is not apparent. Mimicry still a little exaggerated. Good orientation. Conscience, intelligence and behavior intact. Mood slightly sad, but somewhat justified because of her concern about her future. Fair appreciation of her previous pathological state; appreciation of her situation is appropriate.
>
> There are no new pathological ideas or other symptoms and for the most part previous paranoid ideas are primarily gone. That is to say, after the treatment the patient's anxiety and restlessness had declined rapidly with a concomitant marked attenuation of paranoid features.

The report closed with Moniz's final evaluation of the case—a "clinical cure."

The entire evidence presented for the "clinical cure" consisted of one psychiatrist's subjective impression of the woman's mental state. There was no description of any behavior—for example, how she occupied her time—that could justify the conclusion that she was cured. Nor was there any indication that Sobral Cid examined the patient, even though he surely must have been interested in the results of this bold venture. In view of some of Sobral Cid's later critical comments about prefrontal leucotomy, it seems likely that he decided to maintain all possible distance from the project. Moniz, however, conveyed the impression that his colleague had collaborated on the diagnosis and was in full agreement with the conclusion:

We operated on 20 patients, the majority of them considered chronic, from the Manicome Bombarda Asylum, thanks to the great kindness of our friend, Dr. Sobral Cid, renowned professor of psychiatry at the Medical School of Lisbon, who assisted us in the selection of cases and provided us with excellent clinical observations from the asylum.[4]

These "clinical observations" were, however, part of the patient's medical history Sobral Cid made before consideration of leucotomy. Almost all of the postoperation psychiatric evaluations were made by Barahona Fernandes.

Years later, in 1946, in response to an inquiry from Walter Freeman, Moniz wrote that Sobral Cid had selected the first four patients—two women and two men—but refused thereafter to cooperate:

> From this time on in order to obtain a case it was necessary for me to go to the [Manicome Bombarda] Asylum 9 or 10 times to see Sobral Cid and persuade him to send me more mental cases. He gave the excuse that he wanted to send only patients with complete histories; but the observations were not speeded up and I wore out my patience in these peregrinations.*

However strong Sobral Cid's reservations about the operation, Moniz was not to be thwarted: "I did not desist, having also taken advantage of cases from my own clinic and from other asylums, when I recognized that the method could be advised as harmless and capable of benefiting the insane."[6]

The evidence that the first four operations were beneficial was marginal at best. The patients seemed to be able to tolerate the alcohol injections reasonably well, although there were side effects. While the first patient's psychiatric symptoms may have been substantially reduced, the success with the second patient was described as "only partial." The third and the fourth patients were clearly schizophrenic and had expansive and confabulatory ideas which continued essentially unabated after the surgery. The fourth patient, for example, alternated between asserting, after the operation, that he was the president of Portugal or the Marquê de Pompal. According to Barahona Fernandes, neither the third nor the fourth patient showed significant or lasting improvement.

Moniz persisted, getting patients where he could. In the end, he was even able to get fourteen patients from the Manicome Bombarda Asylum and six from other sources. The patients seem to have been selected primarily on the basis of availability. Cases 5 and 6 were also schizophrenic. Case 5 was a very agitated man, whose symptoms were said to have "ameliorated" after the

---

* This information and the quotations are from a letter, dated 9 July 1946, from Moniz to Walter Freeman. A copy of the letter was sent by Freeman to John Fulton and is in the latter's correspondence file in the Yale University Library. After being suppressed for a number of years, the letter was recently published in Portugal.[5]

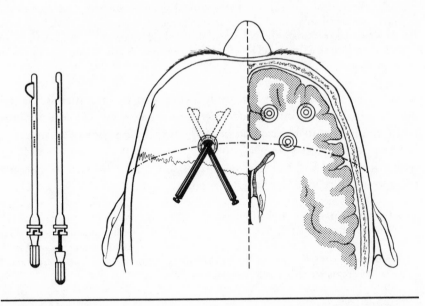

Figure 6.1.   The "core operation." Prefrontal leucotomy as performed by Egas Moniz and Almeida Lima. The leucotome was inserted into the brain at the approximate angles shown. When the leucotome was in place, the wire was extended (*see detail of leucotome*) and the handle rotated. In the first leucotomy, one "core" was cut on each side of the brain. As described in the text, in subsequent operations progressively more "cores" were cut. The right side of the figure depicts a horizontal slice of the brain (parallel to the top of the skull) with Moniz's estimate (without evidence) of the extent of damage. (Drawn by R. Spencer Phippen.)

operation, and whose condition was described as "enduring calmness"—a state that may have been more helpful to the staff than to the patient. Fernandes concluded that in case 6 there was a decrease in delusional "resonance"—a term he commonly used to indicate the intensity of a patient's preoccupation with delusional ideas. Thus, the benefits from the first six operations were far from spectacular, and Moniz found it necessary to increase progressively the number of alcohol injections.

The next few operations were done on patients with "affective" (emotional) disorders; and it soon became apparent that the best results were achieved in such patients rather than in those with thought disorder characteristic of schizophrenics. Case 7 was a forty-six-year-old woman who had severe anxiety attacks and hypochondriacal preoccupations and had been in and out of the asylum during the previous nine years. On 20 December, she was transferred from the asylum to the Santa Marta Hospital. The operation was performed the same day. Two injections of alcohol were made on the right side; but when her pupillary reactions became weak, Lima injected no more on that side but switched to the left side and made four injections there. On 26 December, the patient was returned to the Manicome Asylum. Moniz wrote in his monograph: "The patient is cured of her anxiety and sad ideas. Why does she want to return to the Asylum? It is probably because she has a comfortable place at Manicome and doesn't want to work." On 8 January, only three weeks after the surgery, Fernandes made the last entry described in the monograph: he concluded that the anxiety state was gone, but the hypochondriacal condition was unchanged.

The eighth patient was the first case where the leucotome was used. Moniz never said why he switched from alcohol injections to the leucotome, but the obvious assumption is that he was not completely satisfied with the earlier results—a conclusion he may have arrived at weeks, if not months, before, as the leucotome had to be ordered from a medical instrument company in Paris.*

In any case, when the instrument arrived, Moniz did not try it first on animals to determine the risk of rotating a wire loop within the brain. The leucotome was first used on a patient on 27 December. Only two "cores" were made on each side. Later in the series of operations, four "cores" were made on each side. Shortly after completing the first series of twenty patients, Moniz started making six "cores" on a side, using a modified steel loop that cut, instead of crushing nerve fibers.[8] (See figure 6.1.)

By the twentieth operation, it was evident that the patients who improved most were those suffering from anxiety and depression. Chronic schizophrenic patients with serious thought disorders did not benefit much, if at all. In summarizing his results, Moniz claimed that seven patients, or 35 percent, were cured, another seven patients significantly improved, and six patients were unchanged. All seven of the patients who were considered cured had affective disorders, including "anxiety depression" (melancholia), "anguished-depressed," "involutional depression with anxiety," and "manic-depressive." Five of the seven "cures" had leucotomies, while two had alcohol injections.

On the average, there were only four days between successive operations—an interval much too short to detect postoperative complications, especially as the surgical procedure was changed frequently. Whereas there had been only a single alcohol injection on each side of the brain in the first operation, the number was progressively increased to six on a side. Of the ten leucotomized patients, one received a single cut on each side of the brain, three

---

* Instruments of this type had been used in several countries since the turn of the century. The American neuroanatomist Andrew "Ted" Rasmussen mentions several early physician-investigators using trocars with concealed stilettes that exposed cutting "wings" if a plunger was depressed after the instrument was positioned into the target site. The purpose was to minimize inadvertent damage on the way to the intended site of the lesion.[7]

received two cuts, and six patients were subjected to four cuts on each side. Moreover, three patients were given a leucotomy and subsequently an alcohol injection operation.

The evidence that seven patients were cured was not at all convincing as the description of the patients following surgery was totally inadequate. Their anxieties or delusions were simply said to be gone, but the monograph presented no examples of any behavior indicating how well they were functioning. The most glaring deficiency was the brief postoperative observation period. There were no observations of any of the seven "cured" patients that extended beyond two months after the surgery. Moreover, as several patients had a second operation within the two-month period, they were observed postoperatively for an even shorter time. *In four of the seven "cured" patients, the last observation Moniz recorded in the monograph was less than eleven days after the operation!*

Not only was it too soon after surgery for a valid judgment about any of the patients being cured, but Moniz presented no evidence indicating that any of the patients were capable of living normally outside the hospital. In the majority of cases, the major change was reduction of agitation and increased "calmness." Yet neurologists and psychiatrists commonly described the evidence in Moniz's monograph as "detailed" and "impressive." Apparently they were either repeating what others were saying or accepting Moniz's summary statement without examining the evidence.

Case 9 well illustrates the weakness of Moniz's evidence. The patient was a fifty-nine-year-old widow, whose life had been so filled with adversity that it would have been remarkable had she not been depressed. Following eight pregnancies—the first three stillborn—her husband died, leaving her the sole support of her five children. One daughter became depressed and needed special care from her mother. At the age of thirty-nine, the patient also became depressed and could not continue working. According to the medical records, she stayed in bed for almost six months. Afterward, her son became terminally ill and had to be hospitalized. She felt guilty because she could not afford to bring anything to the hospital when she visited. After her son died, she suffered enormously, crying and moaning for eight days. On 2 November, she went to Mass and said to her daughters afterward that the saints had made faces at her. She was so distressed that her daughters kept her in bed, but she continued to be anxious and agitated, worrying that her son had not made a proper confession before dying.

Soon after her son's death, she began telling one of her daughters that she also had died and that her son had been taken from her by the devil. The patient was hospitalized in the Manicome Bombarda Asylum and, on 27 December—less than two months after admission—was transferred to the Santa

Marta Hospital for a leucotomy. It is not clear how she was selected as a candidate for surgery, but Sobral Cid must have been bypassed as, by this time, he had refused to cooperate with the project. The patient arrived at Santa Marta in an agitated state, either refusing entirely to respond to questions or responding in an evasive and short-tempered way.

The leucotomy was scheduled for the next day, 28 December. The woman was so agitated that she pulled out the intravenous needle that was to be used for anesthesia; the operation was continued with a local anesthetic. Moniz reported that when the leucotome handle was turned, the woman groaned—a response he gratuitously interpreted as evidence that the centrum ovale was able to convey conscious experience. After the operation, the woman was tranquil and spoke slowly; but by the next day, 29 December, she was again agitated and tried to tear the bandages off her head. In the afternoon, she calmed down and talked with ease, but later developed a fever and did not even know she was in a hospital. She was confused, irrational, and disoriented, and thought she was in a "place where they kill people"; surprisingly, her anxiety was said to have disappeared.

On 30 December, her temperature was still elevated. She was excited, sang occasionally, and talked incessantly. She used any word or phrase that someone uttered as an excuse to talk, often using obscene language. Occasionally she was completely incoherent, shouting, "They have killed me twice, but I'm still here." She continued to deny she was in the hospital. The next day she was in much the same state of "maniacal agitation," talking day and night, and had to be given opium and a sedative, but the medication did not calm her.

On 31 December, Moniz considered reoperating, partly because the surgery on the right side had apparently not been complete. The wire loop on the leucotome had broken, and the centrum ovale on that side was judged to have been damaged too little, because Almeida Lima felt no resistance when rotating the instrument. By 2 January, however, the patient was less agitated, spoke less, and had stopped singing. Moniz decided to hold off on a second operation and wrote in the monograph that the patient was now responding to questions with a certain alertness and was more calm.

The last observations of the patient recorded in the monograph were on 4 January 1936:

> The patient is well. She recounts the events of her life normally, but prefers to talk about more recent events such as entering Manicome, etc. One time, she was emotional while speaking about her son, but in an appropriate manner.
>
> No sad ideas, nor persistence of those subjects which dominated her psychic life before the operation.

Results—Cure. The patient was sent back to Professor Sobral Cid, who wanted to examine her in front of her family.

The patient spoke a lot, but the family said that she always had been somewhat talkative.

The inadequacy of this report is obvious, its last observation having been recorded only seven days after the surgery. Moniz provided no information about the quality of this woman's life after her discharge from the hospital. Moreover, she had been hospitalized an extremely short time—less than two months—barely long enough to justify considering her condition to be un-treatable by less radical methods. This is one of the rare cases where Moniz was able to refer to Sobral Cid's opinion about a patient's mental status after the surgery; but even here the psychiatrist only stated that he agreed the patient should be sent home. He may well have been annoyed that such a patient should have been subjected to a brain operation. Later, when he launched a highly critical attack—especially by the contemporary standards of behavior among upper-class Portuguese—this could well have been one of the patients he had in mind when he asked "whether one had the right to inflict so considerable a cerebral mutilation on the patient in order to free him from a psychotic syndrome, which is by its nature, curable and which could have been cured spontaneously in a few months."[9]

Case 14 in Moniz's series is even more difficult to evaluate. The patient was a sixty-year-old street vendor, who had been admitted to the Manicome Asylum in November 1933, after a long history of psychopathology. At the age of forty-one, he had had a depressive-hypochondriacal episode, believing that he had tuberculosis. He stopped working and did little for several months. He had other depressive episodes, often triggered by sexual delusions. When his wife mentioned that a twelve-year-old girl living with them had stopped menstruating, he took this to be an accusation that he had had sexual relations with her. Afterward he spoke little and spent hours with his head in his hands. Later, at the age of fifty-seven, he confessed to his wife that he had "deflow-ered" the young girl who worked for them. The girl denied it, and eventually she was determined to still be a virgin, but he maintained that he had violated her and talked incessantly about wanting to die.

Two months before he was admitted to the asylum, he had set fire to his house and had talked to his family afterward about "collective suicide." At Manicome, he was given opium, which seemed to reduce his anxiety, but he continued to be delirious and to have depressive ideas. He remained much the same for the next two years in the asylum. Periods of calm alternated with periods of agitated crisis. He had a mild outburst of anxiety at the end of December 1935, but was reasonably well oriented when he was transferred to Santa Marta on 7 January 1936. The leucotomy was done the next day.

The day after the operation, the patient was calm and said that he no longer felt that he was "condemned to death." Moniz reported that the patient's two daughters visited the day after the surgery and were surprised by the great improvement. The next day, however, they found him less well and very drowsy. During the first few days after surgery, the patient continued to express hypochondriacal ideas, complaining, for example, that everything he ate made him sick. On 13 January, five days after the leucotomy, he was calm but still had melancholy ideas. Four days later, his daughters visited and were pleased when their father wished to go home with them. Moniz, however, thought that the patient was still apathetic, and knew that he was still having emotional outbursts. Moniz decided that a second operation was called for because he suspected that the centrum ovale had not been sufficiently destroyed. On 24 January, an "alcohol operation" was performed, sixteen days after the leucotomy. The patient was exhausted for the first three days after the second operation. On 5 February, Moniz wrote briefly that the patient was considerably improved during the previous six days, and observed that even though he did not show great enthusiasm about going home, he definitely did not want to go back to the asylum.

Two weeks after the second operation, on 7 February, the patient was discharged from the hospital. When the daughters brought him back for a visit on 15 February they told Moniz that they were very satisfied to have their father home and that they thought he had reverted to the way he was before his illness. They believed he was cured. Moniz, however, wrote at this time that the patient lacked initiative; and more "intensive examination" revealed that the patient was not aware of the value of money, could not give his age, and was confused about time. Moniz wrote in his monograph that the patient "seems indifferent to his situation at home," but concluded the report: "The patient can live at home and has good relations with the family. The patient is 64 years old and has arteriosclerosis; it is an element which must affect his psychic state; but he is cured of his anxiety and his melancholy ideas."

This patient, like the fifty-nine-year-old woman of case 9, was also counted as one of the seven cures. It is difficult to interpret this case because little was expected from this man other than being able to live with his family in peace. It may be true that his anxiety and hypochondriacal ideas were significantly reduced. There was no way of determining from the record whether the changes lasted, as the last report was only twenty-two days after the second operation. Although the daughters were happy to have their father live at home, there was no evidence that he could work or that he showed interest in anything.

The evidence of "cures" in Moniz's monograph could not have convinced a critical reader. The description of the patients' illnesses prior to surgery were only brief summaries abstracted from their medical histories. Postoperative behavior was described equally briefly. Typically, the patients were returned to the Manicome Asylum less than two weeks after surgery, and Moniz was dependent primarily on other physicians—usually Barahona Fernandes—for information about their condition. In a letter to Walter Freeman, for example, Moniz made the astounding comment that he really did not know how two of his earliest patients had fared as they were "returned to the Asylum where I did not follow them."[10]

The many favorable reviews of Moniz's monograph mainly illustrate how receptive people were to prefrontal leucotomy, rather than how convincing the arguments and evidence were. The monograph was published in June 1936; but before the end of July, Freeman had written a review of it for the *Archives of Neurology and Psychiatry.*[11] The epigraph for this chapter, quoted from that review, refers to the "detailed study of twenty patients" and is not sparing of hyperbole. Freeman sent a prepublication copy of his review to John Fulton at Yale on 6 August advising him to get a copy for his library as Freeman considered it "an epoch making work."[12] Fulton, who had received two copies of the monograph directly from Moniz, replied on 12 September:

My dear Freeman,

Many thanks for your thoughtfulness in sending me a copy of your excellent review of Moniz' book. I acquired a copy early in July, and I quite agree with you that he has taken a full and important step forward in the treatment of nervous disease. I have always been struck by the fact that in chimpanzees, after bilateral removal of the frontal association areas, panic and tantrum, as well as experimental neuroses, disappear forever. I have no doubt that Moniz produced a similar effect with his alcohol injection in human beings.

With best wishes and many thanks,

Yours very sincerely,
J. F. Fulton, M.D.[13]

Apparently Fulton had just glanced at the monograph as he was aware only of the initial alcohol injections and not of the leucotomy procedure (see also chapter 8, page 148). Nevertheless, he did not hesitate to speak about an "important step forward in the treatment of nervous disease," while reminding Freeman of the major role of his, Fulton's, chimpanzee experiments in these developments.

Moniz had presented the results of the first twenty operations in Paris, on 3 March 1936, less than four months after the first operation. This report, published several weeks afterward in the *Bulletin de l'Académie de Médecine* of Paris, summarized all the results published in the monograph in June.[14] (A footnote in the *Bulletin* article indicated that the monograph was already in press.) Before the end of 1936, Moniz had published essentially the same data in five more articles, in one of which he coined the term *psychosurgery*.[15] Early in 1937, additional articles of his appeared in the *American Journal of Psychiatry*, the *Medical Bulletin of Trieste*, the *Giornale di Psiquitria e di Neuropatologica* of Italy, and the *Archives Franco-Belges de Chirurgie*.[16] Moniz also sent a manuscript to Professor Reichert in Freiburg asking whether he would translate it into German. Reichert complied, and the article was published during the early part of 1937 in *Der Nervenarzt*.[17] During the fall of 1936, Moniz increased his series of leucotomized patients; and in an article in the January 1937 issue of the *American Journal of Psychiatry*, he reported the results of eighteen additional operations and several more modifications in the operation, including changing the location of the cuts and increasing their number to six on each side. He also changed the construction of the leucotome.

Thus, in a relatively brief period, Moniz had published articles on leucotomy in six different countries. Seven articles in addition to the monograph were published in 1936, and six more and a book appeared in 1937. Clearly, there could be no competing claims for the discovery. As Walter Freeman concluded, "Moniz was taking no chance of further piracy."[18]

One of the few objective summaries of the evidence Moniz presented in his monograph was written by Stanley Cobb of Boston, but it was not published until 1940. Writing about Moniz's patients, Cobb stated that "the reports are so meager that one cannot judge the work. . . . Of 20 cases, he produced cure in 7, amelioration in 7 and no effect in 6. Only 1 of the 20 cases, however, is given in enough detail to allow the reader to judge for himself as to the diagnosis and result."[19] By this time, however, the operation had been changed many times, and Moniz's monograph was only of historical interest.

Never in his monograph did Moniz seriously consider the possibility that any of the patients were worse after the operation. The six patients who did not improve were said simply to have been "unchanged" or "the same as before the operation." Although he conscientiously listed all the complications of the surgery—devoting the last ten pages of the monograph to this topic— these were said to be transient. These complications were discussed under the three headings "General," "Neurological," and "Psychological." The general and the neurological postoperative symptoms were vomiting, urinary and fecal incontinence (severe in two men and one woman), diarrhea, and eye troubles, such as abnormal nystagmus (involuntary oscillations of the eyeball) and ptosis (drooping eyelids).

The psychological symptoms produced by the surgery were apathy, akinesia (retarded movement or inertia), loss of initiative (present in one half of Moniz's patients; and in a few, prolonged), catatonic "attitude," mutism (rapport was difficult to establish, and some patients would only nod yes or no even after much prodding; mutism was observed in eleven of the twenty cases), negativism, disturbed orientation in space and time, puerility (one patient carried around and rocked dolls she had made), kleptomania, and abnormal hunger sensations (experienced by one-quarter of the patients). In all, a serious list of psychological problems, except that Moniz insisted that these symptoms usually lasted no more than several weeks. With absolutely no evidence, he asserted unequivocally that the operation did not impair intelligence or memory.

While Moniz stated in his monograph that the apathy was transient, Sobral Cid, in his critical comments made in Paris in 1937, insisted that the apathy might last much longer and could be permanent.[20] Moniz was not present on this occasion to respond to Sobral Cid, but later seemed to confirm this criticism when in the same year he reported modifying his initial surgical procedure, as "two or three of the patients in my first series have remained somewhat apathetic."[21]

While never clashing directly with his colleague, Moniz, years later, responded to Freeman's inquiry about opposition to leucotomy within Portugal by referring mainly to Sobral Cid:

> After all, this was an old friend with whom I did not wish to have discussions which might give rise to anger. . . . We stopped talking about the subject, each remaining in his position, and continued good friends through life. This, however, was not long, for my unfortunate friend and opposer of leucotomy. Death took him in 1941.

In the same long letter, he speculated about why Sobral Cid had attacked leucotomy so vehemently:

> An almost unbridgeable gulf divides us!
> We in the field of anatomical realities, they [the "functional" psychiatrists] climbing without realizing it, to the inaccessible regions of a kind of metaphysics far from the reality of medicine.
> Sobral Cid was given to an excessive psychiatric phraseology to which clings, in that language of special terms, at times interesting, with literary reflections, an atmosphere of scholastic erudition, but far from medical reality, forgetting completely that psychiatry is a branch of that science which we profess.

Although they had been classmates at Coimbra University and good friends, Moniz wondered whether Sobral Cid might have been jealous of Moniz's fame for his work on cerebral angiography:

Sobral Cid watched, at least with interest, the development of these scientific events, speaking to me of the subject and saying pleasant things.

Nevertheless, there never was, I must confess, a great atmosphere of enthusiasm among the majority of my colleagues, and especially among the professors. In these circumstances a harmful wind of rivalry always blows. Did it also touch Sobral Cid?

Moniz concluded that his own intrusion into the field of psychiatry must have been the main cause of Sobral Cid's opposition:

One day, however, leaving Neurology for the moment, I boldly penetrated in the sphere of Psychiatry. . . . I had the idea, which was more properly that of a neurologist than that of a psychiatrist, of attempting to cure mental cases by surgical treatment. As the operation achieved some results, the Professor of Psychiatry of my faculty could not remain complacent.

Moniz added, unconvincingly, that he "did not blame [Sobral Cid] in the least."

To invade his psychiatric domain and bring to this science some useful contribution, no matter how small, was not calculated to leave the Professor of Psychiatry stoically impassive. The reaction was, even in his subconscious, that of taking a position of moderate hostility, at first hesitant, then made concrete in precise phrases, and finally in an obstructionist attitude.[22]

In 1941, Moniz wrote a brief obituary for Sobral Cid, emphasizing that there had never been "a shadow of rivalry or the whisper of disrespect to disturb our mutual esteem."[23] Walter Freeman later wrote that Moniz revealed his magnanimous personality in writing an obituary for Sobral Cid that included "a generous appreciation of the qualities, intellectual and personal, that had bound these men together over a period of 40 years."[24] A similar opinion was expressed by others.

It takes considerable imagination, however, to detect magnanimity in what Moniz wrote in his obituary of Sobral Cid. One might more readily conclude that with a friend like Moniz, Sobral Cid scarcely needed an enemy. Assuming a patronizing attitude toward his "colleague and long-time friend," Moniz damned him with faint praise, saying that

he was such a perfectionist that his unpublished work is immensely superior to what he published. Many times I asked him for his notes to make them known. . . . At times, he ascended in his psychological conceptions to such heights that I could not follow him. He lived in another world, but we did not collide in the ambit of sterile discussions.[25]

Moniz never, in any publication, answered Sobral Cid's criticism of psychosurgery, choosing no doubt to let the controversy die from neglect. Only in the 1946 letter to Freeman did Moniz discuss the matter. Almeida Lima also ignored Sobral Cid's criticism but later, like Moniz, answered Freeman's inquiry in a letter:

> I am sure that at first he complied with the request of Egas Moniz because he did not believe that a small lesion in the white matter of the frontal lobe would have any influence on psychical manifestations. The first patients were sent without difficulties, but everything changed when it became evident that there was an alteration in the psychiatric picture. Sobral Cid began then to make all kinds of difficulties about sending patients from his hospital. . . .
>
> One of the facts which impressed me most strongly at the time was that Sobral Cid more than once changed the preoperative diagnosis owing to the surgical results. I am convinced that this modification of diagnosis was made with complete sincerity and derived from the conviction that schizophrenia was an illness of the mind and that it was unalterable by means of an operation on the brain. Therefore, if the patients improved or at least showed a noticeable modification of the psychiatrical picture, Sobral Cid preferred to think that he had been wrong in his diagnosis [rather] than to having to confess the influence of the operation on the schizophrenics.[26]

Lima's letter, however, is hardly an adequate answer to Sobral Cid's criticism. Even among those who still practice psychosurgery today, the overwhelming majority do not think that the operation helps schizophrenics—an opinion consistent with Moniz's summary in his monograph. Sobral Cid was thus correct in stating that the patients who appeared to benefit from the surgery were not the schizophrenics but the manic-depressives and those agitated patients who normally have a much better prognosis for recovery.

Sobral Cid was not the only physician in Lisbon who had reservations about the evidence in Moniz's monograph. Diogo Furtado, a Lisbon neurologist, who had presented the paper (written with Moniz) that provided the occasion in Paris for Sobral Cid's attack, later changed his mind about the effectiveness of prefrontal leucotomy. At the First International Congress of Psychosurgery in Lisbon in 1948, he sharply criticized leucotomy—after he had first briefly dispensed with the necessary tribute to Moniz:

> We cannot doubt that the operation imagined by the genius of Egas Moniz resolves certain states of anxiety tension, causes the disappearance of obsessive ideas, diminishes the intensity of delusional activity and of psycho-sensorial error [hallucinations]. *The problem which is more than ever in the forefront, is to know what is the price paid for the percentage of improvement obtained, that is what are the final changes of personality suffered through frontal mutilation.*[27] [Italics added]

Furtado went on to observe that the best evidence of the "price paid" came from intensive studies of a small number of patients such as had been undertaken by the Swedish psychiatrist Gosta Rylander.* By contrasting thorough, long-term observations with Moniz's shallow descriptions of patients, Furtado was implicitly criticizing him. After presenting a detailed description of six recently leucotomized patients, Furtado drew conclusions sharply different from those reported by Moniz:

> The changes of personality were generally serious. In five of the patients could be noticed an affective indifference, a loss of ethical and moral inhibitions, a loss of higher interests, both affective and intellectual, which causes the personalities of lobotomized patients to become much inferior to what they were before lobotomy.... The therapeutic results obtained in our series are very slight, and the operation, by making the patients easier to manage, made them also, in a most definite way, very inferior as personalities, to what they were before.... We conclude, therefore, that lobotomy, as a method of treatment, should only be authorized in special cases, when all other therapeutic means have been exhausted and when there is no longer any hope of a spontaneous improvement.[29]

Furtado not only stressed the impairment produced by the operation, but also suggested that Moniz may have given undue weight to those behavior changes that made patient management less difficult. Furtado seemed also to be in agreement with Sobral Cid's criticism that Moniz had operated before less drastic therapy had been adequately tried.

At the same Lisbon congress, Furtado gave a second lecture based on a follow-up of the 1937 paper he and Moniz had presented in Paris. Furtado explained that the results they had described in Paris were only the immediate results of leucotomies, but he could now report the results of a twelve-year follow-up of these same nineteen patients. Most of the patients, he said, who had improved following the operations had since relapsed. Also, some patients had developed epilepsy, and a suspiciously high number of patients in the group had died. The three remissions that persisted were in patients whose diagnosis of schizophrenia was questionable. Furtado concluded that Moniz had been too hasty in reporting success. When patients were followed for an adequate period, the results should "be considered unfavorable."[30]

Why did Moniz move so rapidly to broadcast his results? There are many obvious reasons, and they probably all contributed. He was, after all, over sixty, and had not many years left to achieve a place in history. It is not an uncommon pattern among scientists approaching the end of their career to take a particularly bold and often controversial step. While these factors probably had an influence, they do not adequately explain his great haste. An

---

* Rylander had become convinced by his studies during the 1940s that many of the defects produced by leucotomy were missed by casual observation and even by most psychological testing. It was necessary, he maintained, to observe patients in their everyday activities in order to appreciate the extent of the impairment.[28]

explanation that has not been adequately considered—mostly because the history of the period is not generally known—is that Moniz was aware that several other physicians on the continent were considering a similar undertaking.

In 1895, Emory Lamphear of St. Louis summarized Burckhardt's experiment, word of which had been brought to the United States by William Keen (Walter Freeman's grandfather), in the *Journal of the American Medical Association*. It was in this article on the surgical treatment of insanity that Lamphear drew the conclusion I used as the epigraph to chapter 5; that is, "the knife may be the means of restoring to reason many cases now considered incurable."[31]

Interest in brain surgery to treat mental illness continued up into the 1930s. In 1932, for example, Maurice Ducosté in France described a technique for treating paretics by injecting blood from malarial patients directly into the frontal lobes. The operation, which involved injections made through holes trepanned in the skull, was a direct extension of Wagner-Jauregg's fever therapy. Whereas the latter had routinely injected malarial blood into the body of paretics, Ducosté injected the blood (up to 5 cubic centimeters) directly into the brain. He performed this procedure on over one hundred patients and reported no subsequent complications or discomfort* and "uncontestable mental and physical amelioration." Ducosté also noted that for several years he had been injecting malarial blood and other substances into the frontal lobes of schizophrenic and manic patients with "encouraging results."[32] Moniz was surely aware of Ducosté's work—although he never cited it in any of his articles—as the two had presented papers, one after the other, at the same session of the Academy of Medicine. Moniz reported recent work in arterial angiography, while Ducosté described results of malarial blood injections into the frontal lobes of paretics. The publications of their presentations were adjacent in the same journal issue.[33]

Ducosté's study was not an isolated event: in 1933, a year after his report, E. Mariotti and M. I. Sciuti in Naples and Ferdiere and Coulloudon in France also reported encouraging results following injections of malarial blood directly into the frontal lobes of paretics; and by 1937, the former were withdrawing blood from schizophrenic and melancholic patients and injecting it into the patients' prefrontal area.[34] In 1933, A. M. Dogliotti in Italy had designed a pointed trocar which could be tapped through the orbital plate behind the eyeball into the frontal lobes.[35] Dogliotti mainly used this approach to inject opaque substances into the ventricles for ventriculography; but by demon-

---

* In some patients, Ducosté actually repeated the injections more than twelve times and observed that injections of malarial blood into the brain avoided infection of the liver and spleen—a problem that often resulted from Wagner-Jauregg's method.

strating how easy it was to gain access to the frontal lobes, he had paved the way for frontal-lobe operations.* In 1936, H. Chavastelon in France reported that he had achieved 80 percent "cures" following injections of blood into the prefrontal area of over four hundred demented paretics—work that he said had been in progress for several years.[37]

In his memoirs, Henri Baruk, an eminent French neuropsychiatrist who had written extensively on the effects of frontal-lobe brain tumors, wrote that ideas about psychosurgery had "floated around for quite awhile." He reported that Thierry de Martel, the French neurosurgeon, had proposed doing prefrontal operations before Moniz had started, but that he (Baruk) had convinced de Martel that such a "serious mutilation" of the brain was too dangerous.[38]

The interest in brain surgery as a possible treatment of mental illness was widespread and is also reflected in the remarks made by the surgeon William J. Mayo in a speech to the Association of American Railroads in July 1935—just before the International Neurology Conference in London. Mayo, who, with his father and brother Charles, had founded the Mayo Clinic in Minnesota, had earlier learned through his good friend William Keen about Burckhardt's experimental brain operations on Swiss mental patients. William Mayo's remarks in this speech were consistent with views he frequently expressed on the subject:

> If necessary we shall perform exploratory operations on the head at an early stage of disease and turn the light of day onto many of the pathologic conditions of the brain . . . Are we not in the same position in the treatment of the mentally afflicted that we were in the surgery of the abdomen, when I began fifty years ago, or as we were in the surgery of the chest? Day by day, I can see the extension of remedial measures to conditions of the brain at earlier stages.[39]

A report by F. Ody, a neurosurgeon in Geneva who had trained in Boston with Harvey Cushing, also illustrates the prevalence of the idea that frontal-lobe operations might benefit mental patients. Although Ody's report of unilateral removal (resection) of part of the right frontal lobe of a catatonic schizophrenic was not published until 1938, he made a special point of establishing that he had done this work before Moniz's monograph was published: "When that publication [Moniz' monograph] reached me, I had already operated several months before (6 April 1936) on the patient who is the subject of this report."[40]† Ody quoted an earlier published account of this case clearly

---

* In 1937, A. M. Fiamberti used Dogliotti's technique to perform transorbital lobotomies for the first time (see page 201).[36] These operations were later adopted by Walter Freeman (see chapter 10).

† D. Bagdasar and J. Constantinesco tried the Ody operation at the Central Hospital in Bucharest. While the operations were not judged to be successful, these physicians were clearly receptive to the general idea and, within only months after Moniz's monograph was published, had performed ten "core" prefrontal leucotomies.[41]

for the purpose of establishing his independent role in exploring neurosurgical interventions to treat mental disorders: "In the cause of objectivity, I will cite here . . . the report presented to the Swiss Society of Neurology by the eminent psychiatrist, Dr. Ferdinand Morel."[42] Ody did not conceal his feelings about Moniz's haste in publishing his results, noting that he himself had "waited to determine if there was a sufficient remission to allow me to draw reliable conclusions."[43]

Ody argued that even though Moniz and he were "both guided by the current ideas about the participation of the frontal lobes in the psychic life," their methods and theories were different. Ody quoted Morel to establish the point that, following his unilateral frontal-lobe operation, there were much fewer adverse consequences than after Moniz's bilateral operation: "It proves at the very least that the operation did not induce either apathy, akinesis, nor the loss of initiative, nor disorientation as followed Moniz' less radical, but bilateral operation." Ody added that his operation did not produce problems in conjugate eye movements, nystagmus, or incontinence, all of which Moniz had seen in his own patients. Moreover, Ody claimed the improvement following his operation was lasting, and implied that he suspected this might not have been the case in Moniz's patients as the latter's "description of the patients treated by his method came to a stop following the operation."

> There is more, our intervention has brought about a clear improvement which has lasted for almost two years.
> On questioning, the mother specified that she had been very pessimistic. . . . She was rapidly struck, however, by the cheerfulness and manageability of her son, who at the present time seeks, to the best of his ability, to make himself useful to her. There is no longer any question of confining him. And that alone justifies our intervention.[44]

Lastly, employing a little sarcasm, Ody referred to Moniz's theory about abnormally stabilized ideas and neural patterns as the basis of schizophrenia as a "very ingenious argument," but quickly pointed out that

> as a matter of fact, one picks up few results that are encouraging for the treatment of schizophrenia in Egas Moniz's statistics. In the seven operated cases, one counts five patients unchanged and in the two whose agitation disappeared, it reappeared— to be explicit—one month later.[45]

Apparently, one had to be in competition with Moniz to see the obvious weaknesses in his claims.

The prevalence of interest in the frontal lobes at the time is also evident from the fact that the organizing committee of the International Neurology Congress in London devoted a full day to a symposium on the subject. The interest is also demonstrated in an unusual way by Eugene Zamiatin's prophetic Russian novel *We* which, written in the early 1920s, is considered to be the forerunner of *Brave New World* and other "dystopian" novels. Zamiatin described a totalitarian state that uses science to control the populace:

> The latest discovery of our State science is that there is a center for fancy—a miserable little nervous knot in the lower region of the frontal lobe of the brain. A triple treatment of this knot with X-rays will cure you of fancy.[46]

What makes this passage particularly interesting in the present context is that, in the original Russian from which the psychoanalyst Gregory Zilboorg made this 1924 translation, there is no reference to the frontal lobes. Zamiatin used the phrase "Varolian bridge"—referring to an anatomical "bridge" or connection in the pons.* In the brain, the pons is a long way from the frontal lobes; but Zilboorg took the liberty of making the substitution, apparently believing that it made more sense to place "fancy" in the frontal lobes.

Thus, Moniz's initial operation on the frontal lobes was an idea whose time had come. He reported his "success" widely and in great haste lest someone else get ahead of him. In less than three months after Moniz's monograph was published, prefrontal leucotomy had been performed in Italy, Rumania, Brazil, Cuba, and the United States.

---

* Named for the Italian neuroanatomist Constanzo Varolio. Contrast Zilboorg's translation with Bernard Guilbert Guerney's 1970 one which, more faithful to the Russian, uses "Varolian bridge."[47]

7

"The Cat That Walks by Himself"

What I aspired to be,
And was not, comforts me.
—ROBERT BROWNING,
"Rabbi Ben Ezra" (1864)

WALTER JACKSON FREEMAN had, like Moniz, a tradition of achievement to live up to. Freeman was born in 1895, the year after S. Weir Mitchell attacked the superintendents of the insane asylums for isolating themselves from the rest of medicine (see pages 14–15). Freeman's maternal grandfather, William Williams Keen, had collaborated with Mitchell on the classic monograph *Gunshot Wounds and Other Injuries of Nerves* (1864), based on Civil War casualties.[1]

Keen was a leading figure in American medicine. A distinguished professor of surgery at the Jefferson Medical College in Philadelphia, he was one of the first brain surgeons in the United States. Among the neurosurgical techniques he is credited with developing are a method for tapping the ventricles of the brain to reduce pressure, a procedure for severing the Gasserian nerve ganglion to alleviate the often excruciating facial pain of tic douloureux (facial neuralgia), and a stimulating electrode used to determine the function of different cerebral cortical regions during brain surgery. Keen's many books on anatomy and surgery and his revision of *Gray's Anatomy* (a standard text by the nineteenth-century English anatomist Henry Gray) were highly regarded, basic texts. In 1906, after he stopped operating, he edited the six-volume *Surgery: Its Principles and Practices;* and, in 1917, at the age of eighty, he added two more volumes to it. He was also active politically in medicine, and his presidential address to the American Medical Association in 1900 probably helped defeat antivivisectionist legislation.[2] Keen attended medical meetings until his death at the age of ninety-five in 1932; the year before, after a stroke, he had himself wheeled onto the platform at the AMA convention.

Keen, who was small in stature (only five feet two inches tall), was a man of inexhaustible energy and broad interests. He had enormous influence on his grandson Walter—first as a model and later in opening doors to professional opportunities. Young Walter and the six brothers and sisters who followed him grew up hearing from their mother tales of her father's accomplishments and honors. In 1893, for example, Keen had been summoned to remove the left side of President Grover Cleveland's cancerous upper jaw—an operation undertaken in secret, on a private yacht sailing between New York and Newport, so as not to add to the public's nervousness during the financial crisis of that year.[3] It was not until Walter was sixteen years old, however, when on a Caribbean voyage with his mother, an aunt, and Keen, that he first fully appreciated his grandfather's eminence. Their ship had stopped in Panama, where the Canal was under construction; and Colonel William Gorgas,* a former student of Keen and the medical officer in charge of the project, obviously went out of his way to please his former professor. Taking the party under his wing, Gorgas arranged for a private sightseeing car and showed them around the locks and through Old Panama.

Again like Moniz, Freeman was born to a life of comparative luxury—and he himself acknowledged having been born with the proverbial silver spoon in his mouth—with servants and good private schools and all the advantages that money could buy or his mother could devise. He grew up in a large house—four stories high with three coal furnaces to heat it—in a fashionable Philadelphia neighborhood, just off Rittenhouse Square.

Freeman's father, the first Walter Jackson Freeman, was also a physician, a specialist in otolaryngology. Though shy and asocial, he had, while a student at the Jefferson Medical College, mustered sufficient courage to court the daughter of the distinguished professor of surgery. Later, as a family man who loved the outdoors and only tolerated the many dinner parties arranged by his socially ambitious wife, he took his two oldest sons with him on vacation trips backpacking in the wilderness—as his son Walter later did with his own children. The elder Walter Freeman was a good surgeon, although he used the knife only as a last resort; but in contrast to his father-in-law, who wrote hundreds of articles, he rarely published or attended medical meetings and read little of the literature even in his own field.

Freeman's mother, Corinne Keen Freeman—at the other extreme from her husband in interests and personality—dominated her children's lives. Intensely interested in her family's genealogy, for years she tried to qualify as

---

* Through his efforts to control yellow fever, Gorgas made possible the completion of the Canal.

a Daughter of the American Revolution, but was not successful in finding a relative who had fought in the war and had to settle for the Colonial Dames of America. Gifted with her father's ambition, energy, and studiousness (both father and daughter were Phi Beta Kappa at college), she worked hard trying to instill in her children the qualities she felt would help them be successful, socially and professionally.* She arranged dinner parties and Sunday afternoon teas for her children; enrolled them in dancing, riding, and boarding schools; and was tireless and efficient in planning "worthwhile" activities.† She provided a succession of governesses—French, German, and Spanish—and at dinner, the children were required to speak the language of the current one. In the big house, one or another of the children could always be heard practicing on a musical instrument: Freeman played the cello.

In spite of all his mother's efforts to develop social skills in her children, Freeman was, while growing up, basically an observer not a participant and, rather than team sports, preferred swimming, cross-country running, rifle shooting, and, later, golf. His mother had observed that, from an early age, he preferred his own company, and compared him to the cat in Kipling's *Just So Stories* "that walks by himself." Although, as Freeman later remarked, he could hardly be with his mother without disagreeing with her about something and resisted her attempts to direct his life, he ended up pursuing his career with all her driving energy and ambition.

After graduating from the Episcopal Academy a year ahead of the normal schedule (he finished first in his class and won the coveted Greek medal), Freeman entered Yale. Younger than his classmates, he was—like many a freshman before and since—overwhelmed by the freedom of the university and unable to take advantage of all it had to offer. He took little science or mathematics and did not really begin to enjoy college until his third year, when he enrolled in poetry courses—Tennyson, Browning, and some of the contemporary poets. He did not like being a junior reporter on the school newspaper—a job his mother urged him to seek as a way to get to know important people on campus—and became the photography editor of the *Yale Courant*. This job led to his name appearing in the *New York Times*—the first of many occasions—when the paper published the picture he submitted of the Scroll and Key secret society singing in front of their "mausoleum" at midnight.

When he began his senior year, he still had no clear plans for his future. His father advised him against medicine—pointing to relatives who had been

* Two of her sons eventually were listed in *Who's Who*: Walter and Richard, a professor of art at the University of Kentucky.

† A typical Saturday morning program arranged by Corinne Freeman for her children involved a trip to: a steel mill, a carpet factory, the Fleer "bubblegum" plant, the Baldwin piano company, the Philadelphia Electric Station, the mint, the navy yard, or the Bell Telephone Company.

successful in business and able to retire at fifty and "enjoy life." The previous summer, his mother had arranged for him to work as a machinist apprentice at the General Electric shop. When she observed that he enjoyed learning to operate lathes, drill presses, and milling machines, she persuaded him to "declare" engineering, which he did without any real enthusiasm. He soon found that he had no idea what descriptive geometry—an engineering prerequisite— was all about and, by the end of his senior year, chose medicine, having decided that the businessmen he knew who had retired early were not enjoying life but were bored with it. His grandfather, on the other hand, though almost eighty, was still publishing books and enjoying a full and exciting professional life.

After graduation from Yale, Freeman was accepted at the University of Pennsylvania Medical School, contingent upon his making up courses in chemistry and biology. To this end, he enrolled in summer school at the University of Chicago and also took a correspondence course in typing and shorthand. Before the summer was over, however, arthritis had developed in his joints, so painfully that he could not complete the courses. Since at this time the tonsils were being blamed for many illnesses, his father performed a tonsillectomy. The dean of the medical school consented to admit Freeman without his completing the courses, and he started medical school in the fall of 1916, when he was twenty-one years old.

He excelled from the beginning, even though he almost killed himself the first year. All alone in the laboratory, working on an "independent research project" on crystal formation, he put his mouth on the wrong end of a pipette containing cyanide. Suddenly his legs started to give way, and he could hardly breathe. In spite of this inauspicious beginning, his project was judged sufficiently interesting to warrant presentation at the Medical Student Research Association meeting. He was the only freshman on the program. Medical school had transformed him from the indifferent student he had been at Yale, and his grades were close to the top of this group of bright students. He joined medical fraternities, although he thought their solemn oaths, secret handshakes, and other rituals were nonsense, preferring the medical societies where students and staff members presented reviews of new developments in science, historical notes, and clinical case reports.

When the United States entered the war in 1917, Freeman and his classmates were placed in the Medical Reserve Corps. Armed with letters from his grandfather to the surgeon general of the army and the chief medical officer of the navy, he got an assignment to work in the hospital at Fort Dix in New Jersey, where he spent most of his time in laboratory work—urinalysis, blood counts, and examinations of the stool samples from men assigned to be cooks or bakers. He tried to look like an officer and designed his own uniform with leather puttees—but the arduous laboratory routine took most of the "spit and polish" out of his uniform and revealed his real status as student assistant. The influenza epidemic was raging, soldiers and civilians were "dying like flies," and he was kept busy assisting at autopsies. During one three-week summer break, he worked at the Philadelphia Navy Yard, reaming out rivet holes on liberty ships.

One summer, he supplemented his medical education with a course in the biochemistry of the brain at the University of Chicago and, the summer before his senior year, Freeman worked at the Mayo Clinic—another position arranged by his grandfather, a close friend of William and Charles Mayo.*

In his senior year, Freeman had to choose a medical specialty. He considered gastroenterology, which his father thought promising, and worked in it at the Mayo Clinic, but did not find the field exciting. He had been bored assisting in general surgery, he resented being an "animated retractor" not able even to see the "bottom of the incision"—and, in spite of his grandfather's eminence in the field, he had at that time no inclination to become a surgeon. He did, however, enjoy assisting in neurosurgery, not because his assignments as a student were any better, but because he was fascinated to see the brain exposed and pulsating.

What excited him most was neurology, and he formed an attachment to William G. Spiller, a dull lecturer but an astute diagnostician, who managed to convey the excitement of systematically eliminating hypotheses while working toward the diagnosis most consistent with all of a patient's signs and symptoms. Freeman was won over to neurology when a patient died, and a tumor was discovered on autopsy to be exactly where he had suspected. This patient had been unable to move his two eyes in parallel. Using his knowledge of French and German, Freeman reviewed all the cases on conjugate eye movements he could find and produced a fifty-page paper on the subject which earned him, at graduation, an award for the best student research.

During his senior year at medical school, it became clear that his father's health was failing and that he was wasting away from a growing cancer. When he became very weak, his eldest son had to shave him, but Freeman admitted later that he often hurried the job and was rather brusque. His father died on 20 December 1920, a few months after Freeman began his internship at the Pennsylvania Hospital.

After completing the required emergency-room and ambulance service portion of his internship, Freeman rotated among the different medical specialties. Temple Fay and Francis Grant, former classmates now headed for careers as neurosurgeons, would switch rotations, giving him extra experience

---

* The three of them had together established the American College of Surgeons. In 1909, William Mayo had operated on Keen for a blocked intestine.

in neurology and allowing them to get additional time in neurosurgery. The internship completed, he took postgraduate courses in laboratory techniques, having decided to specialize in neuropathology. The course was taught by Nathaniel Winkelman, who afterward offered Freeman the opportunity to work in his laboratory. There he learned the fundamentals of cutting, staining, and mounting tissue and, with his knowledge of photography, soon became expert at photographing specimens. Freeman had to do his share of routine blood and urine tests; but on more than one occasion, when faced with what he considered "mindless" routine, he would pour the sample down the drain (a "sink test") to steal the time for more stimulating work. Winkelman, a neuropathologist convinced of the organic basis of schizophrenia, strengthened Freeman's already formed predisposition toward an organic view of psychiatric disorders.[4]

Committed to neuropathology, Freeman applied for permission to end his internship three months early, in order to become assistant pathologist in the Neuropathology Laboratory at the Philadelphia General Hospital. One of the first patients he saw there had an unusual combination of symptoms. He was a middle-aged man with excruciating pain in the axilla (or armpit), extending down the inside of one arm. The man's eye had narrowed, and his pupil was constricted on the same side as the pain. Noting all the symptoms, Freeman hypothesized that a tumor was pressing on the spinal cord at the level of the first thoracic vertebra.* After the man's death, the autopsy revealed a large tumor in the very region Freeman had suspected. Pleased with his successful diagnosis, he wrote a brief note describing the case. It was his first publication.

Freeman worked hard and had little time for social life. On New Year's Day 1923, he was saddled with nine autopsies and could not attend the family dinner at his grandfather's house, even though his excellent visual memory made it possible for him to dissect three cadavers before he had to stop and dictate his observations. He decided to spend several years studying in Europe— a trip that, with his grandfather's help and reputation, was not difficult to arrange. It was common at the time for young American neurologists who could afford it to spend a few years working in Europe. European neurologists were considered to be the intellectual leaders in the field, and time spent abroad could have practical advantages in securing desirable positions later, as well as providing an attractive *Wandernjahre* break after the long hard pull through medical school and internship.

Freeman's grandfather gave him a letter of introduction to Charcot's successor at La Salpêtrière, Pierre Marie, who, though close to retirement when

---

* Nowadays such a diagnosis can usually be made by X-ray myelography.

Freeman arrived, was nonetheless still attracting many promising young neurologists from around the world. The Russian Constantin Trétiakoff was
there at this time. Several years before, at the University of Paris, he had
completed his thesis, in which he demonstrated that the substantia nigra* was
atrophied in the brains of patients with "paralysis agitans," or Parkinson's
disease.[5] This work, which turned out to be a most important clue to the
cause of Parkinson's disease, was being hotly disputed at the time, the French
supporting Trétiakoff and the German pathologists disagreeing.

Loner that Freeman was, he preferred laboratory research to examining
patients, but the chief of the clinic would not let him "hide in the laboratory"
and insisted that he needed more clinical experience. Freeman participated in
the examination of patients in rooms so cold and drafty that, as he remarked,
no matter what else patients suffered, they always had goose pimples. At the
Pitié Hospital, just next door to La Saltpêtrière, Freeman had the opportunity
to observe the elderly Joseph Babinski examine patients. Mostly, however,
Freeman studied patients with movement disorders—chorea and Parkinson's
disease. Having earlier been introduced by his grandfather to Pierre Janet at
a meeting in Atlantic City, Freeman, mainly for his own benefit, translated
the French neurologist's most recent book into English.

After six full months in Paris, Freeman went to work with Giovanni Min-
gazzini† at the Ospedale Psichiatrico in Rome. Freeman knew almost no
Italian but, after three months of tutoring, could get along fairly well, even
reading The Brothers Karamazov in Italian, but mostly reading articles on pathology and neurology. He studied the position of sensory-cell groups in the
brains of children who had died at different ages, and wrote a long paper on
the development of sensory systems in the brain for a special volume honoring
the Estonian neurosurgeon Lodovicus Puusepp.

In Rome, Freeman also had a chance to study the brain of an elephant that
had died at the city zoo. It took two assistants several hours, using crowbars
and pickaxes, to remove the brain from the dead animal so that he could
satisfy his curiosity about the neural innervation (the nerve connections that
activate, rather than convey, sensory information to the brain) of the elephant's
trunk. After Rome, he went on to Vienna to study with Otto Marburg at the
Brain Institute, which had a large collection of brains from different animals,
including elephants. When Freeman confirmed his earlier observation on the

---

* Substantia nigra (literally, "black stuff") refers to a region of cells that appear black because
they contain melanin, a precursor of dopamine (which such patients lack). The substantia nigra
is in the brain stem.

† Mingazzini had founded the laboratory of pathological anatomy at the psychiatric hospital,
where he also was the director. At the same time, he held the coveted position of professor of
neurology and psychiatry at the University of Rome.

huge representation of the trunk in the brain, he added a footnote and a sketch to his manuscript for the Estonian volume.

Although he had planned to spend a year in Vienna, a few months after his arrival Freeman received a letter from his grandfather congratulating him on his appointment as senior medical officer at the Blackburn Laboratory of St. Elizabeth's Hospital in Washington, D.C. Apparently, when Admiral Edward Stitt, the chief medical officer of the navy, had inquired at the University of Pennsylvania about potential candidates, Winkelman had recommended Freeman. Stitt, an authority in tropical diseases, had been a former student of William Keen, so Freeman was soon the top candidate. Eventually, Freeman received Stitt's letter offering him the position at an annual salary of $5,200. Freeman had never heard of St. Elizabeth's Hospital and did not know who Blackburn was, but Otto Marburg, reaching for a copy of Blackburn's *Illustrations of the Gross Morbid Anatomy of the Brain of the Insane*,[6] explained that Isaac Blackburn had been a distinguished pathologist and that the Government Hospital for the Insane in Washington—a name that Freeman did know— and St. Elizabeth's were one and the same. Freeman accepted the offer and cut short his European postgraduate training.

It was July 1924 when Freeman arrived in Washington. President Coolidge's sixteen-year-old son had just died from blood poisoning, which had started with a little blister on his heel, and the White House door was draped in black. At St. Elizabeth's, after a brief chat with the superintendent, William Alanson White, Freeman was shown around the hospital grounds and then taken to the Blackburn Laboratory. It had recently been expanded from six to over twenty rooms, and Freeman was so pleased that not even the red tape necessary to run a government laboratory could dampen his enthusiasm. The requisition forms, budget applications, and progress reports all seemed manageable. He felt he was doing very well indeed, only twenty-nine and head of a large pathology laboratory—and, as White, his immediate superior, was directly responsible to the secretary of the interior, Freeman liked to say that he was "just three steps from the President."[7] Admiral Stitt invited him home and told him that he had also been appointed lecturer in pathology at the Naval School, with the responsibility for demonstrating autopsies to young medical officers and some students from George Washington Medical School.

Washington was an excellent place for a researcher with Freeman's broad interests. When he was curious about the cranial capacity of Incas, he found the world's largest collection of Peruvian skulls at the Smithsonian Institute and a curator more than willing to spend hours talking about them. The Surgeon General's Library—which later would become a major part of the National Library of Medicine—seemed to have everything, including rare medical books and the most obscure periodicals. The Patent Office's files could often supply information about useful techniques, and the Army Medical Museum had a fascinating collection of medical memorabilia—including, he found, many anatomical specimens that his grandfather had contributed. Much later, Freeman would note that his own contributions to the museum were in the thousands, many times more than his grandfather's.

A fortunate coincidence almost immediately established his reputation as an expert in pathology. As a medical student in Philadelphia, he had come across a relatively rare case of yeast mold on a microscope slide of a brain slice. He had determined that the mold was not introduced on the slide, but had been in the living brain. He became interested in learning more about these molds and, having located similar cases, soon was somewhat of an expert on the topic. When a young navy pathologist in Washington handed him a slide sent from California and asked if he could find time to examine it in his laboratory, Freeman simply held the slide up to the light and said it looked like tortulosis—a yeast mold. When his opinion was confirmed a few days later, word spread about the "brilliant young pathologist" at St. Elizabeth's.

Freeman was able to get to work almost immediately. There was plenty of space, and a surprising amount of useful equipment was just lying around. He found a huge Sartorius microtome, capable of slicing a whole brain, which had never been unpacked. He established separate units in bacteriology, biochemistry, serology, clinical pathology, histopathology, photography, autopsies, and "museum work," the latter including nonclinical subjects of interest such as trepanned Inca skulls or the brains of rare animals. Freeman was good at delegating the routine jobs and also the government paperwork. He got a neuropathology appointment for Armando Ferraro, a friend from La Saltpêtrière, but Ferraro was lured away by the New York State Psychiatric Institute.*

Although they would later disagree about lobotomy, Freeman got on well with William White, respecting his ability to handle the administrative task of running a large hospital while still finding time to write books and to be a leader in psychiatry. Even though White had introduced malaria-fever therapy into the United States (see pages 29–31), he was not as committed to an organic approach in psychiatry as Freeman thought at the time. But Freeman did not get along with everyone at St. Elizabeth's, and he and Nolan Lewis seemed to irritate each other. Lewis, who later became director of the New York State Psychiatric Institute, was a neuropathologist turned psychiatrist. An entertaining conversationalist, he always had a crowd around him in the cafeteria at lunch, and Freeman could not resist contradicting him when he was holding forth. Their relationship deteriorated further when Freeman

---

* Like Freeman and Winkelman, Ferraro also searched for the cause of schizophrenia in neuropathology.[8]

produced evidence against Lewis's "pet theory" that schizophrenics have smaller hearts than normal and, as a consequence, a smaller blood supply to their brains.[9] Lewis's resentment appears to have simmered a long time: twenty-five years later, even though a substantial amount of psychosurgery had been done under his auspices at the New York State Psychiatric Institute, he harshly attacked Freeman and lobotomy.

Soon after settling in at St. Elizabeth's, Freeman had made contact with the neurology and pathology staffs at George Washington and Georgetown universities. Both of these medical schools were of marginal quality at the time. Their training facilities were inadequate, the condition of the George Washington Hospital was described as scandalous, and the staff was generally mediocre. (Later, in 1931, when medical schools were being inspected, both schools were put on probation and came close to losing their accreditation.) Freeman was immediately given appointments at both schools. Starting with the 1924–25 academic year, he was made professor of neuropathology at George Washington and associate in pathology at Georgetown. He wryly commented that both schools had more titles than cash.

Because of the lack of facilities and staff, the medical students at George Washington and Georgetown were frequently sent to other institutes for training. Since the senior students at George Washington had been attending William White's lectures on psychiatry at St. Elizabeth's, a precedent had been established for students to go there also for instruction in pathology. Freeman was only thirty at the time, and many of the second-year students, who had had to work to save money before enrolling, were about the same age. He thought that his lectures were not well received; but as soon as he started to do autopsies, the students became fascinated. The Blackburn Laboratory had a dissection room with concrete bleacher seats which could accommodate about one hundred students. He would select different students to assist him, and "suited them up" with rubber aprons and gloves. The dissections were often full of surprises, as many chronic mental patients were not able to talk to the staff about their medical complaints and their case histories often provided no hint of what would be found. Freeman was expert at this work, holding up organs to the class and commenting on his observations and conclusions. He used a motorized saw to remove the skull cap; and as he dissected the brain, he would teach neuroanatomy, using the blackboard to draw diagrams.

He started to look for interesting patients and sometimes, instead of an autopsy, he conducted a clinic, examining a patient in front of the class, coming up with a diagnosis, and discussing what was known about the cause of the disorder. Freeman's students thoroughly enjoyed these sessions, especially in comparison with their professor of neurology, who "lectured" by sitting on a desk and reading chapters from Osler's *Practice of Medicine*. A student committee went to the dean to complain and to request more hours with Freeman. Thus, in the fall of 1926, Freeman was appointed professor of neurology and head of the department. He was only thirty-one and well aware that he was the only one of his former classmates who had become a full professor by this time.

As soon as he became head of neurology, Freeman instituted many changes to improve the training of students. He started presenting neurological patients on Saturday mornings, demonstrating symptoms, explaining with diagrams where an "insult" to the nervous system was likely to be located. He recruited other staff members, especially neurosurgeons, to explain how they would attempt to treat such cases. Younger staff and senior students were given assignments to review the history and epidemiology of a disorder and the relevant physiology. Freeman was always center stage even when someone else was speaking: it was clearly "his show."

Freeman's teaching became increasingly dramatic. He enjoyed the limelight and kept an eye out during the week for patients who would be interesting to present. The way he handled patients at these demonstrations, imitating their gait or discussing a dismal prognosis in front of them, could strike an observer as cold and unsympathetic, but he was always effective and memorable. Students who passed through George Washington often regarded Freeman as their "most unforgettable teacher." Irving Cooper, who later had a prominent, if somewhat controversial, career as a neurosurgeon treating Parkinson's disease, described his experiences as a student at George Washington:

> Dr. Walter Freeman, a goateed, hyperactive, flamboyant antiestablishmentarian, a brilliant eccentric, was professor of neurology at George Washington University. His classes in neurology were electric. His demonstrations of anatomy included on-the-spot dissections of the human brain, which he carried out with a small, sharp-pointed wooden stiletto, neatly demonstrating centers of brain cells and nerve pathways in moments. It would have taken others months to prepare them for demonstration. He drew diagrams upon the blackboard using both hands simultaneously to illustrate the anatomy of the nervous system—a technique that I adopted and spent many months practicing. I have used it in my own teaching ever since. He imitated the symptoms of every neurologic syndrome he described in an unforgettable fashion. Each Saturday morning Dr. Freeman conducted an elective course, a three-hour clinical demonstration of patients with neurologic disease. The demonstrations were brilliant, but Dr. Freeman's treatment of the patient in those clinics was cold and dispassionate, often chilling. However, his elucidation of the anatomy of the brain was inspiring. It was in his classes that I literally fell in love with the brain: its structure, its functions, the possible mechanisms of those functions (at that time largely unknown), its complexity, and the miraculous creation that it is.
>
> I decided during those classes, while I was still in the first year of my medical

education, that the study of the human brain and its diseases was the path that I would follow.[10]

Later these demonstrations became so popular that they were moved to the large "Hall A" auditorium of the Outpatient Clinic. Although they were held on Saturday and attendance was not required, there were often as many as seventy people in the room. Freeman was such a good performer that some of the medical students took their girlfriends along for the show. He also arranged to take small groups of students on Sunday mornings to the Gallinger Municipal Hospital—later to become D.C. General—because it had a different patient population. The nurses complained that he was bothering the patients and managed to get the administration to stop the practice, but when the hospital was enlarged he started again. Students from Georgetown as well as George Washington were included in these sessions. Freeman found ways of getting the students more involved and, in 1933, started the William Beaumont Medical Society, named after the famous graduate of George Washington Medical School.*

Eugene Whitmore at Georgetown University suggested that Freeman get a Ph.D. degree from that university, and agreed to sponsor him. Freeman later described it as a painless degree with few seminar requirements, no term papers or other "academic filigree." With the data he had been collecting at St. Elizabeth's, he wrote a thesis entitled "Biometrical Studies in Psychiatry." The Ph.D. in pathology was awarded in 1931.

Nearly six months after his arrival in Washington on 3 November 1924, Freeman had married Marjorie Lorne Franklin, the eldest daughter of the superintendent of schools in New York. After getting her B.A. and M.A. degrees from Columbia University, Marjorie Franklin had gone on to get her doctorate in economics, also from Columbia, and at the time of her marriage was holding a responsible position at the Tariff Commission. Four of Marjorie and Walter Freeman's six children were born soon after their marriage (four children in forty months of marriage, as Freeman liked to say): Lorne, a girl, in July 1925; twins, Walter and Franklin, in January 1927; and Paul in February 1928. After a seven-year interval, Keen was born in 1934, followed by Randy, born on Christmas Eve 1936.

As soon as his children were old enough, Freeman began taking them on trips—at first, short ones, to explore places close to Washington along the Potomac or Chesapeake Bay, but later trips as far as the West Coast, with camping excursions along the way. In 1931, he and the four children (unlike

* William Beaumont made his classical observation of the changes in the stomach during both digestion and emotional experiences as they occurred in the French-Canadian fur trapper Alexis St. Martin after a fistula opening in his abdomen had been made following a gunshot wound.

her husband, Marjorie Freeman was never interested in coping with nature in the raw) climbed to the top of New Hampshire's Mt. Washington, the highest point in the eastern United States. Freeman led the way, climbing up the trestle of the old cog railway, with Paul, barely four years old at the time, on his shoulders. They had lunch at the top and took a nap, but when a sudden storm came up, they barely made it down, struggling in rain, hail, and snow. It occurred to him that he might have pushed the children a little hard, but in the end he was convinced that such adventures brought them closer together, while teaching self-reliance.

The year before his twins were born, Freeman opened up an office at 1801 Eye Street for private practice. It was a small office, but he barely got enough referrals to meet expenses during the first six months. As his contacts expanded, the practice began to prosper and he moved to a succession of larger offices on Eye Street, Connecticut Avenue, and finally, the one he shared for many years with James Watts, on 2014 R Street.

Freeman always had another physician, usually a neurosurgeon, share the office space with him as there were many advantages to sharing besides saving expenses and covering for each other in emergencies. Theoretically, at least, each physician could refer patients to the other. The relationship between neurologists and neurosurgeons, however, had become asymmetrical. After neurologists have made a diagnosis, there is often little treatment they can offer, so a patient is commonly referred to a neurosurgeon. Many neurosurgeons, especially after gaining experience, come to feel that they can make their own diagnoses. The patients who present the greatest difficulty for a neurosurgeon are those whose problems are suspected of being psychological rather than neurological. As I noted earlier, many neurologists, including Freeman, spent a significant proportion of their time with psychiatric patients.

In keeping with his skepticism about psychiatry, Freeman had neither the interest nor the patience for the kind of psychotherapy that searches for intrapsychic conflicts. When he treated patients with mental disorders, he used drugs or suggested an exercise regime or a change in life style. When the shock therapies were developed in the 1930s, he was quick to adopt them. It was generally acknowledged that he was the first physician in Washington to use insulin, metrazol, and electric shock.[11] On one occasion, when an aunt became very depressed, he decided to give her metrazol shock treatment:

A disturbed relative, aged 70, came under my care in 1937, and with my surgeon brother, Dr. Norman E. Freeman as assistant, I injected 6cc. of metrazol into one of her veins. Within ten seconds she began twitching, then opened her mouth widely, arched her back, stiffened out in a tonic convulsion that lasted about 20 seconds, followed by clonic movements for about 25 seconds. Then she relaxed with no respiratory movements for many seconds. She became cyanotic; her pulse was strong, however. Finally she gasped and then breathed heavily. Gradually, the color returned to her face—also to my brother's. "Jesus!" he said, and wiped his brow.[12]

Curare and other muscle relaxants had not yet been developed to prevent the bone fractures often caused by shock treatment and Freeman would normally ask his secretary to help hold a patient down when he gave such treatment in his office. On one occasion, when he was treating a woman for depression, a new secretary refused to participate, so he proceeded alone. During the convulsion, the patient fractured the humerus bone in both arms. Freeman, who was late for an appointment in the hospital—with another patient—gave the woman an injection of morphine and left her in the office with her husband. When he returned, the woman had been writhing in pain for almost two hours. The husband was furious and contacted his lawyer immediately afterward. Freeman could not possibly justify leaving a patient in distress, and his attorney advised him to be thankful he could settle out of court. Freeman went through a period of mixed feelings—alternately blaming his secretary and himself and feeling humiliated by the condemning eyes of some of his colleagues, but soon put the incident out of his mind.

The incident had occurred because Freeman was overextending himself, striving for recognition on several fronts at once. From his first days in Washington, he had published extensively, but even he recognized that only a couple of his articles amounted to much.[13] Nevertheless, he would purchase, out of his own pocket, as many as a thousand reprints of some articles and send them to colleagues because he considered it "a good investment." Also, to further his ambitions, he followed his grandfather's lead and started to join medical societies immediately after completing medical school. The first ones were in Philadelphia; and for some years, while in Washington, he returned regularly for the monthly meeting of the Philadelphia Neurological Society. (During the Second World War, he was elected president of this group— one of the few out-of-towners to hold that office.) In 1923, he traveled to the International Neurological Congress in London to be with his grandfather, who had just ended his term as president of the society. Freeman later admitted that he rode his grandfather's coattails getting to meet all the important people—to the irritation of his contemporaries, one of whom sarcastically referred to him as "the grandson."

By 1927, the year after he was appointed professor of neurology at George Washington, Freeman had made himself so well known to the influential members of the American Medical Association that he was elected secretary of its Section on Nervous and Mental Disease. Secretaries of AMA sections were usually senior people, but he was only thirty-two at the time. Freeman managed to impress many of the leading figures in the AMA—including the editor of JAMA, Morris Fishbein—and as a result was chosen to serve on the committee mandated to develop a plan for certifying specialists in meurology and psychiatry. After the certification board was established in 1934 (see pages 20–22), he served as one of the examiners for years and later he was elected president of the board. During the late 1930s and early 1940s, any physician certified in neurology had to pass Freeman's critical scrutiny.

As Freeman's reputation grew, he cultivated the image of a brilliant, acerbic, and opinionated physician. He sported a goatee beard in the style of the French neurologists at a time when few American physicians had beards. Just under six feet tall and weighing about 180 pounds, he had a penetrating way of looking at people, with his head cocked to one side (see figure 7.1).* His later partner in lobotomy, James Watts, first saw Freeman in Atlantic City at the 1933 American Neurological Association meeting strutting down the boardwalk wearing a Texas sombrero and carrying a cane.[14]

Freeman was always charismatic, presenting papers in a dramatic, challenging, self-assured, and well-organized manner. At most meetings, he also had an exhibit of his current work and would make it a practice to arrive early to set up his booth as dramatically as possible and also to cultivate newspaper reporters looking for a story. As a result, his name was frequently in the newspapers or magazines. When he put himself on the Section of Nervous and Mental Disease's program at the 1931 AMA meeting, *Time* magazine considered his speech on physiochemical factors in mental disorders newsworthy and ran a picture of him. The magazine referred to him as the "brilliant young chairman of the A.M.A. section on Nervous and Mental Disease," and reported that he said dementia praecox was probably a "deficiency disease, comparable to scurvy or rickets."[15] Freeman made an effort to attract to his booth physicians strolling through the hall and sometimes, in order to draw a crowd, would call out that the show (short films he often made for a meeting) was just about to begin.

Soon after beginning work at St. Elizabeth's, he decided to write a book on neuropathology. He started collecting material, read articles in the French and German literature as well as those in English and wrote almost every evening. It took about six years to complete and was not published until 1933. Toward the end of the task, his mother and grandfather died, and he began

---

* When Freeman was a little over one year old, he developed, on the right side of his neck, about thirty enlarged lymph nodes, which were suspected of being tubercular. His grandfather William Keen removed these nodes, but Freeman's trapezius and sternomastoid muscles were paralyzed afterward, so that, throughout his life, his right shoulder was lower than the left and his head was tilted slightly to one side. He was, when young, sensitive about this tilt and tried to correct it looking at himself in a mirror; as a grown man he came to regard this tilt as a sign of his "built-in arrogance."

Figure 7.1. Walter Freeman, Portland, Oregon, July 1951. (Courtesy of the National Library of Medicine.)

to drive himself relentlessly. Maintaining an extremely busy schedule between his duties at George Washington, private practice, and his activities in many medical societies, he would get up at 4:30 A.M. to start writing and would resume work on the book after dinner. When it was nearly done, he became depressed and could not sleep. He felt his brain was on fire, and was impatient and irritable. He finished the book while recuperating from a car accident in which, while driving through an intersection, he was hit broadside, bruising his ribs.

Also during 1933, the year he finished his book on neuropathology, he wrote "Psychological Plagues," a thirteen-stanza poem that he regarded as the best of the many poems he wrote. Certainly it is the most self-revealing, picturing Man—particularly "brilliant youths"—as driven to the breaking point by the "Master of Evil" and his sons "Hurry," "Flurry," and "Worry." The poem is also prophetic in its reference to the frontal lobes of the brain:

> . . . Then said Worry unto him: "All this of
> which my brothers have spoken can I do,
> And more also in addition.
> For I can work upon Man's imagination,
> That same power with which he develops his
> theories and his machines.
> I can cause him to foresee events, both those that
> will happen and those that will not.
> I can cause him to interpret falsely the actions
> of his friends
> And of his family
> And of the wife of his bosom.
> I can cause him to fear decisions.
> And to brood over mischances.
> And I can cause him to lose sleep in prospect
> of the morrow."
> "'Tis best of all," said the Master of
> Evil, "For sleep would defeat our
> every object.
>
> "Go ye three upon the Earth
> And burn the candle of Man's life at
> both ends.
> Spare not his frontal lobes where intelligence
> And ambition
> And judgment
> And self-control
> Are centered . . ."[16]

Freeman's *Neuropathology: The Anatomical Foundation of Nervous Diseases* was published in 1933 by W. B. Saunders, who had also published Keen's volumes on surgery. The chapter on neurosyphilis was acknowledged to be one of the best summaries of the brain-cell changes that occur during different stages of paresis and following malaria-fever treatment. About schizophrenia, Freeman wrote that in his opinion it is "a deficiency disease"; he reported finding a deficiency of iron in a particular layer of the cerebral cortex—a finding that was never confirmed. Nevertheless, the book was well received and for some years was one of the leading English-language texts in neuropathology. Much later, Freeman was pleased to discover two copies of his book marked "Not to Be Removed" in the library of the Royal College of Medicine in London.

Shortly after the book was published, Freeman became severely depressed.

He would not seek psychiatric help and decided instead to get away for a while. His wife put him on a boat headed for France, where he spent a week at a chateau owned by friends before catching the same boat on its return voyage back to Baltimore. He would not allow himself to think about the future until the last day at sea, when he decided to resign from St. Elizabeth's Hospital, to find some time every week for relaxation, and to spend no more hot summers in Washington. Characteristically, Freeman attempted to turn this breakdown into something positive and, in 1935, wrote a paper entitled "Danger Signals: On the Advantage of a Nervous Breakdown, or a Few Neurotic Symptoms in Certain Men Under Forty Years of Age." He observed that, years before, Osler had commented that a little albumen in the urine (a sign of diabetes) could be a blessing in disguise if the patient heeded the warning, "restricted his appetite, gave up whiskey and champagne, resigned from six or eight boards and started to live a rational life." Freeman, himself just forty, wrote that "it is better to suffer a nervous breakdown at forty, than to suffer a cardiovascular breakdown at sixty." He pointed out that those who might benefit from this "early warning" by amending their life style are the group

> which comprises constitutionally vigorous individuals, hard working, hard playing, capable executives, lawyers, physicians, engineers, brain workers of all sorts. These men get a lot out of life, just as they put in a lot. They drive themselves as well as others, enjoy the battle of stiff competition, and to them the fruits of victory are sweet ... until they have a nervous breakdown. They may realize that they are working too hard, that the tension is great and the rewards dubious, they may lose some sleep worrying over problems and how to meet them, they may become grouchy about the home.[7]

Freeman might have been describing himself, but his temperament would not allow him to follow his own sensible advice to change his pace; and, indeed, he developed diabetes in later life and, at the age of fifty, had a "minor stroke" while pushing himself to complete a paper for the World Congress of Psychiatry meeting in Paris.

In August 1935, Freeman and his wife went to London for the International Congress of Neurology, where he exhibited work he had done on ventriculography using Thorotrast as the contrast medium. It was at this congress, as mentioned earlier, that he met Moniz for the first time. Freeman shared the reservations of other neurologists about some aspects of Moniz's work on cerebral angiography, but was clearly impressed by the man. Thus, seven months later, Freeman was very receptive when he stumbled across Moniz's first report of prefrontal leucotomy.

Just after the London conference, in the fall of 1935, the neurosurgeon

James Winston Watts joined the staff at George Washington Hospital. He had been a fellow in neurosurgery in Charles Frazier's service at the University of Pennsylvania. Looking for a teaching position and a place to start his neurosurgery practice, Watts had sought out an introduction to Walter Freeman. This led to an invitation for Watts to give a talk at a meeting of the Medical Society of the District of Columbia, followed by an appointment as associate professor of neurosurgery. Watts had grown up in Lynchburg, Virginia, attended the Virginia Military Institute, and studied medicine at the University of Virginia, where his uncle was a distinguished professor of surgery. He came with excellent qualifications, being recommended highly by John Fulton of Yale University and Francis Grant. Watts had trained in neurology and neuropathology in Boston; in neurophysiology with Fulton at Yale; worked in Chicago with Percival Bailey; and in Breslau with Otfrid Foerster.

Temperamentally, Watts was almost the direct opposite of Freeman. Slow and careful, he appeared gentle and retiring. He was not the protypical, assertive neurosurgeon. Watts disliked controversy but had firm convictions. When opposed, he could be firm, but he tried to avoid conflict and antagonizing anyone. Tall and slender, with a delicate physique, Watts always paced himself to save his energy. He relaxed on vacations, avoiding strenuous physical activity, and at work routinely took a short nap after lunch. He was not inclined to drive himself by self-imposed deadlines as was Freeman.

Watts came to George Washington in a junior position, but he rose rapidly to be the professor of neurological surgery. He and Freeman started performing prefrontal lobotomy together just one year after Watts came to George Washington University. The two of them soon established the Department of Neurology and Neurological Surgery and would serve for many years as its co-chairmen.

8

Releasing
Black Butterflies

The operation is based, however, on sound
physiological observation.
—*New England Journal of Medicine* (1936)

IN THE SPRING OF 1936, while abstracting French articles for the *Archives of Neurology and Psychiatry,* Walter Freeman came upon Moniz's first, brief article on leucotomy, published in the *Bulletin de l'Académie de Médecine* in Paris. Freeman immediately ordered a copy of Moniz's monograph, which had been cited as "in press" and was published in June. When it arrived from its Paris publisher, he read it and showed it to James Watts. They agreed that Moniz's arguments and evidence were compelling and decided to try prefrontal leucotomy as soon as they returned from their summer vacations. Before leaving Washington, Freeman ordered two leucotomes from Moniz's instrument maker, Gentile et Cie, in Paris, wrote a review of the monograph, and sent off a letter to Moniz mentioning his and Watts's plans. As I have said, by 6 August, Freeman had sent a prepublication copy of his review to John Fulton at Yale and, in his letter to Fulton, called the monograph "epoch making" (see page 112).

By the time Freeman and Watts returned from their vacations early in September, the leucotomes had arrived, as had an inscribed copy of the monograph from Moniz. They immediately started to prepare for the operation. Using "formalin-fixed" brains from the morgue, they practiced inserting the leucotomes into the centrum ovale of the prefrontal lobes as Moniz had described. It did not take long to master the technique, and they started looking for a suitable patient.

Within a week, a patient they judged to be an appropriate candidate was referred to Freeman. A sixty-three-year-old woman from Topeka, Kansas, she was agitated and depressed—just what Freeman and Watts were looking for, as they had noted that Moniz had the best results with agitated patients.* The patient and her husband agreed to the operation, choosing surgery over institutionalization. When the woman was being prepared for the surgery, she changed her mind when she realized that her hair would have to be cut, but she finally consented after Freeman assured her that every effort would be made to save her curls. The operation was performed on 14 September 1936. The curls were not saved—but, as Freeman observed, "she no longer cared."[1]

Freeman and Watts performed the surgery together, following the latest procedure described by Moniz. They cut six "cores" on each side—three medial and three lateral—and rinsed each site successively with a warm saline solution to determine whether there was any internal bleeding. The scalp was then stitched together, and the head bandaged.

Immediately after recovering from the anesthetic, the woman reported that her previous sense of terror was gone. Freeman asked her a few questions:

FREEMAN: "Are you happy?"
PATIENT: "Yes."
FREEMAN: "Do you remember being upset when you came here?"
PATIENT: "Yes, I was quite upset, wasn't I?"
FREEMAN: "What was it all about?"
PATIENT: "I don't know. I seem to have forgotten. It doesn't seem important now."[2]

She greeted her husband with a peaceful smile.

Freeman later wrote that the results seemed spectacular. He also recalled that, a few days after the surgery, "Watts and I were so sure of ourselves that we really rushed into print with our first case." In fact, the patient took a turn for the worse one week after the surgery; she was unable to talk coherently and was repeating syllables. By 24 September, she improved somewhat and at least could say the days of the week, repeating only a few syllables. When writing, she also repeated letters, but "she appeared unconcerned." Several days later, she seemed to have recovered her speech and was "anticipating her return home, without eagerness, but also without apprehension."[3] Seventeen days after the operation, the case was reported at the October meeting of the District of Columbia Medical Society. When published, this report, which had already been sent to the journal, was entitled "Prefrontal Lobotomy in Agitated Depression: Report of a Case." Convinced that nerve-cell bodies as well as fibers were destroyed by the procedure, Freeman and Watts had

---

* Freeman later corresponded with Karl Menninger about this patient. In 1949, according to Freeman's unpublished "History of Psychosurgery," when he was visiting Karl Menninger at the Menninger Clinic in Topeka, the latter located his first letter about the patient and wrote "historical document" across the top.

leucot →?
lobot

decided on *lobotomy* as a more appropriate name for the operation than *leucotomy,* which implies the cutting of nerve fibers only. Freeman's presentation was enthusiastic, but the written report was more conservative: "Whether any permanent residuals of frontal lobe deficit will be manifested is uncertain at the present time, but the agitation and depression that the patient evinced previous to her operation are relieved."[4]

Eager to pursue this work, they performed operations on five more patients within the next six weeks, rushing to complete enough cases to report at the upcoming Southern Medical Association meeting in Baltimore. Although the deadline was passed, Freeman used his influence to get the paper accepted. On 18 November, they reported the results of six operations:

> In all of our patients there was a substratum, a common denominator of worry, apprehension, anxiety, insomnia, and nervous tension, and in all of them these particular symptoms have been relieved to a greater or lesser extent. In some patients there has been amelioration or even disappearance of certain other symptoms such as disorientation, confusion, phobias, hallucinations, and delusions that were present before the operation.

After describing the cases, they concluded with some caveats:

> In making this preliminary report upon a small series of cases we wish again to define our position in regard to the operation. We have undertaken it as a measure of relief from symptoms that were causing great distress to patients and to their families. These symptoms have been worry, apprehension, anxiety, sleeplessness and nervous tension. These symptoms have tended to subside following the operation, and the patients have become more placid, content, and more easily cared for by their relatives. The symptomatic relief has been almost immediate, and has persisted to the present time. We have not yet been in a position to establish controls satisfactory to ourselves or to others, such as simple anesthetization, incision of the scalp, drilling holes in the skull, unilateral operations, or partial operations, in an effort to determine what part is played by the procedures other than the actual reduction in the number of axis cylinders interrupted. That such experimental procedures should be carried out is highly desirable, but our duty to the patients has precluded these, and we have followed to the letter the suggestions of Moniz.
>
> We wish to emphasize also that indiscriminate use of the procedure could result in vast harm. Prefrontal leucotomy should at present be reserved for a small group of specially selected cases in which conservative methods of treatment have not yielded satisfactory results. It is extremely doubtful whether chronic deteriorated patients would be benefited. Moreover, every patient probably loses something by this operation, some spontaneity, some sparkle, some flavor of the personality, if it may be so described. In the patients operated upon up to the present, memory has not been obviously impaired and concentration has been improved, possibly on account of relief of the preoccupation. Judgment and insight are apparently not diminished, and the ability to enjoy external events is certainly increased. Special psychological tests will be used to determine whether learning ability has been

affected, and only a return to the previous occupation will reveal whether these patients are able to function satisfactorily at the [their] socio-economic level.[5]

More cautious than Moniz, Freeman and Watts were qualifying their conclusions and pointing to the need for "experimental controls" before improvement could be attributed to the destruction of specific nerve fibers. In print, they reminded their readers that "we have not mentioned recovery. We have not spoken of cure. These words cannot be used until after a period of five years." Speaking informally, however, Freeman did not conceal his optimism, saying about the first patient, "This woman went back home in ten days and she is cured."[6]

The first comments from the audience were relatively mild and unchallenging. One psychiatrist questioned whether the patients were really psychotic, while a neurologist raised ethical and legal concerns about how consent for such an operation was obtained. At this point, other people overcame their initial shock, and hands went up all over the room. The atmosphere was heating up. Adolf Meyer, however, asked for the floor and captured everyone's attention. The seventy-year-old "dean of psychiatry" (see pages 15–16) said: "I am not antagonistic to this work but find it very interesting. I have some of those hesitations about it that are mentioned by other discussants, but I am inclined to think that there are more possibilities in this operation than appear on the surface." Meyer further observed that the results were really not surprising as they were consistent with what "we are learning concerning the frontal lobes and their role in the functioning of the personality," but cautioned: "I should hesitate to promise that we could remove distraction and worries by operation. To call attention to what is possible might start an epidemic of hasty human experimentation." He stressed the importance of following each case "scrupulously," but added that "at the hands of Dr. Freeman and Watts, I know these conditions will be lived up to."[7] Freeman was pleased and wrote later "that had it not been for his [Meyer's] sympathetic and helpful discussion, the advance of lobotomy would have been much slower than it was."[8]

Meyer's comments blocked criticism not only because of his prominence, but also because his "psychobiological" theory of mental disorders was universally regarded as reasonable and balanced. For his part, it would have been awkward for Meyer to remain silent, allowing the audience to grow more hostile, as he and Freeman had worked together to establish the American Board of Psychiatry and Neurology in 1934 and had been serving together on it ever since.

Despite Meyer's calming influence, Dexter Bullard, a prominent psychoanalyst who headed the private Chestnut Lodge Sanitarium in Washington, could not restrain himself. Although he and Freeman had been friendly up to this time, Bullard was so disturbed by the idea of performing brain operations to treat mental illness that without waiting to be acknowledged by the chairman of the session, he stood up and with a mixture of anger and disbelief of the claims, blurted, "Now, Walter, that just isn't true!" For some years afterward, Bullard hardly spoke to Freeman and stopped using him as a consultant at Chestnut Lodge.[9]

Soon after the Baltimore meeting, Freeman tried to get his former chief at St. Elizabeth's Hospital, William A. White—another prominent psychotherapist—to allow him to perform lobotomies at that institution. White's reaction was similar to Bullard's. Making no effort to conceal his feelings on the subject, White replied, "It will be a hell of a long while before I'll let you operate on any of my patients."[10] Eight years later, when Winfred Overholser became superintendent at St. Elizabeth's, a relatively small amount of psychosurgery was performed there, but Overholser was never enthusiastic about these operations.*

But Freeman and Watts were not to be deterred. By the end of 1936, they had completed twenty Moniz-style operations, averaging about one every six days from 14 September, the date of their first operation. On Christmas Eve, Freeman and Watts were out searching the bars for their fifteenth lobotomy patient, an alcoholic, who had dressed, put his hat over his bandages, and walked out of the hospital. (Marjorie Freeman was at Columbia Hospital giving birth to her last child, and Freeman did not see his son Randy until early the next morning.) The patient was eventually found and returned to the hospital barely able to walk after having consumed quantities of alcohol.

The opposition to Freeman and Watts's report in Baltimore had been relatively restrained. Not only did Adolf Meyer's comments dampen the criticism, but many people were taken by surprise and did not fully react to the presentation until they had a chance to discuss it with others. Quite different was the response to Freeman's lecture to the Chicago Neurological Society on 18 February 1937. He had made certain that his presentation in Baltimore would receive maximum coverage in the press; and thus, when he spoke in Chicago three months later, most of his audience knew in advance what he and Watts had been up to, and several physicians were ready to express their feelings and opinions.

Freeman later recalled that after his presentation in Chicago, there was such "a torrent of adverse criticism that I almost bit the stem of my pipe off trying to retain control of myself." He admitted that his own manner was partly responsible, for by "this time I was feeling my oats and probably pre-

---

* The Harvard psychiatrist Harry Solomon persuaded Overholser to try lobotomy, and eventually about eighty lobotomies were performed at St. Elizabeth's Hospital.

sented in an imperial, dogmatic, and provocative fashion.'"[11] The published discussion after his presentation only partly captures the intensity of the feelings of some of the people in the audience. Dr. Loyal Davis started off his comments with a sarcastic reference to all the publicity Freeman had been getting in the popular press:

> Unfortunately, the only report of the procedure which Dr. Freeman advocates available to me was that in *Time*. I had hoped that the reports were grossly exaggerated and that some physiological facts would be given as the basis for indications for the operation. . . . The only explanation given so far is that of a patient who said that Dr. Freeman removed his "worry center." . . .
> I assume that Dr. Freeman's experience as a pathologist has made him curious to know the exact location and extent of the lesions he produces. As a pathologist, he may not have much fear of looping out pieces of subcortical tissue, but his statement that one person died after the operation of cerebral hemorrhage [see page 149] and that in many instances blood vessels are pulled out with the wire loop gives one's surgical conscience a twinge. The offhand manner in which this surgical procedure is described and discussed is no credit to the essayist as a surgeon, a pathologist, or one who is searching for a scientific truth.[12]

Dr. Harry Paskind added to the sarcasm by saying that he was ignorant about neurosurgical matters and asked whether it would be explained to him why the operation would not have been as successful "if done on the hands, or even on the feet."[13] Dr. Lewis Pollack then stood up and continued the attack stating that it was

> immoral to offer to the public any sort of a procedure which would awaken expectation and hope without possible fulfillment.
> On what does this procedure rest? First, this is not an operation, but a mutilation. Moniz said he was unable to state the exact extent and location of the parts removed by the leucotome. That is evident, for there are different sizes of skulls and of brains, and there is no way of knowing what may be removed. The wire loop may strike a blood vessel.

Pollack ridiculed the assumption that damage to the frontal lobes produced no impairment but only removed the depression or anxiety. He cited a case of a Methodist minister with a frontal-lobe tumor that was removed: some of his symptoms have been alleviated, "but he will never be a functioning minister." Pollack then spoke about the high incidence of "spontaneous remissions" in patients suffering from depression and anxiety: "Many patients with recurrent depression are highly competent and contribute largely to science, literature and art. Ought one to remove the frontal lobe to relieve the state of anxiety or depression which may disappear spontaneously? What will be left of the musician or the artist when the frontal lobe is mutilated?"[14]

The criticism at the meetings was so strong that Freeman decided to call off a scheduled speaking engagement in St. Louis and encouraged Watts to make the next scheduled presentation at the American Medical Association convention in Atlantic City in June.

Although many physicians felt strongly opposed to psychosurgery, little of the criticism was published during the first ten years after lobotomy's introduction in the United States—partly reflecting the long-established tradition in medicine of considering it bordering on the unethical to criticize in public another physician's treatment. Indeed, psychosurgery was given an early endorsement—if slightly qualified—in the 3 December 1936 issue of the *New England Journal of Medicine,* unquestionably one of the most influential medical journals in the United States. An unsigned editorial entitled "The Surgical Treatment of Certain Psychoses" began by pointing out that, for decades, brain tumors had been successfully removed, relieving many patients of mental symptoms:

> These and a few other examples suffice to call attention to neurosurgery as one of the adjuncts of psychiatric practice ... [as] has recently been brought to the attention of the medical profession by the work of Egas Moniz, Professor of Neurology in Lisbon. Moniz is well known for his work on arterial encephalography. . . .
> ... The operation, a relatively simple one in neurosurgery, can be done under local anesthesia. Professor Moniz reports observations on nineteen [*sic*] cases from a varied group of psychoses. Those that did best were the agitated depression type. In patients with this disease the agitation disappeared immediately and the patient became calm and tractable. Patients who formerly were able only to be cared for in a hospital for mental disease could return to their homes under nursing care. To be sure, they had lost something of their initiative and possibly their power of discrimination. They were, however, nearer a normal state of health than they had been before the operation. Six patients have now been so operated upon in this country with equally good immediate results.
> Such a radical procedure is not to be widely recommended at the present time. . . . It may mean a better future for certain patients with chronic mental disease. . . . The operation is based, however, on sound physiological observations and is a much more rational procedure than many that have been suggested in the past for the surgical relief of mental disease.[15]

Although the editorial recommended only limited use of psychosurgery "at the present time," the net impact was to encourage its adoption by suggesting these operations had a sound basis in "physiological observation." As John Fulton was on the editorial staff, he probably saw a copy before it was published, if he did not actually write the editorial himself. Regardless of who wrote it, it reflected the position that Fulton consistently took in regard to psychosurgery. While he carefully maintained some personal distance by advising caution, more research, and, later, modifications in the surgical procedure, he always was in favor of continuing psychosurgery and, indeed, significantly influenced its evolution (see pages 261–67). It is apparent that he glanced only casually through Moniz's monograph and, initially, had the impression that Moniz had used only alcohol injections—when, in fact, he had used the leucotome from the eighth operation on. Fulton's later references to the careful, detailed study of Moniz's patients demonstrated that he had not read the monograph critically. Even as late as 1948—as revealed in a letter to Freeman—Fulton had still not realized that Moniz was *not* the surgeon:

> Henry Viets* directed my attention to the fact that Moniz only proposed the operation of lobotomy, and he states that he did not carry out any of the surgical procedures himself. I have just been perusing the Moniz monograph of 1936 and this would seem to bear Viets out. How well do you know Almeida Lima? It was he, I gather, who actually did the early leucotomies. Has Moniz done any himself as far as you are aware?[16]

As will be seen, there are good reasons for believing that Fulton, despite all his accomplishments and great prestige, was often dangerously superficial in reporting the observations of others.

Although Freeman canceled the scheduled lecture in St. Louis after his strong rebuff in Chicago, it did not take him long to rebound. Within a year he had lectured on lobotomy in New Haven, Boston, New York, Philadelphia, and Memphis, in addition to the June 1937 presentation with Watts at the American Medical Association convention in Atlantic City. Awareness of psychosurgery was increasing around the country.

In spite of their initial enthusiasm, Freeman and Watts were forced to recognize that there were serious problems with the operation. Patients they considered improved had relapsed, and many of them were reoperated. For example, a woman lobotomized on 5 November 1936 was discharged seven days later. Freeman and Watts reported that, a month after the operation, her anxiety was reduced and "she was planning to return to work in two weeks." She never did return to work. Instead, she progressively deteriorated and had to be committed to St. Elizabeth's Hospital one year after surgery. Twelve years later, she was still institutionalized and described as presenting a "typical picture of a deteriorated praecox, fat and frowzy, refusing to speak and spending all her time stretched out on a bed or bench."[7]

Eight, in fact, of the first twenty patients had to be reoperated a second time, and two of these had a third operation. The six spherical "cores" on each side of the brain recommended by Moniz were increased to nine, some of these involving deeper penetrations. Freeman and Watts then found that

---

* Henry R. Viets, neurologist at Harvard University and on the editorial board, with John Fulton, of the *New England Journal of Medicine*.

the effect of a slight error in estimating the angle of penetration was magnified with the deeper penetrations; and in two cases, cerebral arteries were torn and the patients died from the hemorrhage. They started to check their anatomical accuracy by injecting opaque "contrast" fluids—first Thorotrast and later iodized oil—into the destroyed area. X rays revealed that the variability in the location of the damaged areas was unacceptably large.

Furthermore, Freeman and Watts found that the wire stylet in the Moniz leucotome was not stiff enough, but the blade they substituted had a tendency to break off in a patient's brain. When the blade was eventually removed, bits and pieces of blood vessels and brain tissue were usually attached to it. In addition, some patients remained stuporous, and one developed a serious motor problem. After a while, some patients started to have epileptic seizures, which Freeman and Watts attributed to the "cores" of devitalized tissue, like "foreign bodies," that were left in the brain. As a result of these serious complications, they were forced to slow down to develop a better surgical procedure. They performed only twelve more lobotomies in 1937 after the twenty they had done in rapid succession during the fall of 1936.

During the latter part of 1937, they developed what came to be called the "Freeman-Watts standard lobotomy." In this procedure, which they named the "precision operation," the brain was approached from the lateral surface of the skull rather than from the top, as in the Moniz procedure. As shown in figures 8.1 and 8.2, "burr" holes were drilled on both sides of the cranium at points designated by distances in millimeters from "landmarks" on the skull. A 6-inch cannula, the tubing from a heavy-gauge hypodermic needle, was inserted through one hole and aimed toward the hole on the opposite side of the head. The cannula was inserted about $2\frac{1}{2}$ inches into the brain and, if no fluid oozed out (fluid indicating that the anterior horn of the lateral ventricle, which extends into the frontal lobes, had been penetrated), it was lowered to the bony (sphenoidal) ridge at the base of the skull. The cannula was then withdrawn, and a blunt spatula—much like a calibrated butter knife—was inserted about 2 inches into the track left by the cannula.* Care had to be taken to avoid damaging major arteries located near the midline of the brain. After the spatula was inserted, its handle was swung upward so that the blade could be drawn along the base of the skull, and a cut was made as far to the side as possible. The spatula was then withdrawn, and the site was rinsed. That was only the first of four quadrants to be cut—two on each side of the brain. For the second on this side, the knife was reinserted; only this time the handle was lowered so that the upper quadrant could be cut. After some more rinsing with a saline solution, the process was repeated on the

---

* The "knife" used by Freeman and Watts (and later by others) was a relatively blunt instrument called a Killian periosteal elevator.

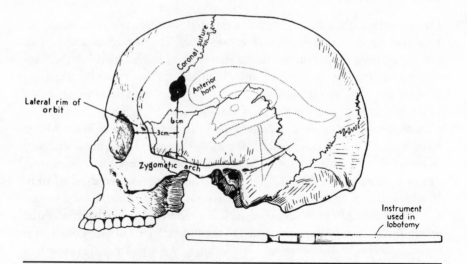

Figure 8.1.   Sketch of the skull coordinates used in the Freeman-Watts standard lobotomy. (From W. Freeman and J. W. Watts, *Psychosurgery: Intelligence, Emotion and Social Behavior,* 1st ed., 1942. Courtesy of Charles C Thomas, Publisher, Springfield, Illinois.)

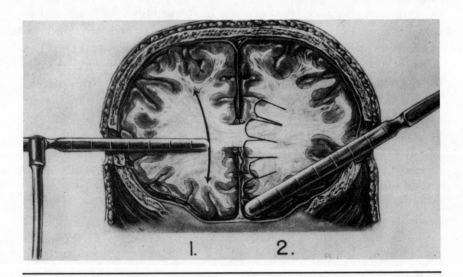

Figure 8.2.   Sketch of the procedure used during the Freeman-Watts standard lobotomy. (From W. Freeman and J. W. Watts, *Psychosurgery: In the Treatment of Mental Disorders and Intractable Pain,* 2nd ed., 1950. Courtesy of Charles C Thomas, Publisher, Springfield, Illinois.)

other side. Although Watts was the surgeon, he and Freeman often operated together, each doing a side while the other crouched down, sighting from the opposite side to make sure the cannula and spatula were inserted at the correct angle (that is, perpendicular to the long axis of the brain and also at the desired anterior-posterior plane).

Having observed that the best therapeutic results seemed to occur when the lobotomy produced drowsiness and disorientation, Freeman and Watts began to use this "disorientation yardstick" to predict success. Whenever possible, they operated under local anesthesia, talking to the patient, asking questions, and getting the patient to perform tasks such as singing or sub-tracting sevens from one hundred. (See figure 8.3.) If they observed no signs of drowsiness and disorientation, they often destroyed a larger area. They reported observing that many anxious patients described a sudden reduction in psychic tension as the final cut was made.

The new procedure was tailored to individual patients: limited and anteriorly placed lesions for those suffering from "affective psychoneurotic disorders"; extensive and posterior lesions for the chronic schizophrenic or the patient who had to be reoperated. The operations were referred to as a "minimal," "standard," or "radical" lobotomy, the latter involving the most posterior lesion. The more posterior the lesion was made, the larger the portion of the frontal lobes removed from the rest of the brain and the more the patient was likely to be debilitated after the operation. By April 1939, when invited by John Fulton to talk at the Harvey Cushing Society in New Haven, Freeman reported that they had obtained good results from twenty-eight standard Freeman-Watts prefrontal lobotomies.[18]

Before the Freeman-Watts standard lobotomy had been used, J. G. Lyerly, a neurosurgeon in Jacksonville, Florida, developed a psychosurgical procedure that made it possible to see the brain area being destroyed (figure 8.4). Neurosurgeons refer to such a procedure as an "open," in contrast to a "closed," operation. Approaching the brain from the top of the skull, Lyerly created an opening into the depth of the brain with a "speculum," a type of long tweezers or forceps. Then, with the aid of a light and a small knife, he could cut nerve fibers under direct observation.

Lyerly first used his procedure in the spring of 1937. By November 1938, he reported results from twenty-one operations.[19] One-half of the patients were inmates in the State Hospital at Chattahoochee, Florida, while the others were private patients in Jacksonville. Lyerly reported that his best results were achieved with melancholic, depressed, and anxious patients, and that many of them were "apparently well" after the surgery. When he finished his presentation, J. C. Davis, a psychiatrist at the Florida State Hospital, stood up and said enthusiastically that he had seen the patients and the results were

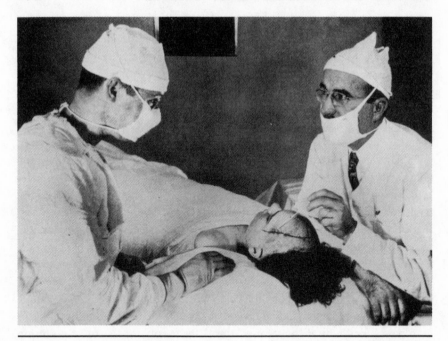

Figure 8.3.   James Watts and Walter Freeman operating. (Harris & Ewing/TIME Magazine, 30 November 1942.)

"nothing less than miraculous."[20] P. L. Dodge of Miami, another psychiatrist in the audience, then commented:

> I have a number of cases now and I am going to write and see if I can possibly induce the relatives and friends to take this chance because the patients are deteriorating, and they certainly will go on like such cases have gone on for centuries past. I was in the New York City Hospital for a number of years and I saw hundreds and hundreds of these patients who just deteriorated. They were absolutely hopeless and helpless.[21]

Lyerly's "open" lobotomy procedure had clear advantages and was adopted by several neurosurgeons; it was later modified by James Poppen of Boston's Lahey Clinic in 1943 (see page 196 and figure 8.4, on facing page).* Although the surgeons did not know what was best to cut, at the very least they were better able to avoid blood vessels and to stop any bleeding that might occur. Typical was the experience of John Love of the Mayo Clinic and the Rochester State Hospital in Minnesota. After watching Lyerly operate, Love concluded that the "open" technique was preferable to the "closed" operation of Freeman

---

* In 1940, Lyerly also performed a lobotomy at Delaware State Hospital.

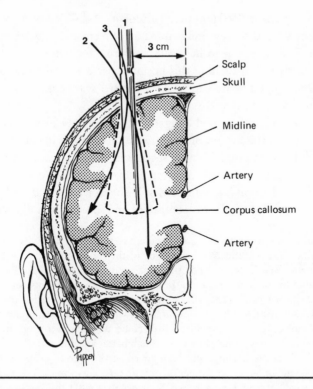

Figure 8.4. The Lyerly "open" lobotomy as later modified by James Poppen: (1) the spatula was inserted and then swung laterally; (2 and 3) a suction tube was inserted to extend the damage more deeply and laterally as well as to clean the area of the wound. (Drawn by R. Spencer Phippen.)

and Watts. Love was also impressed by the cosmetic advantages: "the trephine holes are so placed that they are within the hairline and are not visible." By 6 October 1943, Love reported to the staff of the Mayo Clinic that he had completed a series of forty-two Lyerly operations—forty-one patients at the Rochester State Hospital and one at the Mayo Clinic—with no deaths; and in only one case did convulsions occur postoperatively.[22]

Meanwhile, in the late 1930s, interest in lobotomy kept growing. Many neurosurgeons visited Freeman and Watts in Washington to learn their operating procedure. After hearing Freeman lecture, Mesrop A. Tarumianz, the superintendent at Delaware State Hospital, arranged to visit Washington in the fall of 1937 to observe the operation and the patients. Tarumianz had an M.D. degree from both Moscow and Berlin universities and had received his psychiatric training partly under Emil Kraepelin. The neurosurgeon Francis C. Grant, a former classmate of Freeman's who was practicing in Philadelphia, joined Tarumianz in Washington. Tarumianz and Grant were obviously encouraged by what they saw; and before long, Freeman and Watts were visiting Delaware State Hospital to help select patients for lobotomies. Grant, who was frequently a consultant at this hospital, performed the first of many operations there on 11 March 1938. This pattern was repeated at many other public and private hospitals.[23]

Overcrowding and limited budgets unquestionably played major roles in encouraging superintendents of large mental institutions to consider lobotomy. In a panel discussion on lobotomy sponsored by the Section on Nervous and Mental Disorders at the 1941 meeting of the American Medical Association in Cleveland, Tarumianz, who was still superintendent at Delaware State, described successful results of lobotomies on ten patients and then added:

> From an economic point of view I should like to give some figures as to what this may mean to the public. We have come to the following conclusion with regard to our own cases: In our hospital there are 1,250 cases and of these about 180 would be suitable for such an operation. In our hospital these patients could be operated on for $250 per case. That will constitute a sum of $45,000 for 180 patients. Of these we will consider that 10 percent, or 18, will die, and a minimum of 50 percent of the remaining, or 81 patients, will become well enough to go home or to be discharged. The remaining 81 will be much better and more easily cared for in the hospital. Thus the hospital will be relieved of the care of 99 patients. That will mean a saving of $351,000 in a period of ten years. I believe that, these figures being for the small state of Delaware, you can visualize what this could mean in larger states and in the country as a whole.[24]

This economic argument was frequently raised throughout the period of expansion of lobotomy. In 1948, John Fulton remarked that if only 10 percent of the patients occupying neuropsychiatric beds could be sent home, "it would mean a savings to the American taxpayer of nearly a million dollars a day."[25]

Influential as the economic factor was, nothing played a larger role in stimulating interest in lobotomy than the many popular articles that appeared in newspapers and magazines. Even without any special effort to bring it to their attention, journalists were able to recognize that lobotomy was a "hot topic." Nevertheless, the reporters in the front row when Freeman presented the results of the first six operations in Baltimore were not there solely because they had recognized, in the printed program, a potentially good story. After persuading Watts that, by talking to a science writer in advance, they could prevent the story from being distorted, Freeman called a *Washington Evening Star* reporter, Thomas Henry, who had covered other stories about his work. He asked Henry whether he would "like to see some history made. We've done a few brain operations on crazy people with interesting results." When the reporter expressed interest, Freeman explained a little more and then asked him out for lunch and spent about an hour and a half explaining what lobotomy was all about. He gave Henry a chance to observe an operation and to talk both with lobotomized patients and with patients scheduled for the operation. Henry wrote the story before the Baltimore meeting and gave a copy to Freeman and Watts to check. They made a few changes, mostly removing some fanciful speculations. (Over all, Freeman appreciated the effectiveness of the journalistic style, later remarking that he learned how to "assemble ideas and to present them in a more readable fashion" from such outstanding science writers as Howard Blackeslee, William L. Laurence, and G. B. Lal.)[26]

Henry called the next day to report that the story "was going to make the headlines," and asked if it was all right to show his article to the Associated Press and the United Press as well as to the editors of the Baltimore newspapers. Freeman agreed, with the result that a large group of reporters were clamoring for a press conference before his presentation. Among those present was Alexander Gifford, city editor of the *Baltimore News Post,* one of Freeman's former Yale classmates. To the reporters beforehand, Freeman gave an introductory lecture on the frontal lobes and included the often-repeated Phineas Gage "crowbar story" (see page 90), a summary of the clinical and animal literature on frontal-lobe injury, and a description of the Moniz leucotomy operation.

The day after Freeman's lecture, the *New York Times* carried the story under the headline "Find New Surgery Aids Mental Cases." Quoting Freeman and Watts, the article stated that the operations had been performed on six patients, "all of whom had been materially improved," and continued by noting that "medical men declared that the new operation marked a turning point" in treating mental patients. Dr. S. Spafford Ackerly of Louisville, Kentucky, was quoted by the *Times* as saying that this was a "startling paper that will go down in medical history as another shining example of therapeutic courage."[27] The story was also covered widely by newspapers in Baltimore and Washington and, after being carried by the wire services, was picked up by newspapers around the country. Freeman and Watts started to receive letters inquiring whether the operation would help patients with not only psychiatric disorders but also a wide range of medical problems, including asthma, epilepsy, and mental retardation.

The following spring, when Freeman and Watts presented their results at the meeting of the American Medical Association in Atlantic City, the *New York Times* reported the story the day before, Freeman having given it in advance to William L. Laurence of that newspaper. It appeared as a "Special to the New York Times" on the front page, headed by bold type:

SURGERY USED ON THE SOUL-SICK; RELIEF OF OBSESSIONS IS RE-
PORTED. New Brain Technique is Said to Have Aided 65% of the Mentally Ill Persons
on Whom It Was Tried as Last Resort, but Some Leading Neurologists are Highly Skeptical
of It.[28]

Freeman and Watts also had an exhibit of two leucotomized monkeys at
the same AMA meeting. According to the *Times* article, the monkeys had
been changed from their normally apprehensive and hostile state to behaving
as gently "as the organ grinder's monkey." In their lecture, Freeman and
Watts summarized the results of twenty Moniz-Lima "core" operations. The
Times described the Moniz "core" operation, noting that only two holes needed
to be drilled in the skull, each being sufficient "by proper manipulation of
the instrument, for the separation of six cores." Several brief summaries of
case histories from the "greatly improved" group were presented along with
Freeman's claim that this new "surgery of the soul" could often be performed
under local anesthesia and that it relieved "tension, apprehension, anxiety,
depression, insomnia, suicidal ideas, delusions, hallucinations, crying spells,
melancholia, obsessions, panic states, disorientation, psychalgesia (pains of
psychic origin), nervous indigestion and hysterical paralysis." The *Times* did
report that two patients had died after the operation, and that some neurologists
had "expressed themselves as being very skeptical about the new psychosurgery."[29] The effect of this warning in the *Times* article, however, was probably
minimal because no specific objections were mentioned nor were any of the
critical neurologists identified.

Over the next five years, articles about psychosurgery appeared more and
more frequently in newspapers and magazines, including such widely read
periodicals as the *Reader's Digest,* the *Saturday Evening Post, Time, Newsweek,*
and *Life.* There were also "testimonials" from former patients. One entitled
"Psychosurgery Cured Me," published in *Coronet* magazine, was written by
a former patient, Harry Dannecker, who described his illness and how his
wife had learned about the operation from a newspaper article. They then
went to George Washington Hospital to speak to Freeman. After Freeman
explained what was involved, the wife responded, "Doctor, we have nothing
to lose. Life is not worth living for either of us as it is. If there were no more
than one chance in a million for him to recover, I would still want him to
have the chance. Please—please go ahead with the operation." After describing
his own slow return to normality—having emerged "from that terrible un-
derworld of the sick mind"—and his subsequent success in designing a tool
used in the manufacture of automobiles, Dannecker concluded, "My purpose

in writing this article is a simple one: it may give heart and courage to readers who have afflictions such as I had or who have friends with similar, miserable obsessions."³⁰

Some of the captions to articles that appeared in small-town newspapers are striking:

SURGEON'S KNIFE RESTORES SANITY TO NERVE VICTIMS
WIZARDRY OF SURGERY RESTORES SANITY TO FIFTY RAVING MANIACS
BRAIN SURGERY IS CREDITED WITH CURE OF 50 HOPELESSLY INSANE
FORGETTING OPERATION BLEACHES BRAIN
NO WORSE THAN REMOVING TOOTH³¹

A very widely read article, "Turning the Mind Inside Out," appeared in the *Saturday Evening Post* in 1941, written by Waldemar Kaempffert, the science editor of the *New York Times*.³² Kaempffert had collected considerable information about lobotomy—much of it from Freeman, who had explained that the operation worked by separating the frontal-lobe "rational" brain from the thalamic "emotional" brain. Freeman gave Kaempffert a prepublication manuscript copy of *Psychosurgery*, the book he and Watts had just finished. One horrified physician in New York had asked Kaempffert whether he really was going to publicize that "criminal operation" in that "rag." The science editor replied that he had no intention of wasting all the time he had already spent on the project, and was not going to publish it in *Atlantic Monthly* or *Harper's* and "starve to death."³³

Freeman considered Kaempffert's article a "brilliant exposition of a very difficult subject written by a master of the art of interpreting science to the educated public" and the best description of the role of the thalamus and the frontal lobes in maintaining a balance between "socially acceptable behavior and primitive drives."³⁴ This high opinion of the article was not surprising as Kaempffert simply paraphrased Freeman's ideas. The article was written in clear prose, but was far from "brilliant" and even contained glaring errors, which none of the "experts" consulted bothered to correct.* Its main virtue was that it described the operation accurately and the immediate results realistically. The article's bias, however, was evident in its "journalistic" opening sentence: "There must be at least two hundred men and women in the United States who have had worries, persecution complexes, suicidal intentions, obsessions, indecisiveness, nervous tensions literally cut out of their minds with a knife by a new operation on the brain." After a brief description of knowledge about the frontal lobes, Kaempffert presented Freeman's theory: "It [the thal-

---

* For example, the article stated that the thalamus of the brain is "so old that every animal from the worm up has it."³⁵ The worm, of course, has no thalamus.

amus] is the seat of raw emotion—desire, passion, hate, fear, combativeness, love, appetite. . . . Man must balance emotion and reason. According to the Freeman-Watts theory, the preservation of that balance is a matter of nicely adjusting the thalamic feeling with prefrontal logic." The Freeman-Watts standard lobotomy operation was explained by analogy:

> Imagine a watchmaker confronted with the task of regulating the balance wheel of a fine chronometer, but sternly restricted to boring two holes through the outer case and inserting the tools through these. He knows exactly how the watch works and is designed and where the balance wheel is located. To guide himself he would mark lines and points on the case. You can imagine him drilling holes at just the right marks, inserting tools very carefully to avoid touching little wheels that might be injured and using the guide lines to hold a tool at just the right angle and level. And so it is with Doctors Freeman and Watts. They know the "works" within the skull, they know what arteries, cavities and spots to avoid.[36]

This description, making the operation seem precise and mechanical, was doubtless reassuring to readers who were considering the operation. The validity of the analogy, however, breaks down when one considers the large individual differences in the anatomy of the brain and the position of its blood vessels and also the fact that no one really knew the location of those important "little wheels" that had to be avoided.

Kaempffert continued in this positive tone, reporting that lobotomized patients were "indistinguishable from the rest of humanity." It is difficult to find any mark on their heads, but "worry had been cut out of their minds." Seemingly contradicting himself, Kaempffert explained that, initially, patients were passive, and often confused, and phlegmatic for several days or even weeks after the operation, but that in this period their personalities were being reformed. The new personality is "more than half born" after about five days and discharge from the hospital follows in about ten days. Kaempffert did acknowledge that the social readjustment might take months or even a year, and that there had been three deaths. Nevertheless, he pointed out that 85 percent of the patients showed at least some improvement, even though the results were poor in about 15 percent of the cases.[37]

Lastly, Kaempffert wrote that before consenting to perform a lobotomy, Freeman and Watts considered the patients' social situation, paying particular attention to who would care for them: "The decision is not easily made, and grave responsibilities are assumed when it is made. So there are careful inquiries, painstaking psychological tests, heart-to-heart talks with wives, husbands, business associates, close relatives."[38] Kaempffert was clearly repeating Freeman's orientation message as, in fact, there were no psychological tests useful for this purpose, and Freeman and Watts mainly relied on their own judgment. Also, as many of the patients were referred from out of town, their medical records and what could be learned from relatives who accompanied them were often the only source of information available. Furthermore, in order to minimize expenses, out-of-town patients were usually scheduled for the operation within a day or two after their arrival in Washington.

Kaempffert concluded his article by describing the types of people considered unsuitable for prefrontal lobotomy and what the operation was supposed to accomplish:

> Doctors Freeman and Watts decline to operate on the insane, on actual and potential criminals, on men and women who have in them a streak of cruelty and who glory in it, on those who are so far gone emotionally and intellectually that their lot is not likely to be improved. Nor will the occasional fussers and worriers be touched. "It is good to have some anxiety, some fear of consequences, some degree of self-consciousness," is Dr. Freeman's way of putting it. "These force us to restrain ourselves in our dealings with others." Social responsibility is based on what the world thinks of us. Community life would be impossible if we were not afraid of the police, afraid of upbraidings from our wives and husbands, afraid of the boss and his quirks. The fear, the worry must be grave indeed, must threaten incapacity before Doctors Freeman and Watts decide that the operation is the only hope. "We want a little indifference, a little laziness, a little joy of living that patients have sought in vain for so long," says Doctor Freeman. Though the operation may occasionally transform a morbidly anxious man into a careless happy drone, Doctor Freeman thinks it is better so, than to go through life in an agony of hate and fear of persecution.[39]

A condensation of Kaempffert's article appeared in *Hygeia,* published by the American Medical Association, and was, in turn, reproduced with additional abridgments in the English and Spanish editions of the *Reader's Digest.* With each shortening, the few caveats in the original article that might have given pause to a reader became progressively less prominent, if they were not completely omitted. During this period, almost no articles were published—in either popular, widely circulated magazines or the serious "quality" publications—that would have dissuaded desperate patients or their relatives from considering psychosurgery. The September 1941 issue of *Harper's* contained an article that was enthusiastic about all the new somatic psychiatric therapies, including insulin and metrazol-shock treatment as well as lobotomy: *Harper's* reported that the operations performed by Francis C. Grant on five "malignantly insane" and "apparently hopeless" schizophrenics at the Institute of Pennsylvania Hospital in Philadelphia had produced "substantial" improvement in all patients. Undoubtedly dependent on Grant for the information, *Harper's* described the improvement in one patient as "revolutionary":

> This was a woman who was plagued with voices, voices so torturing in their

persistence that she begged the doctor to puncture her ear drums so that she could not hear the eternal taunts and threatenings. Driven by these hallucinations, she was most unmanageable, given to atrocious conduct, apparently a hopeless case of deterioration. She underwent the operation six years ago, and the transformation was like a miracle. Today that woman is a matron of charm, she has married, has even had a baby (though against the doctor's advice), and is a completely re-oriented, socially attractive, apparently normal personality. It is doubtful if in all the annals of mental disorder a more complete or more dramatic "improvement" can be found.

Results such as these have rendered untenable the old concept of insanity as a mysterious psychological ill that yields only to psychological treatment.[40]

No one received more attention in the popular media than Freeman. He thoroughly enjoyed all the publicity and made a great effort to remain in the limelight. From 1936 to 1946, he and Watts had a "psychosurgery exhibit" at every annual meeting of the American Medical Association. The meetings usually began on a Monday morning, and the exhibits were supposed to be completely set up by 8:00 P.M. on Sunday. Reporters would drift in Sunday afternoon to get a few stories for the Monday-morning papers. Freeman always insisted that their exhibit be ready by noon on Sunday, and he usually stood around ready to collar any reporter who came by. According to Watts, once the meeting had started, Freeman did everything possible to attract attention to his exhibit:

> He would stand by his exhibit, and as people approached, he had a clicker which made a sharp staccato noise. When they stopped he would begin to talk like a barker at a carnival. Then a crowd would gather and before they moved on they had heard the story of dramatic relief of emotional tension, depression, suicidal ideas, had seen pre- and post-operative photographs, x-rays and sketches of the lobotomy operation.[41]

The attention Freeman and Watts were getting from the popular media angered some physicians, who regarded it as a form of advertising and therefore unethical. It was obvious that Freeman in particular had not discouraged this attention and had usually initiated it. It was also generally known that many patients were referred to Freeman and Watts as a result of all the publicity. Motions to censure Freeman were made at the executive sessions of both the American Medical Association and the American Neurological Association. At the latter meeting, Dr. Ernest Sachs of St. Louis walked down the aisle waving a copy of the *Saturday Evening Post* and demanding an investigation to determine whether punitive action was appropriate. There was even mention of expelling Freeman and Watts. John Fulton, however, prevented any action from being taken by suggesting that the matter be tabled for later consideration. Ultimately it was dropped from the agenda.

Freeman was able to take these attacks in stride, considering them only barbs from "small minds." He was convinced that those who disagreed with him would eventually "see the light," and he was not hesitant in using the press again, especially to realize some of its political potential:

> When it is a question of overcoming inertia and taking part in the pressure of the public upon politicians, and particularly upon hospital administrators at least to give the method described a fair trial, then I think that the doctor with a clear conscience can discuss his research with newspaper reporters and undergo the ensuing censure from his colleagues with equanimity. He realizes that something is being done to apply his discovery to the welfare of suffering patients.[42]

As I have noted, prefrontal leucotomy had been adopted in several countries almost immediately after Moniz's monograph was published in June 1936. Although Freeman and Watts were the first to follow Moniz's lead in the United States, they had not been the first in other parts of the world. They were not even the first in the Western Hemisphere, as Mattos Pimenta did four operations in Brazil; and Ramirez Corria operated on three patients in Cuba during the summer before Freeman and Watts got started.[43]

Leucotomy was also performed in Italy and Rumania before Freeman and Watts started. For several years, more psychosurgery was performed in Italy than in any other country, including Portugal. When Moniz traveled to Italy in 1937, several surgeons were already using his procedure extensively. Emilio Rizzatti and his associates, for example, had performed over one hundred operations by this time (see figure 8.5).* In 1939, Rizzatti, the director of the Racconigi Asylum located outside Turin, reported the results of two hundred operations, noting that it would be possible to draw reliable conclusions if his cases were added to those of Moniz and to the operations done by G. Sai in Trieste and by Bagdasar and Constantinesco (who had earlier tried the unilateral prefrontal operations introduced by Ody [see page 119]) in Bucharest.[44]

When Rizzatti described his two hundred leucotomized patients in 1939, he also reported that he had used insulin-coma treatment on three hundred patients, cardiazol (metrazol)-convulsive therapy on two hundred patients, and malarial-fever therapy on several thousand patients—thus letting it be known that the Racconigi Asylum was in the forefront in the use of the new somatic therapies. He introduced his article with a reminder that surgeons at Racconigi were the first to perform radical thyroidectomies and partial destruction of the pituitary gland in agitated patients suffering from "mental confusion and dissociative behavior." He wrote that he was skeptical about removing glands thought to be essential; but when he observed that patients actually improved afterward, he was also encouraged to try leucotomy.[45]

Although Rizzatti was quick to adopt somatic therapies, he did not make extravagant claims about the results. He reported, for example, that only

---

* One of Rizzatti's associates was Lodivicus Puusepp, the Estonian neurosurgeon, who had attempted a form of psychosurgery early in 1900 (see pages 43–44).

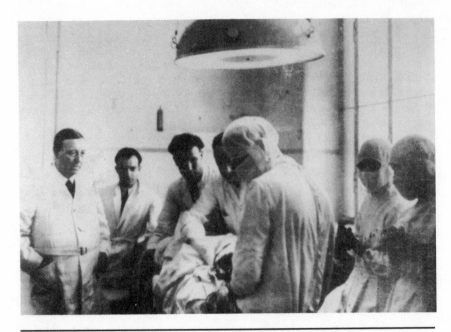

Figure 8.5.   Moniz observing a prefrontal leucotomy when he visited Emilio Rizzatti at the Racconigi Asylum (outside of Turin) in 1937. (Courtesy of the National Library of Medicine.)

about 15 percent of patients were helped after leucotomy. Even in the successful cases, he added, it was difficult to know for certain what produced the improvement, as most of the patients had also been treated with insulin, metrazol, and fever therapy. Speculating that some of the successes, as well as some of the failures, with leucotomy might be due to interactions with some of the other treatments, Rizzatti did not propose to do a "clean experiment" but advised that "it is our duty to try everything and even repeat treatments after reasonable intervals . . . as one can cure schizophrenia when it is least expected."[46]

In 1937, Sai of the Provincial Psychiatric Hospital in Trieste reported his experience with an operation that combined prefrontal leucotomy and alcohol injections in a single procedure. Moniz, as I have said, began with "alcohol operations" and then switched to leucotomy, but later gave some unsuccessful leucotomized patients alcohol injections. Sai reported substantial and lasting improvement in two and transient improvement in a third patient: "In my cases, the Moniz procedure has given encouraging results so that I conclude it is justified and its use should be expanded. If additional results will replicate the results obtained we would then be in a position to apply the procedure with success to diseased states hitherto unassailable."[47]

Even those who did not have much success were not discouraged. Bagdasar and Constantinesco of the Central Hospital of Bucharest reported using the Moniz leucotomy procedure on ten chronically ill patients between 4 November 1936 and 19 January 1937:

> We cannot speak of recovery in the sense of Egas Moniz. There is a decided improvement in certain patients, but there always remains a certain number of symptoms that are refractory to surgical intervention. Perhaps our cases were not chosen as carefully as those of Egas Moniz. Later we will try this method again, and will more carefully select the psychotic patients.[48]

As in the United States, it was not long before several surgeons started to modify the Moniz-Lima leucotomy procedure. Many variations were tried. In the early spring of 1937, E. Mariotti and M. Sciuti reported "good results" on schizophrenics by injecting either formalin or the patients' own blood ("autohemotherapy") into the frontal lobes instead of injecting absolute alcohol or performing a leucotomy.[49] After Mariotti and Sciuti reported that their procedure reduced aggressive behavior, Mario Torsegno in Genoa speculated "whether one could obtain good results with criminals?"[50]

The Italians seemed particularly receptive to psychosurgery during the first few years after Moniz's work became known. If Mussolini could make the trains run on time, the very least psychiatry could do was to keep abreast of new developments in science. By July 1937, Amarro Fiamberti, a psychiatrist in Varese, Italy, had described a much "simpler and faster" technique than the Moniz leucotomy. After putting a patient under local anesthesia, Fiamberti used a technique developed for injecting substances into the lateral ventricles in the frontal lobes of the brain: that is, he entered the brain by inserting a trocar (a sharply pointed hollow instrument) just above the eyeball and forcing it through the bony orbit. (This technique was originally published in 1933 by Dogliotti [see pages 118–19].) Through the hole in the trocar, he then inserted a hypodermic needle about 2 centimeters into the brain. This procedure put the point of the needle into the centrum ovale, Moniz's target, and then either absolute alcohol or formalin was injected. Ten patients operated on in this way were said to "tolerate it very well," experiencing only a moderate headache and sometimes a slight, but transient, temperature. Fiamberti, apparently sharing Moniz's impatience with psychiatric theories, wrote: "In the present state of affairs, if some are critical about lack of caution in therapy it is on the other hand deplorable and inexcusable to remain apathetic with folded hands content with learned lucubrations upon symptomologic minutiae or upon psychopathic curiosities, or even worse not even doing that."[51]

In Japan in the fall of 1938, Mizuho Nakata of the Niigata Medical College began to remove—essentially, to amputate—up to 5 centimeters of one frontal lobe as a treatment for mental disorders. Even though he began this work later, he did not seem to be influenced by Moniz's monograph, citing primarily the clinical reports of brain tumor patients by Ackerly, Brickner, Bucy, and Penfield.[52] Fifty-one operations were performed with this technique, but only seven involved psychiatric patients; the rest were epileptics who were having intractable convulsions. In January 1942, Nakata started to perform the Freeman-Watts standard lobotomy procedure; but, in general, there was little interest in psychosurgery in Japan until after the war ended in 1945. Lobotomy was then reintroduced, partially by Paul Schrader, a neurosurgeon attached to the United States occupation forces (see pages 178–79).

In some countries, psychosurgery came close to being discontinued after early failures. In Brazil, where Aloysio Mattos Pimenta had operated on several patients during the summer of 1936—almost immediately after Moniz's monograph became available—interest dropped off. No further operations were performed in that country until 1942.* Apparently Moniz's report of "hopelessly ill" mental patients being cured was taken literally, and the most chronically ill patients were selected for psychosurgery—with often disappointing results.

Surprisingly, in France, where Moniz had many friends and visited often, prefrontal leucotomy was initially received with only polite skepticism. Henri Baruk's extensive volume on psychiatric illnesses, which appeared in 1938, did not cite Moniz's monograph or even mention psychosurgery.[53] In Great Britain, where insulin-coma and electric-shock treatment were being used widely, the attitude toward psychosurgery was much more conservative than in the United States at this time. Moreover, with the outbreak of the Second World War in 1939, many surgeons were drafted into military service. It was late in 1940 before the first psychosurgery was performed in Great Britain (see page 172). Two medical teams reported the first operations in *Lancet* in July 1941. The two separate reports described eight operations performed at the Burden Neurological Institute in Bristol and four at the Warlingham Park Hospital.[54]

By 1942, many physicians and lay persons were aware of psychosurgery, but the number of operations that had been performed was small compared to what would be done toward the end of the decade. *Time* magazine reported that there had been only three hundred psychosurgical operations in the United States by 1942.[55] By this time, there probably had been fewer than one thousand lobotomies worldwide, with Rizzatti's two hundred being the largest number reported by any one physician.

* Personal communication from Raul Marino, Jr. (director of functional neurosurgery, São Paulo Medical School), 22 November 1982. Before December 1945, 209 lobotomies were performed: the Freeman-Watts operation was used in 48; the remainder were Moniz-Lima prefrontal leucotomies.

In the United States, after Freeman and Watts had performed their first operation in September 1936, Lyerly was second, performing his first lobotomy the next spring. During the following few years, psychosurgery was tried at the Massachusetts General and the McLean hospitals by William Mixter, and by Francis Grant, whose operations at the Delaware State and the Institute of Pennsylvania hospitals have been noted. Early lobotomies were also performed by J. Grafton Love of the Mayo Clinic, who did most of his operations at the Rochester State Hospital in Minnesota; and also by Magnus Peterson (in collaboration with Harold Buchstein, also of the Mayo Clinic), who did most of the operations at Minnesota's Willmar State Hospital; by Louis H. Cohen of the Norwich State Hospital in Connecticut; at the Clarkson State Hospital in Omaha by A. E. Bennett, chairman of neuropsychiatry at the University of Nebraska; and by Paul Schrader of State Hospital No. 2 in St. Joseph, Missouri. As can be seen from this information, most of the neurosurgeons were affiliated with well-known universities and hospitals—Massachusetts General, University of Pennsylvania, and the Mayo Clinic, but the majority of the early operations were performed on institutionalized patients at state hospitals.

At the end of 1940, Freeman and Watts started to write a book summarizing their experiences performing psychosurgery, including a history of the field and a theory of how lobotomy worked based on neuroanatomical and behavioral studies. Freeman undertook the major share of the work, writing the chapters on history, anatomy, pathology, clinical observations, and theory. Driving himself much as he did when working on his neuropathology book, he wrote every morning between four and seven o'clock, before a full day of clinical work and teaching. Watts worked mainly on the sections describing details of the surgery and postoperative care. The psychometric data and personality profiles were done by Thelma Hunt, who had a Ph.D. in psychology as well as a medical degree. Freeman pushed the project to completion, hounding everyone to work faster. While sick in bed, he proofread the galleys; and sick again when the page proofs came, he checked them over and also prepared the index while in bed. It was during this illness that he got the idea for the design of the title page. It was based on the French expression for feeling depressed: "*J'ai des papillons noirs*" ("I have black butterflies"). Freeman drew a skull with a trepanned hole from which black butterflies were emerging (figure 8.6). The publisher, Charles Thomas, liked the idea and had an artist draw such a medallion with a gold background for the title page. In the book's preface, the authors thanked Adolf Meyer for his anatomical assistance in preparing the slides of the autopsied brains; and the dedication read: "To Egas Moniz who first conceived and executed a valid operation for mental disorder."[56]

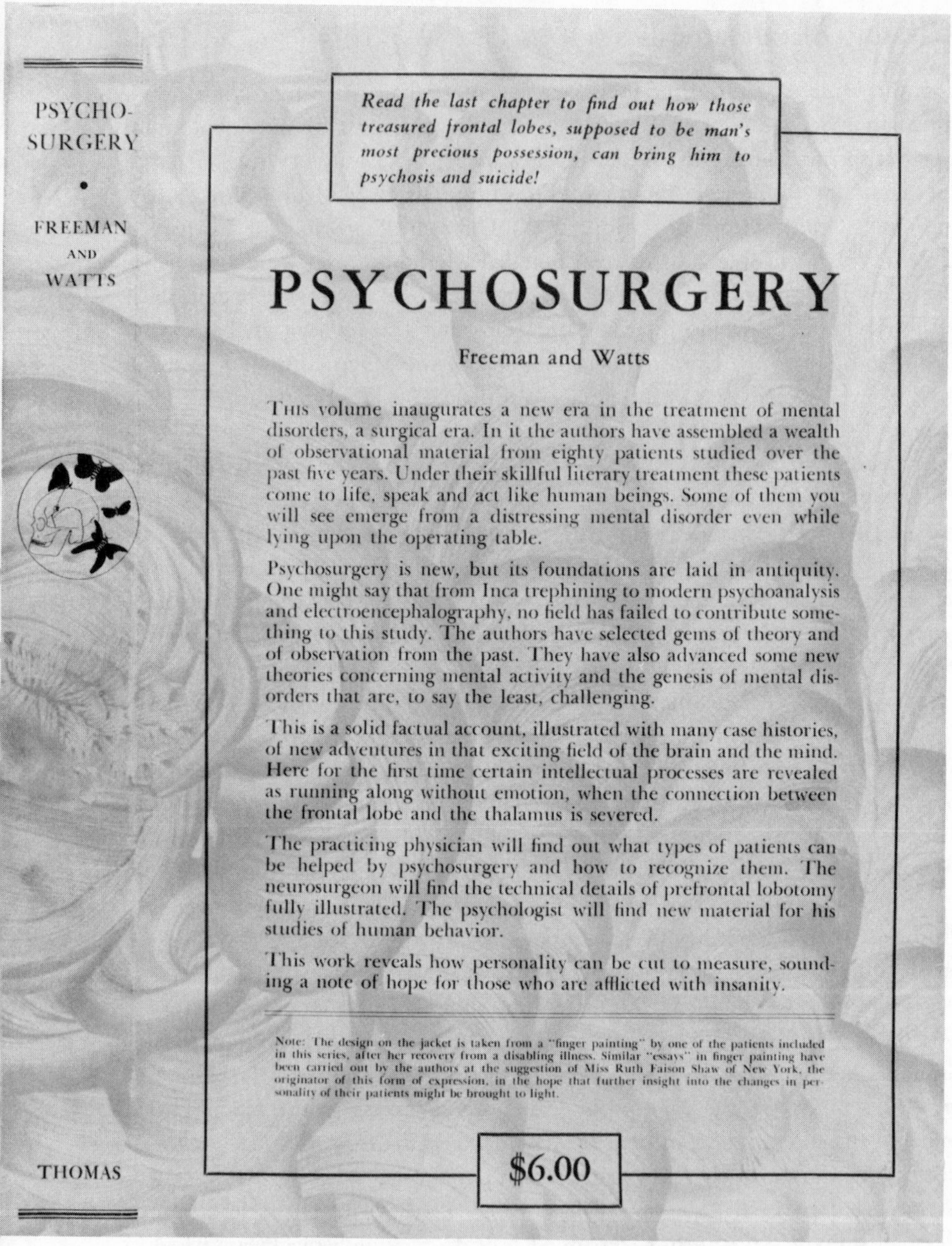

PSYCHO-
SURGERY

•

FREEMAN
AND
WATTS

THOMAS

Read the last chapter to find out how those treasured frontal lobes, supposed to be man's most precious possession, can bring him to psychosis and suicide!

PSYCHOSURGERY

Freeman and Watts

This volume inaugurates a new era in the treatment of mental disorders, a surgical era. In it the authors have assembled a wealth of observational material from eighty patients studied over the past five years. Under their skillful literary treatment these patients come to life, speak and act like human beings. Some of them you will see emerge from a distressing mental disorder even while lying upon the operating table.

Psychosurgery is new, but its foundations are laid in antiquity. One might say that from Inca trephining to modern psychoanalysis and electroencephalography, no field has failed to contribute something to this study. The authors have selected gems of theory and of observation from the past. They have also advanced some new theories concerning mental activity and the genesis of mental disorders that are, to say the least, challenging.

This is a solid factual account, illustrated with many case histories, of new adventures in that exciting field of the brain and the mind. Here for the first time certain intellectual processes are revealed as running along without emotion, when the connection between the frontal lobe and the thalamus is severed.

The practicing physician will find out what types of patients can be helped by psychosurgery and how to recognize them. The neurosurgeon will find the technical details of prefrontal lobotomy fully illustrated. The psychologist will find new material for his studies of human behavior.

This work reveals how personality can be cut to measure, sounding a note of hope for those who are afflicted with insanity.

Note: The design on the jacket is taken from a "finger painting" by one of the patients included in this series, after her recovery from a disabling illness. Similar "essays" in finger painting have been carried out by the authors at the suggestion of Miss Ruth Faison Shaw of New York, the originator of this form of expression, in the hope that further insight into the changes in personality of their patients might be brought to light.

$6.00

Figure 8.6.   Spine and front of the dust jacket of W. Freeman and J. W. Watts, *Psychosurgery: Intelligence, Emotion and Social Behavior,* 1st ed., 1942. Walter Freeman designed the medallion that appeared on the spine (and on the book's title page). He also wrote the jacket copy. (Courtesy of Charles C Thomas, Publisher, Springfield, Illinois.)

9

"Hit Us Like a Bomb": Psychosurgery in the 1940s

> The emotional nucleus of the psychosis is
> removed, the *sting* of the disorders is drawn.
> —WALTER FREEMAN AND JAMES WATTS
> (1942)

FREEMAN AND WATTS's *Psychosurgery: Intelligence, Emotion and Social Behavior Following Prefrontal Lobotomy for Mental Disorders*, published in 1942, vastly extended interest in lobotomy. Indeed, reflecting on lobotomy's early history, the Swedish psychiatrist Gösta Rylander observed that had it not been for the Freeman-Watts book and their standard lobotomy procedure, he doubted "that much would have come of it."[1]

While the successful surgery reported in the popular press had attracted the attention of patients and their relatives, Freeman recognized that lobotomy would be widely adopted only if backed up with a plausible theory. If most neurologists and psychiatrists were going to recommend these brain operations, they had to have at least a reasonable explanation of how it was supposed to alleviate mental illness. An explanation, however, required more reliable data. Freeman frankly admitted that even though prefrontal lobotomy had achieved some excellent results, there were also "a good many failures" that had not been explained: "The indications and contraindications for the operation have to be made more precise, and modifications of the operations attempted both in the direction of more extensive and more restricted incisions. Efforts must be continued to find the critical zone, the important fibers, the necessary areas to be resected, tracts to be cut, and dangers to be eliminated."[2]

Only when the results—both good and bad—could be related to specific brain pathways, Freeman concluded, would it be possible to have a truly adequate theory. He acknowledged that Moniz's theory was not satisfactory because stereotyped thoughts were not at the root of most mental disorders.

Instead, Freeman emphasized the ability of the operation to reduce the intensity of emotions: "The emotional nucleus of the psychosis is removed, the 'sting' of the disorder, is drawn."[3] Later, he referred to the change produced by a lobotomy as "bleaching of the affect."

The theory that Freeman offered was based on contemporary knowledge of the function of specific brain structures. Moniz had written only vaguely about abnormally stabilized synaptic patterns and offered no argument for cutting any particular band of nerve fibers. Freeman, however, made use both of the anatomical studies by the University of Chicago neurosurgeon A. Earl Walker, who demonstrated that the prefrontal area of a monkey's brain is connected to the dorsomedial thalamus;* and of the theory of C. Judson Herrick, one of the foremost neuroanatomists in the United States, that cognition and emotions are joined by the combined action of the frontal lobes and the medial area of the thalamus.[4] Through these two brain areas, Herrick had written, ideas gain "dynamic power" by the addition of emotions. In *Psychosurgery,* Freeman placed this concept at the heart of his explanation:

> The dorsal medial nucleus of the thalamus by its connections with the cerebral cortex may well supply the affective tone to a great number of intellectual experiences and through its connections with the hypothalamus may well be the means by which emotion and imagination are linked. . . . This fasciculus [bundle] of fibers connecting the prefrontal region with the thalamus is unquestionably of great significance, and there is little doubt that its interruption is of major importance in the alteration of the personality seen after frontal lobotomy.[5]

Thus, the theory was based on known anatomical connections between the prefrontal area, the thalamus, and also the hypothalamus—an area known to regulate the visceral responses that accompany emotional states.† They suggested that, for lobotomy to be successful, one had to cut the specific band of nerve fibers—the anterior thalamic peduncle—connecting the prefrontal area and the dorsomedial thalamus. By injecting opaque oil into the area cut by the leucotome, Freeman and Watts had obtained X-ray pictures of what pathways were severed during each operation, and had begun to compare this information with subsequent changes in a patient's behavior.

The most dramatic parts of the book were the transcripts of the conversations between Freeman and patients during the operation itself. The patients, who

---

* In common with many brain regions, its name denotes its anatomical location. Thus, *dorso* refers to the dorsal or upper part, while *medial* refers to the midline; and the dorsomedial thalamus is a region at the top of the thalamus, near the midline of the brain.

† Ignored was Karl Lashley's criticism of the idea that the thalamus is the center of emotions.[6] At present, neuroscientists believe that many brain structures regulate the intensity of emotional experience—among them the hypothalamus and the limbic system (see pages 264–65), as well as medial parts of the thalamus.

were given only a local anesthetic, often revealed progressive changes in mental state as the operation proceeded. As I noted earlier, some of the patients reported feeling a sudden relief from their anxiety or apprehension after two or three quadrants had been cut. Freeman and Watts explained, however, that "this is not enough": "Unresponsiveness or disorientation are usually necessary in order to obtain a satisfactory clinical result. When euphoria, not associated with other intellectual change [unresponsiveness and disorientation], is present within a few days after lobotomy, a relapse usually occurs."[7]

Other sections of the book, written by Watts, constituted a "do-it-yourself" manual. He not only described in detail how to perform the operations, but also provided practical information on caring for patients during the recovery period—what to tell relatives, possible deleterious consequences, and the immediate postoperative indications of whether a patient was likely to improve.

In two separate chapters, Thelma Hunt, an associate professor of psychology at George Washington University, summarized the results of psychometric tests she used to assess intellectual performance and personality. A delayed-response task was modeled after the one Jacobsen had used with chimpanzees (see pages 95–96); the human version, the Coin-under-Cups Test, tested a patient's ability to lift the correct cup in each of five pairs after observing all five cups being "baited" fifteen seconds earlier. Performance on this task, as well as on most of the other tests that were administered, actually improved postoperatively because patients generally were less agitated and more attentive. Speed of performance on some tasks was slower, but accuracy was generally improved. The performance of a few patients suggested that lobotomy might have produced some impairment. The results on the Maze Test, for example, suggested some loss in "the capacity to sustain several parallel lines of activity separately." For most readers, however, Thelma Hunt's conclusion was clear: "there is no impairment of the intelligence following prefrontal lobotomy."[8]

The personality evaluation relied heavily on the Rorschach Inkblot Test and the Word Association Test. No obvious personality changes were found, but Hunt reported a tendency for some lobotomized patients to have a more "constricted" personality because their responses were less varied and individualized than before the surgery. The overall conclusion was that the operation had not produced any devastating impairment in intellect or personality as far as psychological tests could detect.

The last three chapters of *Psychosurgery* presented a highly speculative view of the role of the frontal lobes in normal human behavior and in the psychoses:

> It is well for the individual to have a little fear, a little anxiety, a fairly high ambition, a bit of the perfectionistic spirit, an abundance of foresight and an awareness of himself. . . . He should build a few castles in the air, and people his world of fantasy

with good and bad genii. . . . But all this good can be turned to bad by excessive
elaboration, and especially when indecision brings about rumination, inaction, pure
fantasy and further retreat from reality with feelings of guilt and inferiority. . . .
anything that prevents such a fixation can be of benefit. . . . To the normal individual
the frontal lobes are indispensable; to the sick individual they may be destructive.
Without the frontal lobes there could be no functional psychoses.[9]

Severing the connections between the "thinking" brain and the "feeling"
brain, it was claimed, produces individuals whose emotional reactions are less
intellectualized and whose intellectual reactions are less emotionalized. Such
ideas, while containing some element of truth, were clearly dangerous in their
arrogant oversimplification of very complex processes. "Thinking" and "feeling" are certainly not simple, homogeneous concepts. We think and feel in
many different ways, and the contribution of the frontal lobes and the thalamus
to these processes is poorly understood even today. Moreover, many structures
in the brain are important for "thinking" and "feeling" besides specific portions
of the frontal lobes and thalamus; and these regions have many different
functions, in addition to thinking and feeling. In emphasizing the balance
between emotion and thinking, however, Freeman and Watts had provided
an alternative to Moniz's weak theory, which they now criticized for the first
time, even though gently.

> The theory of Egas Moniz that the operation is successful through the breaking
> up of constellations of neuron patterns will bear further examination. There is no
> doubt about the stereotyped thinking and stereotyped activity indulged in by many
> of the patients. But whether such ideas and activities are actually in relation with
> abnormal stabilization of synaptic patterns in the brain is another matter.[10]

Freeman and Watts did include some warnings in the preface of *Psychosurgery*:

> Operations upon the brain are by no means to be applied indiscriminately in the
> treatment of functional mental disorders. In fact, in view of certain unfortunate
> results, the operation of prefrontal lobotomy is reserved for those patients whose
> outlook for recovery is poor, whose response to other treatment is unsatisfactory,
> and for those who are facing disability or suicide. Not always does the operation
> succeed; and sometimes it succeeds too well, in that it abolishes the finer sentiments
> that have kept the sick individual within the bounds of adequate social behavior.
> What may be satisfactory for the patient may be ruinous to the family. It is in part
> from a desire to instil caution that our bad results have been emphasized, and it is
> hoped that our experience will be a guide to the correct choice of individuals to be
> operated upon.[11]

However, as I will describe, in the absence of any alternative treatment, this
warning was disregarded not only by most physicians but especially by Walter

Freeman, who came in time to believe that lobotomy should not be considered the "last resort" as it was dangerous to wait too long.

For Freeman—and it was he who wrote all the theoretical and speculative sections of *Psychosurgery*—the prefrontal area also makes it possible to project oneself into the future. This part of the brain, Freeman believed, is critically involved in the ability to anticipate the effect of one's actions on oneself and on others. It regulates the involvement "of the self with the self."*

> Consciousness of the self is an undeniable blessing; representing man's highest endowment. But let self-consciousness become elaborated until the individual begins to hear his name mentioned, to see total strangers follow him and watch him, to see people in conversation look at him covertly . . . it can not but have a grave effect upon his ability to meet his fellows upon a plane of equality.[13]

What was most influential in *Psychosurgery*, however, was the theory that specific brain pathways between the frontal lobes and the thalamus regulate the intensity of the emotions invested in ideas. This was widely and uncritically accepted as the most promising scientific justification for psychosurgery.

In the United States, the attention the popular media gave *Psychosurgery* was truly unusual for a medical book: not only was lobotomy inherently interesting, but Freeman did everything to attract a wide audience. He designed an eye-catching dust jacket; and above the title, ran the legend, written by Freeman himself: *"Read the last chapter to find out how those treasured frontal lobes, supposed to be man's most precious possession, can bring him to psychosis and suicide!"* (See figure 8.6, page 166.)

There were many reviews in newspapers and magazines. Waldemar Kaempffert, who had earlier written the widely read *Saturday Evening Post* article on lobotomy, noted in the *New York Times* that the book is for everyone: "It is not too technical even for a layman. . . . No novelist ever had a more thrilling subject." *Psychosurgery* was said to be "practical and realistic" and would probably "be used by every progressive neurologist and psychologist." Kaempffert observed that all the pertinent literature was reviewed:

> the trephining of savages, the accidents to the brain that have changed human character, the operations performed to remove brain tumors . . . the work done by such surgeons [*sic*] as Dr. John F. Fulton on the brains of monkeys and anthropoid apes . . . the brain injuries that soldiers sustained and finally the successes and failures of psychosurgery.[14]

The favorable review of *Psychosurgery* in *Time* magazine influenced great

---

* Over the ensuing years, Freeman became ever more convinced that self-consciousness was central to the changes produced by lobotomy, and he elaborated on this theme in a book written with the psychologist Mary Robinson.[12]

numbers of people to consider lobotomy as a possible treatment for intractable mental disorders. Although only 80 cases were described in their book, *Time* determined that Freeman and Watts had completed 136 operations by the time it was actually published: "Drs. Freeman and Watts regard 98 as greatly improved, 23 as somewhat improved, twelve as failures. Only 13 patients are still in mental hospitals; most are back at their jobs or housekeeping after one to six years of psychotic incapacity."[5]

The wide coverage and enthusiastic response to Freeman and Watts's book had an enormous effect. Freeman had made a special point of persuading the publisher to ship copies of the first printing to England, but the boat was sunk by a German submarine. The next shipment got through and had an immediate impact, as the neurosurgeon Eric Cunningham Dax recalled: "The book by Freeman and Watts hit us like a bomb. With over 300 pages and about the same number of references, we felt that the thoroughness of their enquiries and the amount of work they had done was so great that there was hardly anything to say."[6] In the manuscript sent to the publisher, Freeman had written that "we have been unable to find even hearsay evidence of operations carried out in England or in the Scandinavian countries." A footnote added to the galley proofs noted that the first report of psychosurgery in England had just been published in *Lancet*.[7] Clearly, the book had arrived at a time to have the largest possible influence. Writing in 1949, Maurice Partridge of St. Andrew's Hospital in Northampton described the influence of Freeman and Watts's book:

> They brought some order to a confused and chaotic subject by the publication, in 1942, of an exuberant and brilliant monograph *Psychosurgery* which crystallized six years experience.... This was a considerable achievement. The results were not only a tribute to the moral courage and perseverance of the authors, but formed a unique contribution to knowledge, as the first attempt at systematization of this bewildering subject.[18]

Kevin Walsh later observed that, after *Psychosurgery* arrived in England, "the classical lobotomy of Freeman and Watts became the standard procedure in most centres."[19] Although this was an exaggeration—several British surgeons were already developing their own operations—there was indeed great interest in the Freeman-Watts procedure in Great Britain. The detailed instructions and illustrations made it possible for someone to learn to do the surgery directly from the book.

The number of operations started to increase rapidly in England. In March 1943, the Royal Medico-Psychological Association, the predecessor of the Royal College of Psychiatrists, sponsored a symposium in London on prefrontal leucotomy. Five separate medical groups in Great Britain reported the results of over two hundred operations. The Freeman and Watts procedure and "fronto-thalamic theory" were discussed at length during the symposium.[20] In France, where there had been considerable skepticism and opposition to lobotomy, several psychiatrists and neurosurgeons were persuaded that it was no longer possible to ignore the accumulating evidence.

Because of the war, it took a few years before the full impact of *Psychosurgery* was felt, but shortly after the hostilities ended in 1945, lobotomies increased rapidly. There had been no serious attempt to survey the amount of psychosurgery done in the United States before 1942, but *Time* estimated that there had been about 300 operations by this date.[21] In 1943, Lloyd Ziegler's survey of eighteen hospitals in the United States and Canada revealed that at least 618 lobotomies had been performed by mid-1943.[22] At the end of 1946, *Time* reported that 2,000 operations had been performed in the United States.[23]

The interest in lobotomy was accelerating rapidly as attested to by the results of C. C. Limburg's 1949 survey of 855 institutions in the United States, "including all mental hospitals, general hospitals with psychiatric wards and all medical schools." His data indicate that there had been an almost exponential increase in psychosurgery between 1945 and 1949. In 1945, only 150 operations were performed; but there were 496 in 1946, 1,171 in 1947, 2,281 in 1948, and 5,074 in 1949. Between 14 September 1936, the date of Freeman and Watts's first operation, and 15 August 1949, when Limburg's survey ended, there had been 10,706 lobotomies in the United States.[24] Freeman and Watts were responsible for about 10 percent of the total, as they reported over 1,000 operations when the second edition of *Psychosurgery* was published in 1950. James Poppen modified Lyerly's operation and completed 470 prefrontal lobotomies by 1948 (see page 196).*

Morton Kramer conducted a later survey and reported that 18,608 lobotomies had been performed in the United States by 30 June 1951. He found that, in 1945, there were only 49 hospitals in which any psychosurgery had been performed, but by 1950 these operations were being done in 286 hospitals in forty-three of the forty-eight states and also in the territory of Hawaii. Idaho, Maine, New Mexico, North Dakota, and Utah were the only states where psychosurgery had not been done by this time. During 1950, 56 percent of the total lobotomies in the United States were performed in state hospitals, 16 percent in medical school hospitals, 12 percent in Veterans Administration hospitals, and 6 percent in private hospitals. The remaining 10 percent were done in "psychopathic," city, or federal hospitals.[26] The staff at medical schools was more involved than these figures might suggest, however, as neurosurgeons affiliated with universities often did the surgery in state hospitals.

* Poppen was affiliated with the Lahey Clinic, but the majority of the surgery was done at the Boston Psychiatric Hospital.[25]

While its influence can hardly be overestimated, Freeman and Watts's book was not solely responsible for the rapid increase in psychosurgery. Had there not been a problem in desperate need of a solution, the arguments and evidence presented in their book would have had little impact. The desperate problem, of course, was the huge number of mentally disturbed people around the world who were receiving little or no help and were, as a result, deteriorating in overcrowded and understaffed institutions.

A major part of the tenfold increase in lobotomies between 1946 and 1949—from five hundred to five thousand per year in the United States—took place in state mental hospitals, where approximately 56 percent of all psychosurgery was performed. These hospitals, lacking the funds to care adequately for their patients and any therapy to cure their illnesses, had become ugly "warehouses," where hopeless patients were stored for years. According to the Census Bureau data of 1904, nearly 40 percent of mental patients in all hospitals had been there five years or more.[27] The figures were even more discouraging for state mental hospitals, where the duration of confinement had been steadily increasing. In the state mental hospitals of Massachusetts, the average duration of confinement rose from 8.9 to 9.7 years between 1929 and 1937.[28] Nathan Allen's description of mental asylums in the United States in 1875 applies equally well to conditions in the 1940s: "Unless there are some means besides death of eliminating and removing the incurable and the harmless insane from our lunatic hospitals, these institutions become filled up with a class of patients, very few of whom can ever be benefited by curative treatment."[29]

Jacob Norman's 1947 report on Massachusetts state hospitals was depressing enough, but the conditions he described were actually better than in most states.[30] In Massachusetts, Norman reported that from 40 percent to 45 percent of patients in state hospitals were suffering from psychoses caused by organic brain disease, mainly senile dementia, cerebral arteriosclerosis, and neurosyphilis. The second largest group were the chronic, "burned-out" schizophrenics, most of whom had been living in their own world within the hospitals for many years. There were also alcoholics who usually considered the hospital a jail, and psychopaths sent for diagnosis by the courts. It was a hopeless, depressing atmosphere; and psychiatrists themselves had to struggle not to be engulfed by it. A series of exposés in the 1930s and 1940s describing the "ugly," "crowded," "incompetent," "perverse," "neglectful," "callous," "abusive," and "oppressive" conditions in state mental hospitals effected little change.[31]

Patients were beaten, choked, and spat on by attendants. They were put in dark, damp, padded cells and often restrained in straitjackets at night for weeks at a time. Life magazine's article "Bedlam 1946" vividly described the deplorable conditions that existed in most of the 180 state mental institutions.

The conditions were said to have degenerated "into little more than concentration camps on the Belsen pattern." A photograph taken at Philadelphia's Byberry Hospital showed nude male patients on concrete floors: they were given "no clothes to wear [and] live in filth" (see figure 9.1). A handcuffed patient in a state hospital in Iowa was said to have been kicked repeatedly in the back and genitals by an attendant.[32]*

The lack of trained personnel in state mental hospitals often made it impossible to use even the few therapies that were available. Psychotherapy was totally impractical and, besides, most psychiatrists seriously doubted that it could help the gravely deranged patients in these institutions. According to the report of the West Virginia Board of Control, there was insufficient staff at the Huntington State Hospital in 1943 to use insulin therapy because it required staff both to administer intravenous injections and to monitor patients closely.[34] This hospital had many more patients than beds, and a large number of inmates had to be housed either in the jail or in the old dormitory on the hospital's farm. While shock therapy had helped some patients, it had not "proved to be as successful as it was originally believed." Fever treatment was still used for paretics, and a strain of malarial mosquitoes was maintained at the hospital. Hydrotherapy—mostly wet packs—was given to the more disturbed patients and was considered "the most useful single form of treatment available in the hospital."[35]

In the Northern State Hospital in Sedro Valley, Washington, another institution where Freeman later introduced transorbital lobotomy (see chapter 10), only $2.21 was allocated for the daily care of each patient even in 1954, and then only after the deplorable conditions in mental hospitals had been described in numerous books and magazine articles.[36] Out of this small amount, $1.33 went for staff salaries, and 87¢ was used for "general operations," leaving only about 16¢ a day for food.[37] The annual expenditure for patients in Washington was close to the median figure, as twenty-two states had a lower average per-capita expenditure.[38] Charles Jones, one of the Washington State psychiatrists whom Freeman would train to do transorbital lobotomies, asked: "What sort of medical care will that buy? It is quite obvious that such appropriations have been made with the supposition that only custodial care was to be given to the mentally ill. It certainly does not appear that anyone had had the idea that a great deal of treatment would be forthcoming on appropriations of this size."[39]

The American Psychiatric Association estimated overcrowding in mental hospitals even in 1948 to be in excess of 50 percent. More than 230,000 hospital

---

* Problems persist to this day. In 1984, the *New York Times* reported that a mental patient in a straitjacket at Creedmoor Psychiatric Center in New York died following an alleged beating by a therapy aide.[33]

Figure 9.1. These men, patients at Philadelphia's Byberry Hospital in the 1940s, are representative of thousands of other mental patients, both men and women, who were consigned for years to the wretched conditions of many state mental institutions across the country in the first half of this century. In overcrowded and unsanitary buildings, these people lived, naked and surrounded by their own filth, with neither exercise nor any other activity or therapy to relieve them. This photograph appeared in a *Life* magazine exposé (6 May 1946) entitled "Bedlam 1946." (Courtesy LIFE Picture Service; photo by Charles Lord.)

beds for the mentally ill were judged to be substandard, and many more people needed hospitalization. William Menninger reported that there were 700,000 beds for mental patients in the United States; but that hospitalization of everyone needing intensive psychiatric care would require 1,500,000 beds. The Central Inspection Board of the American Psychiatric Association found that twenty-nine mental hospitals had no psychiatrist at all; and that of the 300,000 registered nurses in the country, only 1,200 were directly involved in psychiatric nursing, even though 55 percent of all hospital beds were occupied by the mentally ill.[40]

Almost any proposed treatment was considered worth trying, as long as it did not require much money or large numbers of skilled personnel. Certainly almost anything could be tried with impunity, as malpractice suits and ethics committees were rare. This attitude was reinforced by the many books and articles that exposed the conditions in state hospitals, while waxing enthusiastic about the new somatic therapies. One of the most influential books was Albert Deutsch's classic *The Shame of the States,* which had first appeared as a series of newspaper articles.[41] After vividly describing the horrendous living conditions in state asylums, Deutsch advocated more scientific treatment of patients, referring to the "promising new somatic therapies." Most psychiatrists were ready to explore these new therapies, as the hopeless condition of mental patients seemed to justify any risk; also, by using these somatic techniques, psychiatrists felt they were brought closer to the mainstream of medicine, from which they often felt isolated. The psychoanalyst Roy R. Grinker commented on the eagerness with which psychiatrists in large institutions seized on the new somatic treatments:

> With avidity of interest aroused, and the rapidity with which the use of insulin and metrazol treatment spread to every corner of this country—attests a certain preparedness and eagerness of the rank and file psychiatrists for an organic approach to psychosis. Shock treatment has livened up the mental hospitals and psychiatrists once again. . . . But interest in the uncovering of basic psychological causes has decreased, for the busy psychiatrist now hardly waits for the patient to undress in his hospital before shocking them into insensibility.[42]

Experience during the Second World War had intensified interest in new ways to treat mental illness by increasing awareness of the magnitude of the psychiatric problem in the United States. William Menninger, who headed psychiatry in the U.S. Armed Forces during the Second World War, reported that—of the 15,000,000 men examined prior to admission to the armed forces—1,846,000, or 12 percent, were rejected for psychiatric reasons. An additional 632,000 were discharged after admission for "mental breaks."[43]

In 1943, the seriousness of the situation prompted the Veterans Administration to issue a memorandum encouraging neurosurgeons affiliated with its neuropsychiatric installations to obtain training in prefrontal lobotomy. The memorandum specified that patients suitable for psychosurgery were those "in which apprehension, anxiety, and depression are present, also cases with compulsions, with marked emotional tension," after shock treatment and other available therapies had been exhausted.[44]

After the Second World War, psychosurgery increased in virtually every country where there were neurosurgeons. In England the results of over 1,000 prefrontal leucotomies were reviewed in 1947.[45] Charles Burlingame, the psychiatrist-in-chief of the Institute of Living in Hartford, attended the annual meeting of the Royal Medico-Psychological Association in July 1949. It was the second year of socialized medicine in England, and Burlingame reported that "the interest in psychosurgery in England is acute."[46] By 1954, the outcome of 10,365 operations had been described—a figure that did not include all the psychosurgery done in Great Britain during the period.[47] One particularly active neurosurgeon, Wyllie McKissock, at St. George's Hospital and the National Hospital in Queens Square, London, completed his five-hundredth leucotomy by April 1946.[48] The number of articles appearing in the medical literature in Great Britain also reveals the growing interest in psychosurgery. Between 1936 and 1941, the Library of the Royal Society of Medicine listed fewer than twelve publications a year on the topic of psychosurgery published worldwide; by 1951, there were about three hundred articles a year.[49*] Great Britain had made up for its slow start, and its influence spread to India, where leucotomy had also been begun during the war.[51] Prefrontal leucotomy was started in Czechoslovakia during the German occupation.† Psychosurgery was started in Hungary in 1946 after Freeman had sent a copy of *Psychosurgery* and some reprints of articles to T. De Lehoczky the year before.[52]

In Japan, where psychosurgery had first been performed in 1938, there was little opportunity to pursue this practice until after the war. In 1946, Dr. Paul Schrader, a surgeon attached to the U.S. Armed Forces in Osaka, demonstrated the Freeman-Watts lobotomy to Japanese surgeons. The following year he was invited to give a lecture on the subject at the Japanese Medical Congress in Osaka. Schrader had worked in the State Hospital in Farmington, Missouri, where, in spite of his relative isolation, he had performed over one hundred lobotomies before being called up for military service. Schrader's lecture and

---

* In the 1950 edition of *Psychosurgery,* Freeman and Watts noted that they had found over three hundred titles of articles on psychosurgery published in 1949 alone.[50]

† According to a personal communication (dated 20 December 1983) from Pavel Nádvornik, professor of neurosurgery at the Neurosurgical Clinic in Bratislova, the Moniz-Lima operation was performed in Prague and Bratislova in 1942: the surgeons, Knobloch and Kukura, collaborated with two psychiatrists, Matulay and Vondracek.

demonstration accelerated Japanese interest in psychosurgery. By April 1950, two thousand operations had been performed in twenty-eight different hospitals in Japan.[53]

Little surgical skill was required to perform psychosurgery, and by 1946 one could learn it at several hospitals. All of the residents supervised by Freeman and Watts at George Washington Hospital became familiar with the operation, and many of them started performing psychosurgery at other institutions when their training ended. Some of these residents came from foreign countries, primarily from Latin America. After returning home, they spread the interest in psychosurgery; they included Humberto Fernández-Morán, a Venezuelan, Abraham Moscovich from Argentina, and also Manuel Velasco-Suarez, who had great influence on the practice of neurology and neurosurgery in Mexico.[54]

The popular press continued to stimulate interest in prefrontal lobotomy during this period. The 1947 article in *Life* magazine—"Psychosurgery. Operation to Cure Sick Minds Turns Surgeon's Blade into an Instrument of Mental Therapy"—reached millions of people around the world. *Life* explained that the operations were started in Portugal by Egas Moniz, a "daring and imaginative neurosurgeon [*sic*]," and that two thousand prefrontal lobotomies had already been done in the United States. Describing psychosurgery as "the most direct and dramatic attack ever made on mental illness," *Life* stated that the operations had been performed

only on "hopeless" patients who had failed to respond to other methods of treatment, people who had little to lose and everything to gain. The results were spectacular: about 30% of the lobotomized patients were able to return to everyday productive lives. Another 30% benefited considerably, finding relief from the painful anxiety and profound depression that their psychoses inflicted on them. The rest were mostly unaffected. A very few deteriorated mentally. Only two or three percent died.[55]

Clearly, *Life*'s summary of the results of lobotomy was uncritically enthusiastic and misleading. The statement that 30 percent of "hopeless patients" were returned to normally productive lives was not true. Moreover, while *Life* acknowledged that the operation was not a "cure-all" and did not replace all other therapies, of these other therapies the magazine mentioned only "somatic therapies," primarily insulin and electric shock treatment. A series of photographs showed psychosurgery being performed at the Boston Psychopathic Hospital, where two hundred lobotomies had been done. *Life* explained that the surgical procedure illustrated (the Lyerly-Poppen procedure) differed from that used by Freeman and Watts in that larger skull openings were made in order "to enable the surgeon to see what he is doing." An accompanying cartoon explained in psychoanalytic language how the operation liberated an agitated and depressed patient from a tyrannical "superego" that had the "id" all tied up in knots (see figure 9.2).

The article in *Life* was typical of many of the simplistic and misleading articles that appeared in popular magazines. While *Life* was picturing the operation as weakening the superego, the affiliated news magazine *Time* described a "criminally insane" woman who had been arrested in Detroit on charges of robbery and arson and had apparently acquired a superego after a lobotomy. She was said to have spent fifteen years in reform schools and prisons, been thrown out of bordellos for injuring patrons, and boasted that she had committed two murders. *Time* reported that after her lobotomy she was "presented" at a meeting of the Michigan Society of Neurology as demonstrating an "amazing transformation." She was quoted as wishing "to go home and lead a normal life"; and the hospital staff now considered her to be "friendly, cooperative" and "seriously interested in the future"—a prediction later events did not fulfill.[56] Thus, while *Life* described the effects of lobotomy as destroying the superego, *Time* implied that the operation created a superego where apparently there had been none before. The medical literature also contained much discussion of the effect of lobotomy on the superego.[57]

Although psychoanalytic concepts were often used to explain the changes produced by lobotomy, most psychoanalysts were opposed to these operations. However, as most physicians realized that psychoanalysts had no practical solution for seriously ill mental patients, their criticisms tended to be discounted. Freeman had several heated arguments with psychoanalysts, but as most of these were face-to-face encounters and were not published, their influence was limited. The analyst Harry Stack Sullivan, who had been trying to treat a small group of schizophrenic patients with psychotherapy, however, did publish an editorial critical of lobotomy: "When the treatment of victims of severe mental disorders by diffuse decortication and destruction of some of their human abilities is sanctioned on the grounds of social expediency, and all other methods of therapy are discarded as impractically difficult, one may well be concerned for psychiatry."[58] Sullivan's opposition, however, had relatively little effect on the practice of lobotomy. Neither *Psychiatry,* the journal Sullivan edited, nor the variation of psychoanalysis he developed— which emphasized an intensive "working through" of a patient's current interpersonal relationships—offered a practical alternative for psychiatrists working in institutional settings. Freeman, who did not bear grudges, later greeted him at a cocktail party, extending a hand, "How goes it, Harry?"

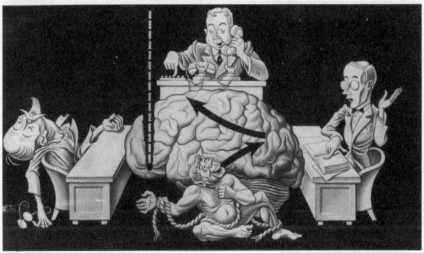

Figure 9.2. Boris Artzybasheff's cartoons explaining, in psychoanalytic language, how lobotomy works. The caption was: "In agitated depression (*top drawing*) the superego becomes overbearing and unreasonable, unbalancing the whole mind and making impossible demands on the individual. The victim, always goaded by an overdeveloped conscience, lives in agony of guilt and deep anxiety. Impulses are stifled, intelligence is overridden; even actions are repressed. The surgeon's blade, slicing through the connections between the prefrontal areas (the location of the superego) and the rest of the brain, frees the tortured mind from its tyrannical ruler (*bottom drawing*). Intelligence is not affected (patients have done well in college and business). Lobotomy, however, should be performed only on those patients whose intelligence is sufficient to take over the control of behavior when the moral authority is gone." (These cartoons appeared in *Life,* 3 March 1947. Used by permission of the estate of Mina Turner.)

Sullivan lost control of himself and, shaking his fists over his head, shouted, "Why do you persist in annoying me?"[59]*

In January 1949, Walter Freeman organized a postgraduate course in psychosurgery at the George Washington School of Medicine.[62] Although the majority of the contributors to the proceedings were performing psychosurgery or actively involved in its practice, also invited were Dexter Bullard, David Rioch, and Winfred Overholser, who were known to be opposed to lobotomy. Harry Solomon was asked to moderate a discussion on the pros and cons of psychosurgery. Bullard, the medical director of the Chestnut Lodge Sanatorium in Rockville, Maryland, opened with brief remarks about past travesties perpetrated under the guise of psychiatric therapy. He went on to state that his "major criticism of psychosurgery is what it is doing to the medical profession itself, and what it will do to the medical students, the doctors of tomorrow."

> When psychosurgery is prescribed, the emphasis is on symptomatic relief, not on etiology, not on insights to facilitate living, not on efforts to understand the process involved, not on their potential modifiability by psychotherapy. All that is known of the dynamics of psychotherapy is neglected. All that is known of mental hygiene is discarded. Perhaps this is because the psychosurgeon is ignorant of them.[63]

David Rioch, an eminent neuroanatomist turned psychiatrist, was disturbed by the permanent destruction of brain tissue. He said that prefrontal lobotomy was a therapy that "amputated functions," and characterized it as "partial euthanasia": "Born of the frustration which every physician feels in working with the psychotically ill, the enthusiasm for cure results in the application of this procedure to patients who have a 1 in 4 chance in the worst state hospitals and a 4 in 5 chance in our best institutions of putting themselves together." Speaking metaphorically, Rioch stated that "the psychosurgeon is indeed treading on dangerous ground when he decides that a patient without a soul is happier than a patient with a sick soul."[64] At this same meeting, Winfred Overholser summarized the experience at St. Elizabeth's Hospital, pointing out that the condition of many of their patients was worse after lobotomy. The successful cases could be sent home; but, in general, few of them could hold jobs, and psychological studies had shown that their imagination and ability to solve problems were much reduced. Lobotomy, Overholser concluded, substitutes "an organic syndrome for the pre-existing psychosis."[65]

---

* In spite of this encounter and their very different views of mental illness, Freeman's description of Sullivan in his 1968 book on psychiatrists was well balanced.[60] Freeman had written a letter of recommendation for Sullivan when the New York Certification Board discovered that Sullivan had gone to a medical school that was not accredited.[61]

The other side of the debate was argued by Walter Freeman, Lawrence Pool (at the time a neurosurgeon at Columbia University's New York State Psychiatric Institute), and James Watts. The last answered the criticism by describing the suffering of patients who were candidates for lobotomy:

> patients with involutional or agitated depression with feelings of guilt who pace the floor, wringing their hands, mourning and crying. Still others with schizophrenia are confined to strong rooms because of their abusive assaultive behavior. If given a bed, they pull it apart and assault attendants with the pieces. They are often naked, refusing to wear clothes, urinate and defecate in the corner.... Food is poked through a crack in the door like feeding an animal in a cage....
>
> Following lobotomy, even the worst improve enough so that they live on an open ward, wear clothes, go to the dining room with other people, and get out in the sunshine.
>
> Some members of the panel have suggested that we wait until they develop some form of therapy more to their liking. Perhaps I am a bit impatient at times. A woman 65 years old with agitated depression does not have but so long she can wait; and a man with dementia praecox confined to a strong room for 5 years has already waited long enough.[66]

Thus the lines were drawn as they had always been. The neuropsychiatrists and neurosurgeons in favor of lobotomy were not likely to be impressed by the arguments of psychoanalysts and were convinced that psychotherapy in any form had little practical to offer for treating the types of patient considered a candidate for lobotomy. The dramatic cases—violent patients who had been in locked wards for many years—were always brought up in such debates, but they were not truly representative of most lobotomized patients. As will become evident later, many patients had not been ill for very long and were neither old nor violent.

For the most part, the criticisms of psychoanalysts were ineffective. Few of them, as noted, were published and, moreover, at least some of the criticism was embedded in psychoanalytic language that was either incomprehensible or offensive to ordinary practicing psychiatrists. Consider the comment by Smith Ely Jelliffe, more than a decade earlier, following a presentation by Freeman and Watts in 1938. Jelliffe criticized not so much the concept of lobotomy as the fact that those doing the operations had neglected psychoanalytic theory:

> If we accept the general Freudian formula, and I see no reason not to, then the compulsion neurosis uses the mechanism of displacement from early erotic fixations.... Psychoanalytic psychiatry has pointed very clearly to a highly important organ libido investment in the "compulsion neurosis." This is the cathexis of the anal sadistic with its massive hostility drive. This anal sensory perception elaborating area, Dr. Penfield I think partly localized at our 1934 meeting.

Jelliffe continued in this vein, loosely combining psychoanalytic concepts with some diffuse ideas about brain function. Asserting that the prefrontal brain area is where many of the different "fixations" identified by psychoanalysis are located, Jelliffe argued that what was needed was a more selective lobotomy severing only the fibers connecting the frontal lobe with the brain area that receives sensory information from the anus:

> I question the general policy of the Charles Lamb Roast Pig Formula of burning the house down to roast the pig. . . . If there is any sense at all in hobbling a horse to prevent his running away it would seem to me that if there could be an isolation of the frontal association wires of these anal sensory perception areas, one might do some definite cutting instead of putting the whole instrument out of commission in order to correct a difficulty that is possibly localized in only a part of it.[67]

Jelliffe did not argue against lobotomy so much as suggest refining it to conform with psychoanalytic theory.

When Jelliffe finished, Abraham A. Brill, another prominent analyst, added his comments to the discussion of Freeman and Watts's presentation.* He made no attempt to conceal his feelings about both lobotomy and also Jelliffe's bizarre comments:

> I cannot understand why such ideas are brought here, and much less the fantastic conclusions drawn from them by some of the discussers. I suppose we are so polite, that anything may be ventured with impunity. I know two of the four cases which furnished the material for the paper under discussion. One of the cases, Miss S. . . . telephoned me and said: "Dr. Brill, I now feel worse than I ever did. . . . I am only sorry I did not write to you before I went to Washington. I had two operations and I am worse than before." . . .
>
> The whole presentation reminds me of a time when they used to treat homosexuality by prostatic massage. I recall such a patient who had been treated . . . for two years, but he was still homosexual. . . . As the physician in question was a man of the highest type, I was sure he must attribute some virtue to such treatment, so I asked him how he expected to cure homosexuality by prostatic massage. He thought the massage might rub out the homosexual cells and that they would be replaced by heterosexual cells. . . .
>
> I feel that there is absolutely no reason why we should be in any way impressed by the seriousness of these presentations, in spite of the fact that I highly regard the readers. I know that they are seriously trying to contribute something to science, but what they showed us is nothing but interesting experimentation. The material they presented this evening is no more conclusive of the validity of their claims than the putative cure of homosexuality by prostatic massage.[68]

Brill's strong attack was not, however, representative of the criticisms of lobotomy most psychoanalysts made at professional meetings. They usually

---

* Other discussants were R. Brickner and Kurt Goldstein.

were skeptical, but ended on some middle ground, recommending a limited use of lobotomy—only on patients in whom all other less "heroic" measures had been completely exhausted. More representative was the encounter that took place in Cleveland at the 1941 meeting of the American Medical Association. The session on lobotomy was chaired by Paul Bucy, an eminent neurosurgeon at Northwestern University. Bucy opened the discussion with an explanation about the composition of the panel:

> I think it is no secret that in some quarters this procedure has met with considerable opposition, on occasions violent. It may appear to the audience that this group [the discussants] in the main has been stacked in favor of the procedure. I assure you that it is not true and I assure you that any apparent stacking has not been done by those who organized the discussion, but those who have opposed the procedure have, with amazing uniformity, refused to give public utterance to that opposition.[69]

Roy Grinker, the Chicago psychoanalyst, was the only panelist critical of lobotomy, and he constantly attacked, or reinterpreted, the favorable results presented by the other panelists. Thus, when it was reported that improvement after lobotomy was correlated with weight gain, Grinker turned this statement around: "I think there is a very strong possibility that those excessive gains in weight are due to some disturbance in the cerebral centers. Particularly one might imagine traction on the hypothalamus or bleeding in the neighborhood." He raised the specter of delayed adverse consequences of lobotomy and speculated that brain "scarring" could produce effects that might not be evident for years. He raised the need for "controls"—unoperated patients with whom to compare the results after lobotomy—and asked whether all the special attention from the extra nurses and the enthusiasm of the surgeons might not, rather than the surgery, be the cause of improvement. Grinker doubted the argument of lobotomy's proponents that it could save enormous amounts of money, and noted that when the patients were returned home, the costs were simply "borne by another agency, namely the home rather than the state."[70]

Grinker also questioned whether many lobotomized patients actually had long intractable illnesses. He called attention to the surprisingly short duration of illness of some of the patients, who "have been operated on after only a few months of mental illness." Taking the position of an elder statesman above the argument, he observed:

> On both sides of this question there is obviously a great deal of preoccupation and emotional bias. I think that those who have taken up this procedure have been interested in physical therapy in psychiatry, which has been based largely on a partially expressed and somewhat unexpressed feeling that the psychoses have a definite organic basis. On the other hand, those who are opposed to the method

have again an emotional attitude toward a mutilating operation that destroys brain tissue. As long as we realize that on both sides there is some bias, perhaps listeners may conclude for themselves which is fact and which is bias.

In the first place, one tries to find out what type of case is selected for this operation. The statement is categorically made that those cases that are untreatable by conservative methods are selected, and the question that naturally enters my mind is, What are those conservative methods? Surely, by conservative methods one does not mean metrazol or electric shock.

We are dealing with a large number of patients who have been chosen from state hospital populations. There have been some private cases, it is true, but when one thinks of what conservative treatment means, of what actual, thoroughgoing psychological study and treatment mean, it is difficult to imagine that the patients who come from state hospital populations are getting the benefit of that type of treatment.

Grinker insisted that neurosurgeons must assume some of the responsibility to make certain that all conservative therapies have been exhausted, and added, "I mean thoroughgoing psychologic treatment." It is not sufficient, he asserted, for a surgeon to point to his "neurologic and psychiatric friends while saying 'I did it because they told me to.' "[71] Apparently everyone agrees that, even in the best cases, the surgery produced defects, Grinker pointed out. Besides the defects in synthesis and capacity for abstraction, he said that there were losses in initiative, ambition, productivity, and creativeness, and that they should not be ignored just because such traits are difficult to measure.

In spite of all this criticism, Grinker said, in concluding, that he would probably disappoint many in the audience who hoped that he would take a stronger stand—a more "iconoclastic position"; but "I think there is no question that the operation has a usefulness" even though the limits of the usefulness have not been clarified: "Older patients, perhaps people who have no chance whatever except to continue in state hospitals, are those on whom further experiments, since I think this is still an experiment, should be made."[72*]

However appropriate his criticisms, Grinker obviously avoided responding to this challenging question from a member of the audience, "What success can be anticipated with psychotherapy in involutional melancholia?" It was one thing to criticize lobotomy, but criticism that could not offer a practical alternative had a hollow ring. The problem was stated clearly by E. A. Stephens, medical director of the Territorial Hospital in Hawaii:

With the advent of interpretive psychiatry as formulated by Kraepelin, Bleuler,

---

* Grinker's conclusion was the same as that suggested by the Boston neuropsychiatrist Stanley Cobb, who wrote: "In my opinion this is a justifiable procedure only when the patient is old and the prognosis hopeless. Specifically, I can recommend the operation only in cases of prolonged agitated depression in persons over 60 years of age."[73]

Freud, and others, it was hoped that cures could be effected by psychotherapeutic measures, but, alas, this hope has rarely been fulfilled. Knowledge of mental mechanisms has, to be sure, increased remarkably in recent decades, but the application of this knowledge is too often frustrated by the regression or inaccessibility encountered in the major psychoses. Since earliest days, a feeling of futility has pervaded all institutions caring for the more serious psychoses.

It is no wonder then that the more drastic methods first introduced by von Meduna with the use of metrazol in 1934 and by Sakel with insulin about the same time were welcomed by workers in state institutions who had become resigned to the use of measures that were, in most instances, no more than palliative. More recently, stimulation of the cerebral cortex by an electric current has been added to the armamentarium of the psychiatrist, and this method, too, has received wide use and acclaim in the fight against mental disease.

Although all these approaches have given encouraging and, in some instances, spectacular results, it must be acknowledged that the earlier promises of success have not been fulfilled. Consequently, state hospitals still have many challenging psychoses that present a serious socio-economic problem because of their chronicity.

About two years ago, it was decided to attempt to meet this challenge in the Territorial Hospital by the use of psychosurgery as first introduced by Moniz of Portugal in 1936 and later by Freeman and Watts of America.[74]

Not one group took an effective stance against psychosurgery, even though many individuals were strongly opposed to the practice. Psychologists, many of whom had a less biological bias than medically trained psychiatrists, did not, as a group, oppose lobotomy, and many participated in studies reporting favorable results. Indeed, the highly respected psychologist Donald Hebb wrote that "prefrontal lobotomy is a landmark in psychiatric therapy" in spite of his reservations about what it was teaching us about "what goes on in the normal frontal lobes."[75] Only a few published criticisms of lobotomy during this period failed to end with a "middle of the road" recommendation to proceed cautiously until more research was done. A 1941 editorial in the influential *Journal of the American Medical Association* exemplifies the typical "establishment" position:

> An emotional attitude of violent unreasoning opposition to this form of treatment [lobotomy] would be inexcusable. True it is a mutilating operation and it does result in certain defects in personality and behavior. However, much surgery is mutilating in the sense that some ordinary normal tissue is removed in order to achieve a beneficial result. . . . No doctor can yet assert that this is or is not a truly worthwhile procedure. The ultimate decision must await the production of more scientific evidence.[76]

A rare exception to this attitude appeared as an editorial, "The Lobotomy Delusion," in the May 1940 issue of the far less influential *Medical Record:*

Lobotomy for frontal lobe malignant tumor we can understand, but this is extended lobotomy, one is supposed to be able to pluck from the brain a hidden sorrow. . . . If the cutting off of one set of frontal lobe association fibers is not sufficient try cutting both sides, i.e., analogously, if cutting off one leg of a paraplegic does not cure the individual cut off both legs. . . . To us the whole procedure . . . is radically wrong. Its advocates overlook entirely the functions of the brain in the make up of the personality of the individual. . . .

In the name of Madam Roland who cried aloud concerning the many crimes committed in the cause of liberty we would call the attention to these mutilating surgeons to the Hippocratic oath.[77]

Although church groups expressed concern over ethical questions that might arise from these operations, and also recommended caution in the use of lobotomy, they nevertheless generally conceded that it was justifiable. Typical was the conclusion of the French Cahiers Laennec, a Catholic-sponsored group of theologians, physicians, and scientists concerned with Christian principles in medical practice. A retreat was held to consider some of the questions raised by lobotomy for the Church and for Catholics in general. It was decided that a person should not be accepted into the priesthood after a lobotomy, and that a lobotomized priest should not hear confessions or administer the sacraments as "here the safer course must be followed." A lobotomized priest, however, might be allowed to continue teaching at a university. Much weight was given at the retreat to an encyclical issued by Pope Pius XII, which implied that an operation that made a Catholic an "automaton" by removing "free will" would be unlawful, but did not object to lobotomy if "free will" was retained, even if there was some diminution of personality. If the soul could survive death, it could probably survive a lobotomy, one member of the Catholic hierarchy observed. Although there were the usual caveats against excesses and statements about the incompleteness of our knowledge of the brain, the group assembled at the Centre d'Etudes Laennec in France concluded: "At present it would be possible to count in thousands the number of leucotomized people who owe to this operation the fact that they lead a normal socio-professional life."[78]

Lobotomy was not readily accepted at all mental institutions. Some approached it cautiously and with considerable skepticism; but in the absence of any alternative treatment, there was a constant pressure to try lobotomy. Harry Caesar Solomon, the highly respected professor of psychiatry at Harvard, described how lobotomy was started at the Boston Psychopathic Hospital in November 1943. He had been skeptical, as "the number of cases reported in the literature in the preceding half dozen years was not great nor was there agreement as to the benefits that occurred." Opinions on the subject were quite strong, Solomon noted: at one extreme, distinctly emotional attitudes "based upon a conscious or unconscious belief that the frontal portion of the brain is a holy of holies, or upon some archaic thoughts concerning the brain as the seat of the soul"; at the other, the conviction "that the disorders of the mind are fundamentally due to pathological lesions of the brain and that the removal of diseased tissue was theoretically a proper approach."[79]

Solomon had been inclined to be opposed to lobotomy, as the few lobotomized patients he had seen had not done well. After reviewing a more representative sample of patients, however, most of the staff agreed that lobotomy should be tried cautiously on some chronically ill schizophrenics. Dr. Frank Lahey of the Lahey Clinic was very cooperative in this undertaking and made available surgical facilities and the services of James L. Poppen and Gilbert Horrax, two neurosurgeons practicing at his clinic. Solomon adopted the strict policy of not operating on any patient if, in the opinion of the staff, there remained "a reasonable expectancy" of improvement without psychosurgery. Where there was doubt, a longer period of hospitalization was required and "further somatic treatments were given."[80]

Before long, Solomon recalled, the staff members, who "had great skepticism, if not prejudice, against the lobotomy procedure," became "thoroughly convinced that the operation had a potentiality for producing benefits as measured in human happiness and contentment." In a few cases of lobotomy, "spectacular improvement resulted"; but others, even though the patient may have improved enough to be sent home, were "not well enough to be completely self-sufficient" and were "quite a load on their families." In considering whether such results could justify the operation, Solomon reported that the general consensus was that it was the responsibility of physicians to produce whatever improvement was possible and "not to become nihilistic because cure is not complete."[81]

In judging the results against "the experience of decades with chronically-ill mental patients," Solomon observed that lobotomy greatly increased the improvement rate, but he presented no convincing evidence that the patients during previous decades were comparable with those selected for lobotomy. While the evidence, without comparable nonsurgical control subjects, had little scientific value, the staff was apparently convinced; and by 1950, more than five hundred Lyerly-Poppen "open" lobotomies (see pages 152–53) had been performed at the Boston Psychopathic Hospital. Forty percent of the chronically ill patients were sent home after the operation, and 20 percent of these were described as relatively self-sufficient and self-supporting. A large proportion of the remaining 60 percent, Solomon reported, "live a more satisfying and contented life in institutions after the operation than before."

Based on his experience, Solomon persuaded his friend Winfred Overholser, the superintendent at St. Elizabeth's Hospital, to try lobotomy at that institution for the first time.

In England, Maurice Partridge reached a similar conclusion, after personally following three hundred lobotomized patients for one-and-one-half to three years after the surgery. He visited them several times during this interval and tracked them down from "east-end tenements to the stately homes of England, from council houses in urban districts to remote farmhouses in the Forest of Radnor and miners' hostels in Ebbw Vale. Some patients had been followed to the Channel Islands and to France. Two were followed to Johannesburg, and a third to Mombasa."[82] Partridge's remarkable 1950 book describing these follow-up studies was dedicated to the memory of Adolf Meyer, who had declared in 1936 that lobotomy should be undertaken only by those who were willing "to follow up scrupulously the experience with each case" (see page 144).

Partridge had no doubt that a lobotomy usually produced some emotional impairment, and commented that these losses always made him hesitate before recommending the operation. Patients were likely to be reduced to a simpler level, and their spontaneity was commonly reduced:

> There is less activity and more inertia. There is blunting of affect, due to a reduced complexity and intensity of feeling, and therefore there is less variation. . . . The intellectual processes are simple, with attention to the immediate rather than to the remote, to the factual rather than the theoretical, with decisions that are simple rather than deliberative, and with a restriction of the intellectual range. The total pattern of reaction is simple, marked by an essential tendency towards avoiding discomfort and courting pleasure, with lowered standards of criticism, reduced self-awareness, and diminished self-control.

Partridge observed that, in some respects, lobotomized patients seem to have followed the physician's common advice to relax and take life easier, but "the difference is that post-operatively the patient has no option, not the same potentialities."[83]

In spite of a patient's reduced potential after a lobotomy, Partridge was convinced that his follow-up study would persuade anyone "that the operation has been abundantly justified":

> When one sees a patient who has been insidiously ill for 8 years and severely so for between 3 and 4 years, with an increasing deterioration which culminates in the eating of faeces while in the padded cell, and sees him just over a year after operation, physically robust, and singing excellently in a well-known choir which is participating in a choral competition, having taken the bus from his home to get there: when one sees a patient with contractures of the semi-tendinosus and semi-mem-

branosos from having spent the better part of 7 years in fixed attitudes of prayer between bouts of being violent with double incontinence, and sees him within a year of operation making 50 at cricket, having taken the afternoon off work to play for a local side, one feels bound to make some inquiry as to how it is done.[84]

Articles in popular magazines continued to stimulate many inquiries from patients, their families, and even their physicians. The case of Millard Wright provides a striking, if not representative, illustration of such a sequence of events. Wright, thirty-seven years old, had been imprisoned repeatedly for robberies. While in prison, he read the 1947 *Life* article described earlier in this chapter, and he and his attorney requested a lobotomy to cure his criminal tendencies. The deputy district attorney also agreed to this course of action and Judge MacDonald was persuaded to grant permission for the operation. The court hearing after the operation was described in *Time* magazine:

> Wright looked like a new man. He was cheerful, sociable, and relaxed. Dr. Koskoff [Yale David Koskoff, senior neurosurgeon at the Montefiore Hospital in Pittsburgh] thought that there was a good chance that he had been cured of the urge to steal. But to complete the cure the prisoner would have to be set free and given a chance to live in a normal environment.[85]

Judge MacDonald had no precedent to help him decide whether a criminal tendency was a "disease" that surgery could cure, or whether Wright should be exempt from punishment for a past crime even if he were cured. The judge was appropriately cautious, noting that, if Wright were absolved from punishment, the courts might be overwhelmed by similar requests from other prisoners. Judge MacDonald said that he personally had "no confidence in such surgery" but would give Wright a light sentence out of a "desire to help medical science." Wright served a little time in prison and was then set free; but within a year, he resumed his life of crime, was apprehended, and eventually committed suicide in prison.[86]

In their desire to increase circulation, magazine editors often competed with each other in trying to be the first to report new developments in science and medicine. Many of the articles were poorly researched, and apparently the primary consideration was how much an article would boost circulation. There is little evidence that magazine editors took any account of how an article would otherwise affect its readers. The results of each new psychosurgical operation were usually described by simply repeating the claims of those promoting the innovative procedure. In August 1948, only seventeen months after *Life* had stated that patients not helped by prefrontal lobotomy "were mostly unaffected," the magazine reported on the damage to brain tissue resulting from these operations: "Patients are sometimes left with serious aftereffects, such as childishness, lack of responsibility and even attacks of

Figure 9.3.   The Wycis and Spiegel stereotaxic device for making electrolytic lesions in the brains of patients. This type of device was first described by Sir Victor Horsley and R. H. Clarke in 1908 for use with animals (*Brain* 31 [1908]:45–124) and was first used on patients in 1947 by E. A. Spiegel, M. Marks, and H. T. Wycis (*Science* 106 [1947]: 349–50). (From E. A. Spiegel and H. T. Wycis, *Stereoencephalatomy*, Part 1. *Methods and Stereotaxic Atlas of the Human Brain* [New York: Grune & Stratton, 1952]. Reprinted by permission of Grune & Stratton, Inc., and the author.)

epilepsy." However, this report of "serious aftereffects" of prefrontal lobotomy was presented together with a description of a new psychosurgical operation performed on the thalamus by Henry T. Wycis, a neurosurgeon, and E. A. Spiegel, an experimental neurologist, of Temple University in Philadelphia. For over fifteen years, Spiegel had been using a stereotaxic instrument to place electrodes into discrete brain regions of experimental animals (figure 9.3). A stereotaxic instrument, *Life* explained, made it possible to produce small and well-localized electrolytic lesions in the thalamus rather than the extensive damage produced by prefrontal lobotomy. Although it was noted that the thalamic operation was still experimental, "of 16 patients—all extreme mental patients—14 have improved, some considerably. Troublesome aftereffects to date have been negligible. The patient was actually able to return to his former trade five weeks later."[87]

About two weeks before this second *Life* article was published, Wycis and Spiegel were reporting in Lisbon, at the International Congress on Psychosurgery, that one of their sixteen patients had died after the operation, another had developed a long-lasting memory impairment, and four cases showed a "diminution of initiative." Moreover, three of the five schizophrenic patients initially thought to have improved had relapsed and were now being considered candidates for a second operation.[88] Hardly a record justifying *Life*'s statement that "troublesome aftereffects to date have been negligible."

The troublesome side effects of psychosurgery apparently did not stop neurosurgeons from continuing to perform these operations, as it was generally claimed that most impairments did not last long. Many patients did experience such problems as urine and fecal incontinence after the surgery, but these usually disappeared after a few weeks. Any epileptic convulsions could usually be controlled with medication, although often at the cost of producing some sedation. Neurosurgeons learn to expect that the surgery they perform will produce some impairment, but justify the operation because it may often save a life; moreover, in time a patient may be able to compensate for the impairment. The adverse consequences of lobotomy generally were less severe than those seen after surgery on brain tumors and were, therefore, considered to be within acceptable limits. A patient who became extremely subdued or passive after a lobotomy was likely to be considered to have improved substantially in comparison with his or her previous state of agitation and anguish. Moreover, confusion, apathy, and puerile behavior had been described by Freeman and Watts as good predictors that a healthier personality was emerging. Patients were sent home to convalesce with the explanation that, helped by a sympathetic family, they would improve. The surgeons and psychiatrists, however, often had only minimal contact with these patients afterward; the fact that they were sent home was sufficient reason to consider them "improved." As the family were told that recovery would be slow, they usually did not complain.

Even a surgeon who was convinced that he was not obtaining good results seldom gave up lobotomy. It was difficult to admit that the effort had been completely wasted, especially when other surgeons were reporting success. Rather than abandoning psychosurgery, neurosurgeons much more commonly introduced some change in the operation in the hope of increasing the success rate. During the five years following the end of the war, there was a great proliferation of different psychosurgical operations—motivated partly by the

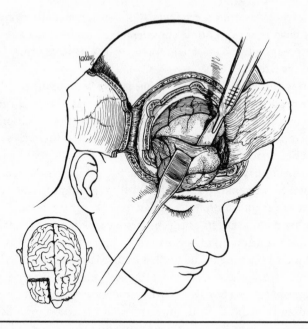

Figure 9.4.   The prefrontal lobectomy operation recommended by W. T. Peyton, H. Noran, and E. W. Miller from the University of Minnesota. As illustrated, this operation involved the removal of the anterior portion of the frontal lobes. It was performed on both sides of the brain. Peyton discussed the advantages of this procedure only from the viewpoint of the surgeon. (Drawn by R. Spencer Phippen.)

desire to improve the results following lobotomy and partly by the natural inclination of each surgeon to develop his own procedure. Had the lobotomy procedure not been modified, the number of operations performed would probably have decreased rapidly. The hope held out for these new procedures was, however, the main reason interest in lobotomy remained high for another five years—until chlorpromazine was introduced in 1954 (see page 272).

One of the more extreme innovations was proposed by W. T. Peyton, H. Noran, and E. W. Miller from the University of Minnesota and the Anoka State Hospital.* They described, in the *American Journal of Psychiatry,* a prefrontal lobectomy operation in which the entire anterior portion of the frontal lobes in both hemispheres was undercut and completely removed, as shown in figure 9.4. No new rationale for this bold step was given except that it was "a more perfect surgical procedure." All of the advantages listed had to do

* Peyton—the neurosurgeon of the team—spoke about the advantages of prefrontal lobectomy at the International Congress of Psychosurgery in Lisbon in 1948.

with surgical considerations: "all dissection is carried out under direct vision"; "the field of operation may be inspected to make sure that a hematoma does not remain"; "masses of cerebral tissue deprived of blood supply are not left to undergo necrosis"; the removed tissue "may be weighed" and "submitted to microscopic study." Peyton, Noran, and Miller, who were members of the departments of neurosurgery and neurology and neuropathology at the University of Minnesota, list as the final advantage of their procedure the fact that "if the expected result does not follow, one is not in doubt about the extent of the operation."[89]

Wilder Penfield, an eminent neurosurgeon at the Montreal Neurological Institute well known for his surgical treatment of epilepsy, did not perform much psychosurgery but experimented briefly with a procedure he called a "gyrectomy."[90] This operation also involved removal of brain tissue, but Penfield excised only selected portions of the cerebral cortex in the frontal lobes. He never made the rationale for this operation explicit and discontinued it because he did not consider it successful. The basic technique, however, was modified by Lawrence Pool, who called his operation a "topectomy."* The topectomy operation was the main psychosurgical procedure used in the frequently cited Columbia-Greystone Project, a collaborative study between the staff at Columbia University and the New Jersey State Hospital at Greystone.[92] The study attempted to relate the area of the cortex removed to the patient's improvement after the operation. In spite of the fact that the results of the study were completely inconclusive, an editorial in the *Psychiatric Quarterly* praised topectomy and stated that, because it makes it possible to relate changes in a patient with removal of specific parts of the brain, "the advance which topectomy represents may be a leap across many centuries."[93]

Although topectomy was not successful in relating the results of lobotomy to specific parts of the cerebral cortex, there had been for several years suggestive evidence that the best results occurred following damage to the deeper areas in the frontal lobe, especially that close to the midline.† The first study designed to evaluate the effectiveness of lobotomies restricted to the so-called ventromedial prefrontal area was undertaken by Leopold Hofstatter and his colleagues at the Missouri Institute of Psychiatry in St. Louis, who reported that the same benefits were achieved, but with fewer adverse consequences,

* The topectomy operation was never widely used by neurosurgeons in the United States, but Jacques Le Beau in France used it extensively for a time. As Pool did not attend the 1948 International Congress of Psychosurgery in Lisbon, Le Beau described the topectomy operation at the session where the different procedures were discussed.[91]

† As early as 1938, Freeman had stated, "We think we are getting closer to the ideal [lesions] when we place them in the base of the frontal lobe, fairly far back, and close to the midline."[94] This conclusion, while not emphasized by Freeman and Watts in the 1942 edition of *Psychosurgery*, was included in the 1950 edition.

when the operation was restricted to this region.[95] Hofstatter had been influenced by early clinical reports—one dating back to 1888—and a more recent study by Karl Kleist, which suggested that injury to the ventromedial region of the frontal lobes produced primarily emotional rather than intellectual changes (see pages 91–92). Interest in psychosurgery had been stimulated in St. Louis, partly by Carlyle Jacobsen of the Fulton-Jacobsen chimpanzee study (see pages 95–97), who had moved from Yale to the Washington University Medical School.* Following Hofstatter's report, other investigators provided additional evidence supporting this conclusion.[96]

The ventromedial (or orbital, as it is also called) area of the frontal lobes, located just behind and above the orbit housing the eyeball, soon became a prime target of new operations developed by several neurosurgeons. James Poppen of the Lahey Clinic modified the Lyerly "open" procedure to restrict the area destroyed to the medial areas (see figure 8.4). By 1949, the Lyerly-Poppen procedure was the most commonly used lobotomy operation in the United States. Poppen pointed out that his procedure appealed to neurosurgeons "because of its simplicity and the rapidity with which it can be executed." The neurosurgeon's only role, Poppen added, should be "to perform the operation as safely and accurately as possible" and not to decide who should be subjected to the procedure, as surgeons do not have the proper training or "the time to devote the many weeks and perhaps months of intimate contact with the patient to weigh justly the merits for and against operative interference."[97] In practice, however, many psychiatrists referred patients to neurosurgeons on the assumption that the latter would further evaluate the appropriateness of the operation for each case.

William Scoville of Hartford, Connecticut, influenced by the animal experiments under way in John Fulton's laboratory at Yale, developed another approach to the orbital area of the frontal lobes. This also was an "open" operation, with suction being used under direct visual observation to destroy brain tissue confined to the ventromedial area.[98] Scoville's "orbital undercutting" operation (figure 9.5) was later modified by Sadao Hirose in Japan.[99]

The use of lobotomy to treat patients suffering from intractable pain contributed to the growing interest in the ventromedial prefrontal area. Freeman and Watts had earlier observed that some psychiatric patients, who also suffered from pain, reported relief from pain after their lobotomy. In 1946, Freeman and Watts reported a case of lobotomy being performed solely to relieve pain.[100] During the same year, Poppen reported a similar case.[101] Before long, several surgeons were performing lobotomies for intractable pain. As these patients had no psychiatric problems—other than those caused by the pain—

* L. Hofstatter, personal communication, 1 August 1983.

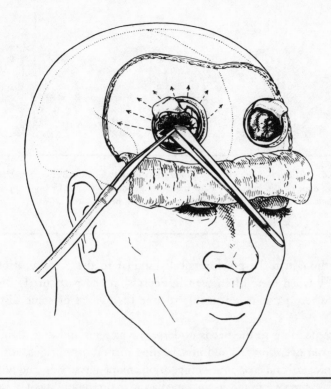

Figure 9.5.   William Scoville's "orbital undercutting" operation which selectively targeted the ventromedial (orbital gyrus) area of the frontal lobes. (From A. Asenjo, *Neurosurgical Techniques,* 1963. Courtesy of Charles C Thomas, Publisher, Springfield, Illinois.)

any adverse personality changes caused by the lobotomy were more obvious than those following lobotomy on psychiatric patients.

To minimize the personality impairment of "pain patients," a few surgeons tried unilateral (one-sided) operations, but these generally produced only partial and temporary relief from pain.[102] Everett Grantham of Louisville, Kentucky, compared the results of his attempts to alleviate pain with unilateral operations and the Lyerly-Poppen "open" lobotomy (figure 9.6). He reported that he achieved most lasting relief of pain, with minimal personality changes, after he further modified the Lyerly-Poppen operation to confine the damage to a more medial prefrontal area. At first, Grantham used a knife, but he later inserted electrodes into the brain, using electrocoagulation to destroy a more restricted target.[103]

When lobotomized patients suffering from intractable pain were asked if they still were experiencing pain, they typically reported that they did, but no longer thought about it all the time. This reduction in preoccupation with

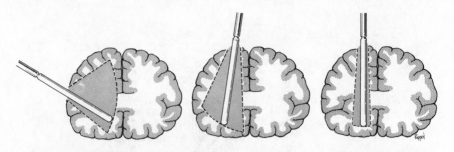

Figure 9.6.   Composite drawings modified from Everett Grantham illustrating extent of damage done by the Freeman-Watts standard lobotomy (*left*), the Lyerly-Poppen operation (*middle*), and Grantham's further modification of the latter (*right*). (Drawn by R. Spencer Phippen.)

pain was thought to be psychologically similar to the preoccupation with emotionally laden ideas often seen in depression and particularly obsessive disorders where patients are unable to free themselves of some disturbing thought.

Thus, responding to the needs of desperate patients and their families, to pressure from overcrowded and understaffed mental institutions, stimulated by the ideological and economic competition among physicians, and promoted by enthusiastic reports in the popular media and in prestigious medical journals, lobotomies increased enormously during the five years following the Second World War. Having gained momentum, the practice of lobotomy reached the stage where, in much of psychiatry, it was accepted uncritically. Whenever electroshock proved ineffective, lobotomy was considered as a possible treatment. In this climate, neurosurgeons, sometimes with little or no scientific rationale at all, proliferated the variety of psychosurgical procedures available, making the problem of evaluating lobotomy increasingly complex. If the results left something to be desired, a new procedure could be tried.

10

"A New Psychiatry": Transorbital Lobotomy

A new psychiatry may be said to have been
born in 1935, when Moniz took his first bold
step.
—*New England Journal of Medicine* (1949)

FREEMAN had for some time been growing dissatisfied with prefrontal lobotomy. In spite of some apparent improvement immediately after the operation, many of the chronically ill schizophrenics relapsed and additional
attempts to help them often produced even more impairment. Thus, when a
standard lobotomy proved ineffectual and a more extensive radical lobotomy
was performed (see pages 150–51), prolonged inertia, apathy, and epileptic
seizures were often the results. One study at St. Elizabeth's Hospital reported,
for example, that after a radical prefrontal lobotomy, chronic schizophrenic
patients were less of a problem on the ward, but for this limited gain a substantial price was paid: the incidence of postoperative seizures was above 30
percent.[1] Freeman became convinced that once patients deteriorated, little
could be done to help them. The only possible way to help them, he concluded,
was to perform a lobotomy before it was too late. Thus, while most psychiatrists and surgeons had insisted on calling lobotomy a treatment of last resort
(even when this was not always their practice), Freeman began to assert that
there was a danger in postponing the surgery.*

Questions of when the operation should be performed aside, Freeman also
realized that conventional prefrontal lobotomies could never cope with the

---

* For some years, Freeman continued to search for some form of lobotomy that might provide
effective treatment for chronic, hallucinating schizophrenics. In the 1950s, with the collaboration
of the neurosurgeon Jonathan Williams, he performed amygdalectomies (temporal-lobe operations)
on such patients, including one blind patient said to hallucinate in braille.[2] The idea for these
operations came from the reports that it was not uncommon for patients with temporal-lobe
brain tumors to experience hallucinations.

magnitude of the problem in the state hospitals. Prefrontal lobotomy required a neurosurgeon and a large staff to care for patients during the long recovery period; neither was available at state hospitals. Florence Ewald, the nurse responsible for Freeman and Watts's patients, described the effort required to care for patients after a prefrontal lobotomy:

> Patients are completely disoriented as to time and environment, and confused as to actualities for the first few days. Vomiting usually occurs during the first twenty-four hours and a liter or two of 5 percent glucose in saline may be given intravenously. . . .
>
> Occasionally there is a tendency for patients to choke due to inertia. . . . A patient should never be left alone while eating. . . . Patients must be taught to eat again. . . . Teaching them to eat correctly takes considerable practice.
>
> At times the patient urged by nature will get out of bed to go to the bathroom. He may or may not find it. One of our patients said of her wastebasket, "My that was the biggest toilet seat!" Consequently, the patient should not be left alone during the first few days.
>
> Incontinence is present and necessitates retraining the patient as one does a child. . . . Ambulatory patients can be seated on the toilet hourly. . . . Occasionally patients have to be diapered with turkish towels. . . .
>
> Nurses, aside from giving professional care, should literally become mothers to these "children" and teach them the rudiments of a normal child's life. . . . Many times a hug or a kiss from a nurse will do much toward developing these emotional children into lovable grownups.
>
> Inertia is probably the most difficult obstacle to displace. Patients are willing to sit for hours doing nothing. . . . They have absolutely no regard for time.
>
> The span of concentration is short and must be developed. To have a patient stop in mid-stream while getting dressed and start doing something entirely irrelevant is not unusual.
>
> At times the patient may become amorous. The nurse should meet such advances with tact. . . .
>
> Many lobotomy patients eat too much. They raid the icebox and snatch from the table or even from their relatives' plates. . . . Lobotomy patients should gain, but not more than twenty-five to forty pounds in the first year after the operation.
>
> We have stressed the need of treating the patient as a child . . . a spanking is good for the patient though it may distress the family.
>
> Usually when a stubborn streak develops, a "roughhouse" is the best way of breaking it up. Pulling, pushing, grabbing around the neck, tickling in the ribs, or wrestling in an undignified way will bring the patient around panting and laughing, and the resistance will drop from him. It is exhilarating if unconventional, this type of nursing![3]

Such intensive nursing care was obviously not possible in state hospitals.

Freeman knew that many patients who did not respond to electroconvulsive shock—used by this time almost routinely in most mental institutions—were not given a prefrontal lobotomy either because they were not considered sufficiently ill to justify it, or, in state hospitals, because there were no funds to pay a neurosurgeon and there was an insufficient nursing staff. He was convinced that some intermediate treatment was needed for patients who did not respond to "more conservative methods of treatment, and yet who [were] not considered sufficiently intractable" to justify the more extensive brain damage imposed by even a "minimal" prefrontal lobotomy.[4]

By the fall of 1945, Freeman decided that this need might be filled by transorbital lobotomy, the operation Amarro Fiamberti had developed in 1937 (see page 163). Freeman adopted the Italian psychiatrist's procedure of gaining access to the frontal lobes by driving a trocar through the bony orbit (the eye socket) behind the eyeball; but rather than injecting absolute alcohol or formalin to destroy brain cells and nerve fibers, as Fiamberti had, Freeman chose to cut the nerve fibers because he feared that any fluid injected might spread farther than intended. He was familiar with Fiamberti's basic procedure, because he had once planned to use it for ventriculography and had experimented on cadavers in the morgue at St. Elizabeth's.[5]

This was not the first time that Freeman had been attracted to a simple and rapid procedure that many neurologists and neurosurgeons rejected as too risky. At St. Elizabeth's Hospital, for example, he had used what he called his "jiffy spinal tap" procedure.* The usual method of obtaining a sample of spinal fluid required several people to hold a patient in a bent-over sitting position so that a needle could be inserted between the vertebrae. Working alone, Freeman instructed patients to sit backward on a chair with their arms folded around its back and their heads bent way down so that their chins touched their clasped hands. He then performed a "cisternal puncture" by inserting a needle through the foramen magnum (the large opening at the base of the skull) into the cisterna magna (a space filled with cerebrospinal fluid located just inside the skull). In this procedure, the needle has to be inserted very close to the base of the brain, where a slight error can produce a life-threatening injury (see figure 10.1).

Although Fiamberti had performed about one hundred transorbital lobotomies, his technique was never popular outside Italy. He had never described his results in sufficient detail to convince more than a few others that the procedure was worth their overcoming their aversion to entering the brain through the eye socket. Fiamberti continued to argue, however, that the simplicity of the procedure made it ideal for many psychiatric institutions where it was often not possible to perform conventional lobotomies.

Once having decided to modify Fiamberti's procedure, Freeman began to practice on cadavers, inserting an ice pick over the eyeball and piercing the

* This "occipito-atlantoid puncture procedure" was first described by others in 1919.[6]

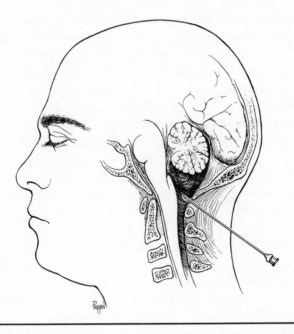

Figure 10.1.   The cisternal puncture technique that Freeman called "my jiffy spinal tap." The technique was originally described by Paul Wegeforth and his collaborators in 1919. (Drawn by R. Spencer Phippen.)

orbit. After he withdrew the ice pick, Freeman injected a methylene blue dye into the brain through the hole in the orbit. When he dissected the brain, he could then determine where the tip of the ice pick had penetrated, and could estimate which nerve fibers were severed when he swung the handle from side to side. Before long, he was sufficiently confident to start operating on patients.*

Looking back on the first transorbital lobotomy, which he performed in his office on R Street in Washington, Freeman reflected:

> When I think back upon the chances that I ran at the time it makes me shiver a bit. I could not go to any recognized hospital in the City and ask for permission from the Superintendent and the surgical staff to perform such operations. It was out of the question to carry it out except in private patients who were willing and whose families were willing to take a chance.[8]

Freeman did the first operation in January 1946, using an ice pick (the name

* When Freeman and Watts first started lobotomy operations, Freeman performed some of the surgery: he and Watts each did one side of the brain. Objections were raised to this procedure, however, because Freeman was not certified in surgery. According to Watts, Freeman enjoyed the surgery and always seemed pleased when introduced as the neurosurgeon of the team.[7]

of the Uline Ice Company could be seen on its handle) as a leucotome:

> It was with some trepidation that I operated on my first patient, Ellen I. I explained
> to her and her husband what I proposed to do and how Fiamberti and his colleagues
> in Italy had been performing this operation for nine years with very few accidents.
> Since the operation required only a few minutes, I believed it could be carried out
> in the unconscious phase after the electroconvulsive shock. Also there would be no
> cutting or sewing so the operation could be done without the need for an operating
> room. And the patient would presumably be able to go home after an hour or so.
> With what may have been an excess of caution, I operated on one side only and
> had her come back a week later for the second side. All went well.[9]

Electroconvulsive shock was the only anesthetic he used; he operated quickly
while the patient was still unconscious after the shock. For some years, Freeman
would argue—without any evidence—that the shocks increased the effectiveness of the operation:

> I prefer three electroconvulsive shocks given at intervals of two or three minutes.
> Stated succinctly and much too simply, I believe the shock treatment disorganizes
> the cortical patterns that underlie the psychotic behavior, and the [transorbital]
> lobotomy, by severing the fibers between the thalamus and the frontal pole, prevents
> the pattern from reforming.[10]

Although he later determined that the operation was just as effective if a
conventional anesthetic was used instead of electroconvulsive shock, he continued to recommend using shock, because it was readily available to psychiatrists in state hospitals, whereas general anesthesia performed by an anes-
thesiologist often was not. Freeman also recommended shock for unruly patients because it rendered them unconscious immediately; and afterward, they
usually remembered nothing of any struggle before the lobotomy.[11]

After Freeman completed the third transorbital operation, he wrote to his
son Paul, then a student at Stanford:

> I have also been trying out a sort of half-way stage between electroshock and pre-
> frontal lobotomy on some of the patients. This consists of knocking them out with
> a shock and while they are under the "anesthetic" thrusting an ice pick up between
> the eyeball and the eyelid through the roof of the orbit actually into the frontal
> lobe of the brain and making the lateral cut by swinging the thing from side to
> side. I have done two patients on both sides and another on one side without running
> into any complications, except a very black eye in one case. There may be trouble
> later on but it seemed fairly easy, although definitely a disagreeable thing to watch.
> It remains to be seen how these cases hold up, but so far they have shown considerable
> relief of their symptoms, and only some of the minor behavior difficulties that
> follow lobotomy. They can even get up and go home within an hour or so. If this
> works out it will be a great advance for people who are too bad for shock but not

bad enough for surgery. This method was tried by some Italians before the War, with poor results, probably because their patients needed something more. At least my studies have shown that there is a nerve pathway where I am making the cut, and it may turn out to be just what is needed for some people.[12]

The fourth transorbital lobotomy, however, turned out to be a disaster. During the operation, a blood vessel was torn, and Freeman could not stop the hemorrhaging. The patient was rushed to the hospital where the bleeding was eventually stopped; but afterward, the patient began having frequent epileptic convulsions. Freeman could not obtain permission to complete the surgery on the other side of the brain, and his later notes indicated that the man was never able to function independently and "aside from selling newspapers on a street corner," he had to be supported by his family.[13] The next three operations were completed without any mishap, and Freeman decided he could safely and more efficiently operate on both sides of the brain during a single session. After completing the ninth operation, he invited Watts to observe the next one—thus triggering a conflict between the two men that had apparently been brewing for some time.

Freeman and Watts had shared office facilities on R Street in Washington for about twelve years. Watts's office was on the first floor. According to Freeman, who had been doing the transorbital lobotomies in his second-floor office, he had previously discussed the operation with Watts, who had declined to work with him and "had been otherwise engaged when I invited him to the laboratory to review the cadaver studies." When he asked Watts to observe the tenth operation, the latter objected so strongly to the surgery being performed in the office that Freeman, as he recalled, "cancelled further operations."[14]

Watts has given a different account of these events. He reported that he did not know about the operations "until one of our secretaries told me confidentially what Freeman was doing in his upstairs office."[15] Watts said that when he walked into the upstairs office, he saw Freeman leaning over an unconscious patient slumped in a chair with an ice pick sticking out from above his eye. Freeman looked up and, without any hesitation, asked Watts to hold the ice pick so that he could photograph the patient. Watts refused and made it clear that he was strongly opposed to performing a brain operation as an office procedure.* Although Freeman stopped performing transorbital

---

* Watts probably knew earlier that Freeman was performing transorbital operations upstairs as he must have been aware of the emergency brain operation at George Washington Hospital when the fourth patient was rushed there with a hemorrhaging brain. Also, a letter from Dr. Jonathan Williams to me (21 September 1983) describes a conversation between Williams and Watts in which the latter acknowledged that he knew all the time that Freeman was operating in the office.

lobotomies in his office, he had no intention of abandoning transorbital lobotomy.

He continued to try to get Watts involved in the transorbital lobotomy project, showing him the results of the first ten patients. Watts remained adamant, however, and insisted that he would oppose any request to use the operating suites at George Washington University Hospital for that purpose. Freeman later compared Watts's attitude to that of the dog in the manger: "I won't and you shan't."[6] Watts's objections, of course, had merit, but much more was involved in their disagreement: this was also a clash between the conflicting interests of two medical specialties. Not only had Freeman intruded onto the turf of neurosurgeons, but he was talking about training psychiatrists around the country to do the same.

Although their relationship was strained, Freeman and Watts kept their disagreement private, remaining cordial and continuing to serve as co-chairmen of the department of neurology and neurological surgery. When Freeman thought it might be useful, he conveyed the impression that the two of them were cooperating on the transorbital lobotomy project. In October 1947, for example, when John Fulton wrote asking: "What are these terrible things I hear about you doing lobotomies in your office with an ice pick? I have just been to California and Minnesota and heard about it in both places. Why not use a shotgun? It would be quicker!" Freeman replied: "Jim and I have recently become interested in lobotomy by the transorbital route. . . . The operation has merits, but it is not recommended as an office procedure. It is much less traumatizing than a shotgun and almost as quick."[7] Thus, Freeman implicated Watts in his activities, even though it was more than a year since the latter had refused to be associated with transorbital lobotomy. Freeman was also being less than completely frank about doing the operation as an office procedure, as he had stopped this practice only after Watts's strong objections.

Unable to persuade Watts to cooperate, Freeman proceeded on his own. He obtained permission to do surgery at two hospitals in the area—the Washington Sanatorium in Takoma Park, Maryland, and the Casualty Hospital, located close to the Capitol Building in the District of Columbia. The superintendents of these hospitals were willing to allow Freeman to operate, and apparently they were not troubled by his use of electroconvulsive shock as an anesthetic, a practice that Watts had found particularly objectionable.

Freeman also made arrangements to try transorbital lobotomies at state hospitals, where—from the beginning—he had thought these operations would prove most useful. During the summer of 1946, he made a trip to the state hospital in Yankton, South Dakota. The following summer he returned to Yankton and also, as I will describe, visited state hospitals in Washington and California. Although most of these early transorbital lobotomies produced no lasting improvement in the patients, Freeman became convinced that he could teach psychiatrists to perform transorbital lobotomies safely. He also hoped that, with a better selection of patients and some modification in the procedure, the operation would be more successful.

By the spring of 1947, Watts had learned that Freeman was going to propose, at the Southern Medical Association meeting, that psychiatrists be trained to perform transorbital lobotomies. Deeply opposed to the idea of psychiatrists doing brain surgery, Watts threatened to move out of their shared office space if Freeman went ahead with his plan. Freeman went ahead anyway; and before the end of 1947, Watts had located another office and moved out. The next few years were the low point in the relationship, Watts even deciding in 1948 not to attend the International Congress of Psychosurgery, which Freeman had been mainly responsible for organizing.

After Watts moved out, Freeman asked the young neurosurgeon Jonathan Williams to move in with him. Williams had just started his neurosurgery practice in February 1947 and was attracted by Freeman's offer to refer any neurosurgical cases to him. The two of them got along well, especially after Freeman consented to call Williams his "associate" rather than his "assistant," and they soon became good friends. Their personalities were well matched, Williams being much more outgoing than the quiet and reserved Watts; moreover, he was willing to be called in an emergency if something went wrong during one of Freeman's transorbital lobotomies.

Freeman was investing ever more energy and time in transorbital lobotomy, having convinced himself that he was on the way to developing a simple procedure that could preserve a mental patient's personality before the illness had produced irreversible deterioration. Not too many psychiatrists would have allowed patients who were not seriously ill to undergo the extreme regression and slow convalescence that usually followed a prefrontal lobotomy. Freeman did not hesitate to recommend the operation, even before he had accumulated much experience with it himself. With Freeman's encouragement, Humberto Fernández-Morán, a Venezuelan who had previously received some training at George Washington Hospital, had performed his first transorbital lobotomy by July 1946.* By December of that year, his article summarizing the results of twenty-five such operations was published:

> The great differences that exist between the lobotomized cases and those operated on by this method [transorbital lobotomy] are evident in their state and conduct after the operation. While the lobotomized patient is almost always disoriented and disturbed during the first stages of the postoperative period, one who has been

---

* Fernández-Morán cut fibers and also injected alcohol and formalin, combining the techniques of Freeman and Fiamberti.

leucotomized by the transorbital route, returns to himself in a few hours with lucidity that is extraordinary and he scarcely shows the effect of the operation except for the disappearance of the state of agitation and emotional tension.[18]

Certainly the transorbital approach was gruesome, but it nonetheless did less damage than the standard lobotomy.

To the hard-pressed hospital superintendents, an offer by Freeman to visit and demonstrate transorbital lobotomy seemed a godsend. In addition to the possible benefit to patients, a visit by a man of his prominence was a welcome relief from the dreary routine of most state hospitals. Furthermore, an enterprising superintendent could publicize Freeman's visit and possibly parlay it into increased appropriations for his hospital.

In the spring of 1946, Freeman wrote to Dr. Frank Haas, the superintendent of the State Hospital in Yankton, South Dakota, explaining that he would be driving that way in the summer and wanted to demonstrate the new operation he had discussed with him earlier at the American Psychiatric Association meeting in Chicago. Freeman asked Haas to identify patients who had been ill for less than a year and had been given electroconvulsive shock therapy without lasting improvement. When Freeman arrived in Yankton that summer, however, he was presented with twelve chronic, deteriorated schizophrenics. Although these patients were not really suitable for his purposes, he operated on them anyway.

The night after he operated at Yankton, he left the hospital and headed toward San Francisco for a medical meeting. With him were his sons Keen and Randy, going on ten and twelve, respectively, and two of his nephews. Freeman, as was his custom then and later, was combining professional and camping trips. On the way west, they stopped at Yosemite, and there a tragedy occurred from which Freeman never fully recovered. As he recounted it in his unpublished autobiography:

> There had been plenty of snow, so the streams were full. Then disaster struck. We climbed to the top of Vernal Falls (on the Merced River) on a hot day, forgetting to fill the canteen. Keen went to the water's edge to fill it, lost his footing and fell in. I was some distance back and was as if paralyzed. I thought later that if I had vaulted the railing and extended my cane I might have saved him. As it was, a young sailor . . . did vault the railing and plunged after Keen. Locked in each other's arms they were carried over the brink and dashed to death three hundred feet below. The last look I had was Keen's face as he went over the edge. . . . Keen was marked for a surgeon . . . a bright mind, an independent spirit, with a touch of ruthlessness. But we lost him.[19]

The river was dragged for two days—while Freeman watched, overcome with grief and guilt—before the bodies of Keen and the sailor were recovered.

After burying Keen in California, Freeman returned to Washington and began to "wall off" his sorrow while at the same time drawing on it as a source of energy for the transorbital lobotomy project, forever linked in his mind with the memory of Keen. He preferred not to talk about the tragedy but, when the topic could not be avoided, would sometimes chill colleagues by his cold, clinical description of the "clean hole in the posterior region" of his son's skull.[20] While Freeman felt the loss deeply, he seemed to regard it as a weakness or as self-indulgence to display emotions.

During the summer of 1947, he revisited the hospital in Yankton and then went on to Western State Hospital in Fort Steilacoom, Washington, at the southern end of Puget Sound. Fort Steilacoom, the first settlement in the state of Washington, was the site of a fort used during the Indian fighting days; the young Ulysses S. Grant had once been garrisoned there. William Keller, the superintendent, had graduated from Rush Medical College in 1899 and had worked as a surgeon for several railroad companies before being given the appointment at Western. Like many superintendents at state mental hospitals, Keller had no psychiatric training. He was well suited for his administrative duties, however, having been active in the social and business community of nearby Tacoma, where he served on governing boards of banks, corporations, and service organizations and was heavily involved in real-estate development. His contacts were said to be legion.[21] He and Freeman had been friends since 1941, when Keller had obtained funds to bring several prominent psychiatrists to Western for a symposium. On that occasion, Freeman—using his experience with journalists—had helped Keller promote Western State as a center of psychiatry in the Northwest.

On the staff at Western during the summer of 1947 was the young psychiatrist Charles Jones, whose background was typical of many of the staff at state hospitals during this period. He had been introduced to psychiatry while serving as a medical officer during the Second World War, but had virtually no training in this field. After his discharge, he had gone to Western for what was supposed to be a training program in psychiatry, but when Freeman arrived, Jones, who had been there only about seven months, was already considering leaving. He had found the staff to be mediocre, there was no training program, and the atmosphere, as in most state hospitals, was depressing. He had been thinking seriously of going into the air force, but Freeman's arrival made Jones change his mind. For the first time, there was something stimulating to look forward to, and he was eager to get involved. Later, Freeman and Jones became friends, several times traveling together to meetings or on camping trips.

It was midmorning of 19 August 1947 when Freeman arrived at Western, after a fast drive from the campsite he had left at dawn. He had written in advance describing the ideal candidates for transorbital lobotomy and also requesting that several cadavers be saved for demonstration purposes. (There were always unclaimed bodies available in large hospitals that had many old, chronic patients.) Immediately after arriving, he began screening the records of about eighty patients who had been selected as possible candidates. Following lunch, he started operating. By 4:30 P.M., he had completed thirteen transorbital lobotomies, explaining the procedure as he went along. During two days at Western, Freeman also gave two lectures, attended a dinner in his honor, and taught several psychiatrists to perform the operation on their own (see figure 10.2).

Charles Jones learned how to perform transorbital lobotomies during this visit. After he and the other two psychiatric residents watched Freeman operate, they went down to the morgue, where each of them practiced on cadavers while being "talked through" the procedure by Freeman.* The next morning, after performing three operations on patients with Freeman standing by, the three residents were considered ready to carry on by themselves. This was the pattern Freeman almost always followed during his first visit to a state hospital. When, during the previous summer, he visited the Yankton State Hospital, he had also demonstrated the procedure to several psychiatrists; and soon afterward, D. B. Williams and Gisela Ebert did twelve operations on their own.[22]

The first thirteen patients given transorbital lobotomies at Western recovered from the surgery without any apparent complications. When the patients' eyes were less swollen and discolored, Keller invited reporters from the leading Seattle and Tacoma newspapers to a press conference. All the newspapers in the area featured the story, and the *Seattle Post-Intelligencer* actually had a front-page article describing every one of Freeman's trips to Western. The lead story in the *Intelligencer* on 28 August 1947 was headed by bold type declaring:

AT STATE HOSPITAL—SURGERY MAY FREE 9 MENTAL PATIENTS. DOCTORS HOPE FOR "MIRACLE" IN NEW OPERATIONS

As accuracy gave way to regional pride, the new surgical technique was said to have been "attempted only in Washington, D.C. before it was tried here last week on eight women and five men suffering from schizophrenia."

> One half hour after the operation, two appeared perfectly normal and will be among the nine who are shortly to go home. The other four responded to the surgery in no way at all. They were neither better nor worse than they had been before....

* A description of Freeman's training procedure was provided by Charles Jones, personal communication, 10 January 1983.

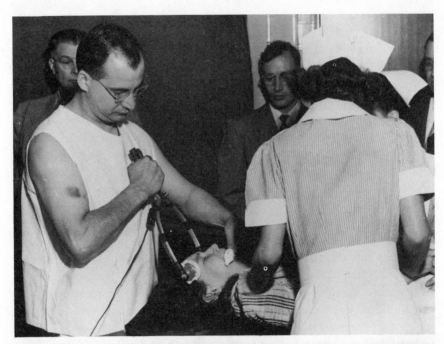

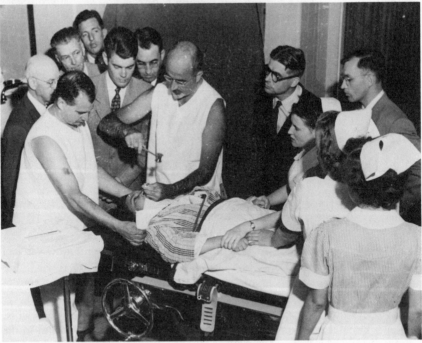

Figure 10.2.   Walter Freeman demonstrating transorbital lobotomy at Western State Hospital, Fort Steilacoom, Washington, on 11 July 1949. (*Top*) Dr. James G. Shanklin, staff psychiatrist at the hospital, is administering the electroconvulsive shock (ECS), used as an anesthetic. (*Bottom*) Immediately following the ECS, Freeman began the operation. Dr. Charles Jones can be seen behind Dr. Freeman's arm, wearing a jacket and tie. (UPI/ Bettmann Newsphotos.)

Three of the women patients who underwent the short, painless surgery came into the operating room in straitjackets. They had to be restrained for fear they would attempt homicide or suicide. . . .

The doctors here do not say that the new technique—transorbital lobotomy, a technique devised by Dr. Walter Freeman, Georgetown University neurology professor—had effected permanent cures. It is too early to say that. They say the results have been excellent. And Dr. W. N. Keller, hospital superintendent, expresses frank amazement at the results.

Dr. L. S. Durkin, Staff Physician, points out that the transorbital lobotomy is not for everyone. "We can perform the operation only on carefully selected cases," he stressed. . . . The only persons on whom the hospital has—or will attempt—the operation are schizophrenic victims whose psychosis is less than six months old, who before the disease struck them were of outstanding intelligence and personality. They must also be persons on whom repeated electric shock treatments standard in schizophrenia have failed to produce any results. . . .

Advantages of the new technique over the prefrontal lobotomy, Dr. Keller asserted, are many. With the old technique there is a 3 percent mortality rate. . . . The mortality of the new technique is zero. . . . Immediately regaining consciousness after the operation, the patient who has undergone a transorbital lobotomy has attained all the benefits of the surgery, he said. With the old technique a retraining and rehabilitation of many months is necessary before the patient is ready for normal living. With the new technique, the operation takes 15 to 20 minutes to perform, with the old, hours. . . .

As Dr. Keller described the operations, a calibrated instrument resembling a thin icepick, designed by Dr. Freeman, is inserted between the patient's eyelid and eye. . . . Dr. Freeman uses electric shock to anesthetize his patients, Dr. Keller continued. The operation is performed when the patient has passed from the electric-shock-produced convulsion to coma. . . .

One of the women—a girl really—on whom the operation was performed by Dr. Freeman, cracked up in her sophomore year at college. Before the operation, she suffered from persecution delusions, heard voices which told her how badly she was treated and was given to clawing and scratching at hospital attendants because she was suspicious of everyone.

Yesterday, even to the trained eyes of the neurologist and psychiatrist here, she appeared normal. Asked if she expected to go home soon, she answered pertly: "That's up to the staff and I never debate with the staff."[23]

The *Seattle Times Herald* article of 20 August 1947 was even more enthusiastic and unreserved. The bold type declared:

BRAIN OPERATIONS ON 13 PATIENTS ARE SUCCESS!

The article noted that the operations were performed by Dr. Walter Freeman, president of the American Board of Neurology and Psychiatry, who came at the "special invitation of Dr. Keller, a personal friend of many years, interrupting a vacation in California with the family in acceding to Dr. Keller's request." All the regional newspapers described the success of the operations as "spectacular" or "amazing." The statement in the newspapers that the patients appeared normal after the operation could be very misleading. To a layman, who might expect only "raving lunatics," a subdued patient might appear normal, but more thorough questioning and observation would probably have revealed that much of the psychopathological thinking was still present. Most patients, however, were not questioned extensively; and it is not surprising that they were subdued after electroconvulsive shock.

During August, local radio stations in the state of Washington aired a program on transorbital lobotomy in which Charles Jones and James Shanklin, another psychiatrist at Western, along with Donald Sergeant, the supervisor of all of the public institutions in the state, described the simplicity and safety of the operation and the rapid post-operative recovery compared with the "older lobotomy."

Not long afterward, Jones and Shanklin reported that 80 percent of Western's forty-one transorbital cases had shown substantial improvement one month after surgery.[24]* The first thirteen patients were the original group operated on by Freeman, but the remainder had been done by Jones and Shanklin. Later, when Freeman commented on the first thirteen patients at Western, he was less enthusiastic—"long follow-up showed several relapses and two suicides"—and added that "the operation was not extensive enough for these state hospital cases."[26] On the same trip, he had also demonstrated transorbital lobotomy at the State Hospital for the Insane in Stockton, California; he later commented that "even worse material was presented at Stockton, with poor results."[27]

Back in Washington, D.C., Freeman was able to perform transorbital lobotomies on his private patients, who were not as deteriorated as those in the state hospitals. Many of the private patients were referrals from out of town, and the operations were usually done within twenty-four hours of a patient's arrival in Washington to save the expense of an extended stay in a hotel or at the hospital. The patients were usually not even "admitted" to the hospital, as they generally were able to leave within a couple of hours. Frequently there was considerable swelling and later discoloration around a patient's eyes—of which, with his customary bluntness, Freeman made light, remarking that "some of the black eyes were beauties, but I usually asked the family to provide the patient with sunglasses rather than explanations. . . . Most patients deny having been operated on."[28]

---

* It has been claimed, in books and a film based on her life, that the movie actress Frances Farmer underwent a transorbital lobotomy at Western State Hospital;[25] but Charles Jones, who was at Western during her confinement there, insists that she never had a lobotomy (personal communication, 10 January 1983).

Freeman's new associate, the neurosurgeon Jonathan Williams, recalled one incident involving an out-of-town referral:

> It was at this time that Freeman achieved his pinnacle of showmanship and bravado. He had an electrical shock machine about the size [of] but a little bit thicker than a cigar box which contained a timing device in seconds and fractions thereof, as well as a transformer which reduced a wall current of 110 volts to a more acceptable level. As time went on, the device started to fail him, and he opened it up, first bypassing the timer and later bypassing the transformer, so ultimately 110 volts came into the box and went out of the box. Freeman judged the time of application by a small toggle switch that he would snap on and off. One day he received word that a patient had come from out of town and was staying at a Silver Springs motel with his relatives, . . . but the man became extremely unruly and the police had to be called. The police, however, said they had no permission to enter the room. Freeman drove out to the motel and, upon sizing up the situation, decided that he could calm the patient down with a few bursts of ECT. The patient was grabbed by his relatives and held down on the floor while Freeman administered the shock. It then occurred to him that, since the patient was already unconscious and he had a set of leucotomes in his pocket, he might just as well do the transorbital lobotomy right then and there, which he did. I guess the family were happy because it saved them the expense of hospitalization and left them only with a motel bill (and Freeman's bill, of course). The patient did well, and the family and patient departed for home a day or so later. Freeman then submitted a claim to Blue Cross for transorbital lobotomy and was advised by them that, since he was not a surgeon, and since the procedure was not performed in a hospital, no payment could be made. He then wrote them back, detailing these events, and Blue Cross was so astonished that they paid him! I know this to be true, because Freeman himself told me this, and I have heard from others that he told them as well.*

Although at first patients were usually not hospitalized, Freeman later became convinced that to make the operation more effective, especially for the more severely incapacitated patients, he had to perform a more extensive procedure destroying nerve fibers in the ventral and medial areas of the brain (see discussion on pages 196–97). Thus, some of the original rationale for a less severe operation was dropped. Nevertheless, because the modified transorbital lobotomy was capable of destroying more specific targets than the standard Freeman-Watts prefrontal lobotomy, the former proved to be less debilitating. By October 1948, he had developed the "deep frontal cut" transorbital lobotomy. Whereas initially he only moved the handle of the leucotome medially and laterally, he now forced the handle upward, driving the tip into the deep parts of the frontal lobe. Before performing the deep frontal cut, he first completed the original procedure:

> Very little preparation is necessary. . . . The patient is given from two to five suc-

* J. Williams, personal communication, 21 September 1983.

cessive electro-convulsive shocks in order to maintain a state of coma for about five minutes. When the last convulsion subsides, the nurse places a towel over the nose and mouth to prevent contamination by saliva or nasal secretions. I pinch the upper eyelid between thumb and finger and bring it away from the eyeball. I then insert the point of the transorbital leucotome into the conjunctival sac, taking care not to touch the skin or lashes, and move the point around until it settles against the vault of the orbit. I then drop to one knee, beside the table, in order to aim the instrument parallel with the bony ridges of the nose . . . and tap the handle to drive it through the orbital plate [the back of the socket containing the eyeball] to a depth of 5 centimeters from the margin of the upper lid [the leucotome had centimeter markers]. The handle is then pulled as far laterally as the rim of the orbit will permit. . . . I then return the instrument half way to its previous position and drive it further to a depth of 7 centimeters. . . . Again I sight the instrument . . . and take a profile photograph. This is the nearest approach to precision of which the method can boast. Then comes the ticklish part. Arteries are within reach. Checking to make certain that the instrument is still parallel to the bony ridge of the nose . . . the handle is moved 20 degrees medially and 30 degrees laterally. . . . The "deep frontal cut" followed: Standing behind the head of the patient [I] strongly elevate the handle of the instrument in this oblique plane [it is still 30 degrees lateral] until the shaft lies as nearly as possible parallel with the orbital plate . . . and then the handle is returned to the mid-position. The instrument is withdrawn, applying pressure on the upper eyelid to control escape of blood and cerebrospinal fluid. . . . I then proceed with the opposite side, using an identical instrument, but freshly sterilized because some germs might be carried away from the lashes or skin of the first side operated upon. . . . In some instances, when the transorbital leucotome is thrust through the orbital plate with a mere tap of the knuckle, there is no difficulty at all; in other cases, however, the plate is so thick and heavy that the operator risks bending or breaking the instrument. Quite often there is a sudden "give" or even an audible crack as the orbital plate fractures. It is remarkable how little reaction there is to such an occurrence. . . . Within an hour after [the original transorbital lobotomy] the patient may be sitting up or even able to walk. . . . The effect of the deep frontal cut upon the patient . . . is somewhat more marked. However, within a day or two orientation is regained and the patient is usually discharged to his home on the day following operation. Relaxation and inertia are evident from the time the patient wakes up. Vomiting occurs in half the cases, incontinence seldom. . . . Swelling of the orbital tissue is marked in some cases, absent in others. . . . Penicillin in oil . . . is used in prophylactic fashion.[29]*

Freeman had started operating on patients with an ice pick, then switched to a "transorbital leucotome," an instrument he developed which closely resembled an ice pick with gradations on it. The "deep frontal cut" modification, however, placed extra strain on the leucotome. Occasionally, when Freeman was forcing the handle of the leucotome up toward a patient's brow (in order to drive the tip down close to the base of the brain), the tip of the

* This quotation is a composite of Freeman's own words to provide a clear description for the layman.

leucotome actually snapped off when pressure was exerted against the bony orbit. Once when this happened, Williams received an emergency call and had to rush to the Casualty Hospital

> where Freeman in his enthusiasm to achieve a deep frontal cut, snapped off the leucotome, leaving two or three inches in the brain. I had to scrub and extract it for him, approaching intracranially through a transfrontal craniotomy. Freeman often was impatient with sterile procedure and what he called "all that germ crap." I often had to assert myself by insisting "Walter, at least let me drape the patient."*

Following several such accidents, Freeman had a Washington machinist, Henry Ator, make several models of a new instrument until, in Freeman's own words, he could use the instrument to "practically lift a door off its hinges without it either breaking or bending." He called the new instrument an "orbitoclast." It was about eight inches long and, as even he had to admit, looked like "a savage instrument."[30] Usually he started an operation with the standard leucotome, but if he sensed too much resistance, he withdrew it and inserted the orbitoclast (these instruments can be seen in figure 11.2).

The publicity in the Seattle and Tacoma newspapers and radio stations had created a demand for transorbital lobotomy at other state hospitals in Washington. Psychiatrists from Northern State Hospital in Sedro Woolley and from Eastern State Hospital at Medical Lake came to observe the operation. Freeman also gave a lecture in Seattle to a meeting of northwest psychiatrists, including Canadian psychiatrists from British Columbia. The *Seattle Post-Intelligencer* reported on 30 October 1947 that fifty-four operations had already been performed at Western on "hopeless patients doomed to a lifetime in a mental institution," and that Freeman planned to complete one hundred transorbital lobotomies by the spring of 1948 in time "for a report to an international medical conference in Portugal."[31] In June 1948, *Newsweek* reported that the one hundred operations had been completed, and described how a "virtually hopeless case" was almost immediately transformed:

> First, Dr. Freeman gave the patient two electric-shock treatments. Then, while she was still unconscious, he drove a slender steel instrument through the bony part of the eye socket into the front of the brain. With a swift turn of the steel pick he severed certain of the brain connections. The instrument was withdrawn and the process repeated on the other side. The operation took only ten minutes.
>
> Within an hour the patient was wide awake, with no memory of what had happened. Later she was heard chuckling to herself. When asked why she was laughing, she replied: "All those foolish ideas I had. How did I get them anyway?"
>
> ... In patients who have been ill for more than a year, Dr. Freeman thinks transorbital lobotomy should be looked upon as a test, not as a last resort. If the

* Personal communication, 21 September 1983.

patient improves, but the improvement does not last, then he thinks the standard lobotomy operation should be used.[32]

Freeman returned to Western periodically to check on the progress of former patients and to get more experience with the "deep frontal cut" procedure. Keller kept the newspapers informed. The Seattle Post-Intelligencer's front-page story on 8 July 1948 was headed:

NEW BRAIN SURGERY USED ON 9 PATIENTS. DOCTORS OBSERVE EASTERN SPECIALIST'S METHOD AT STATE HOSPITAL.

The story went on to explain that the new operation took only seven minutes to perform and opened up treatment "to vastly more patients." Quoting Charles Jones, the article stated:

Transorbital lobotomy, with the new refinement in technique, is adapted to any mentally sick person whose outstanding symptoms are emotional tension, worry, anxiety and fear.

It takes three or four years to train a good neurosurgeon and such a man is needed to perform a prefrontal lobotomy. The limited number of surgeons available necessarily restricted the number of patients who would have lobotomies. A psychiatrist can be trained quickly to do a transorbital lobotomy.[33]

Freeman did not spare himself in his crusade to promote transorbital lobotomy. His trips to state hospitals became ever more extensive and frequent. He would write in advance, giving an approximate date of his arrival, and then telephone when he was close enough to know more precisely when he would arrive. The schedule was always tight, making it necessary to "break camp" early in the morning in order to reach the next hospital in time to get in a full day's work. His daughter Lorne once asked him if he had "to turn every trip into an endurance contest."[34] For anyone less committed, this schedule would have been exhausting, but Freeman could not slow down. Being treated as a distinguished guest at the hospitals did not lighten his load: he was expected to give at least one lecture in addition to the surgery, and was continually "put on display" so that his host could get maximum publicity from the visit.

Freeman's pace eventually took its toll. Never diplomatic, his behavior during this period was often inappropriate and sometimes offensive. Rather than winning people over, he often turned them against him. Watts recalled an occasion at this time when Freeman was demonstrating transorbital lobotomy in a nursing home to some Baltimore surgeons who had come to observe the operation:

The patient was wheeled into an attic room. Freeman took off his coat, leaving on his vest, and instead of using a surgical mallet, he used a carpenter's hammer to drive the modified icepick through the orbit. Freeman always did things in a dramatic way. He liked to shock people. Actually, it doesn't make any difference—the hammer is just as effective and safe, but I liked to drape the patient and make the procedure look like a conventional operation on any part of the body. I am a traditionalist, I guess, and I felt appearances were important especially if there were students present. We used to tug and pull about things like that and I told him: "Walter, you're going to kill the damn thing if you demonstrate with a hammer and an icepick."*

In 1948, on his way to the First International Conference of Psychosurgery in Lisbon (see chapter 11), he stopped to give a lecture and demonstration of transorbital lobotomy at the Burden Neurological Institute in Bristol, England.[35]† Afterward, when he was asked to lecture to a high school audience in Bristol, he showed a film he had brought along to demonstrate the Freeman-Watts prefrontal lobotomy procedure. The particular operation shown was unusually bloody and certainly not appropriate for high school students. Later, after the inevitable happened—five students fainted and had to be carried out—he enjoyed retelling the story, often saying that he was as good as Frank Sinatra in getting young people to faint.[37]

Most of the young psychiatrists watching Freeman demonstrate transorbital lobotomy in state hospitals felt more than a little queasy during the surgery— as can be seen in figure 10.3. Even experienced physicians often had a strong emotional reaction to the transorbital lobotomy procedure. Edwin Zabriski, a seventy-four-year-old professor emeritus of neurology at Columbia University, had already observed Lawrence Pool perform several topectomies before attending one of Freeman's demonstrations. This was at the Greystone Park State Hospital in New Jersey in July 1948. After watching the electro-convulsive shock, followed by the leucotome being tapped into the brain over the eye, and hearing the sound of the orbit fracturing when the handle was forced up toward the brow, this experienced clinician fainted—providing Freeman with another story.[38]

In 1949, the German physician Trangott Reichert invited Freeman to demonstrate transorbital lobotomy in Freiburg. Present was Professor Karl Kleist— probably the most eminent biological psychiatrist in Germany—who did not hesitate to express his strong reservations about the operation, asserting that no one with that much injury to both frontal lobes would ever be able to work again. Freeman, by his own description, almost had to push Kleist out of the way in order to perform the surgery. Actually, no lobotomies of any

---

* J. Watts, personal communication, 22 June 1981.
† There was little interest in the transorbital lobotomy procedure in Europe. One trial in England was reported to have been unsuccessful.[36]

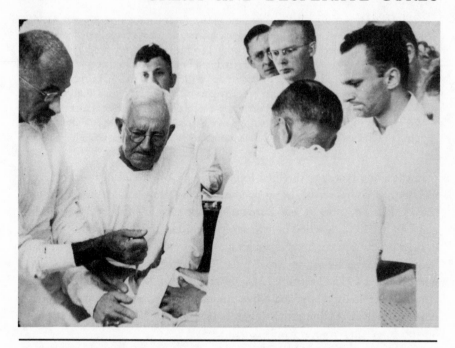

Figure 10.3. Walter Freeman demonstrating transorbital lobotomy at the Green Gables Sanitarium, Lincoln, Nebraska (6 August 1951); note his dispassionate manner. Dr. Paul Royal is holding the patient's head. (Photograph by Charles Jones.)

type had been performed in Germany until 1946, and all psychosurgery was still controversial. The Nuremburg Trials had exposed the atrocities committed by Nazi physicians, and Kleist asserted that lobotomy is "mutilating surgery and Germans should avoid further world criticism for experimental procedures." The "ice pick operation" was not the sort of medicine with which Germans wanted to be identified at that point in history. Even though Freeman was aware of the prevailing attitude in Germany to such operations, he could not restrain from responding to Professor Kleist's criticism with a remark that he knew was insulting by its flippant nature: "If a person had a strangulated hernia the only cure was surgical. What these psychotic patients were suffering from was a strangulated Oedipus complex."[39]

He was not always able to get permission to demonstrate transorbital lobotomy. State hospitals seldom refused, partly because of the desperate conditions of these institutions and also because the staff were usually flattered by a visit from such a well-known figure. In London, however, when he offered to demonstrate the "deep frontal cut" procedure after his 1949 lecture to the Royal Society of Medicine, the neurosurgeons refused to grant permission even though he had demonstrated the standard transorbital lobotomy in Bristol to the neurologists the year before. In 1949, he tried, also without success, to use his long association with John Fulton to advance his transorbital crusade into Connecticut. Fulton refused to cooperate:

> You mention the transorbital procedure and your willingness to demonstrate it at some state hospitals in Connecticut. I am a neurosurgeon by training and I have the greatest misgivings about the justifiability of the procedure because I don't think you can possibly estimate the extent of the interruption which takes place in the blind transorbital approach. Perhaps I am over conservative but I should not like personally to sponsor you or anyone else in carrying out the procedure in this state. I realize that a certain proportion of your results have been good but until your cases have been studied more closely before and after the procedure, I don't think it can be confidently recommended. Here again I am conservative but I counter by taking my hat off to you for your courage and also for the pioneering effort which you and Jim Watts have made in the lobotomy field. At the moment we are trying to find out exactly what tracts should be severed to deal with any particular type of mental disturbance. This necessitates an open approach and a precise knowledge of what is being interrupted in any given case.[40]

Freeman persisted and succeeded in giving several demonstrations in Connecticut without Fulton's sponsorship. As usual, when refused an invitation, Freeman sought another way of reaching the same population. He did not take refusals personally, convinced as he was that it was only a matter of time before his opponents would accept transorbital lobotomy.

In spite of the strong opposition by the neurosurgeons, Freeman tried, mostly in vain, to win people over to transorbital lobotomy. He told his audience at the Burden Neurological Institute that, at the recent meeting of the American Psychiatric Association, "speakers from half a dozen different hospitals emphasized the safety and effectiveness of transorbital leucotomies." These "testimonials" came from his collaborators at state hospitals. Freeman reported that there had been no deaths as a result of a transorbital lobotomy; and that, after two years, the follow-ups of the first ten patients on whom he had operated early in 1946—these were the operations in his office—revealed that six were "fully restored and only one was in the hospital."[41]

By 1949, he had completed 350 transorbital operations, with four deaths ("two fatal hemorrhages and two deaths from other causes").[42] When the second edition of *Psychosurgery* was published in 1950, Freeman reported having performed more than four hundred transorbital lobotomies between January 1946 and December 1949, still with only four deaths. Over all, he wrote, the number of complications "had been small, 10 patients out of four hundred."[43]

In spite of the opposition, Freeman's commitment to his transorbital lobotomy crusade certainly did not weaken. In fact, his visits to state hospitals became increasingly extensive. In 1949, he not only returned to Yankton, Fort Steilacoom, and Stockton, but he also made the first demonstration visits to state hospitals in Galveston and Rusk in Texas; Little Rock, Arkansas; Rochester and Hastings, Minnesota; Columbus, Ohio; Ogdensburg, New York; Sedro Woolley, Washington; St. Joseph, Missouri; and Sykesville, Maryland. Only eight months after his visit to Texas, he received a letter from psychiatrist C. L. Jackson of Rusk, Texas, informing him that he (Jackson) had already performed four hundred transorbital lobotomies, and claiming that there were "only three deaths" and 20 percent of the chronic patients had been discharged.[44]

11

Lobotomy at Its Peak

> Hypochondriacs no longer thought they
> were going to die, would-be suicides found
> life acceptable, sufferers from persecution
> complex forgot the machinations of imagi-
> nary conspirators. Prefrontal lobotomy, as the
> operation is called, was made possible by the
> localization of fears, hates and instincts. It is
> fitting, then, that the Nobel Prize in medicine
> should be shared by Hess and Moniz. Sur-
> geons now think no more of operations on
> the brain than they do of removing an
> appendix.
>
> —*New York Times* (October 1949)

BY THE summer of 1947 psychosurgery had spread around the world, and preparations were under way for an international meeting. Walter Freeman, who was responsible for the planning and for selecting the site, was the driving force that made the meeting a success. As program chairman and general secretary, he wrote to people around the world and was able to persuade two hundred neurosurgeons, neurologists, and psychiatrists, from twenty-eight countries, to come to Lisbon for the First International Conference on Psychosurgery during August 1948.*

Lisbon was selected in honor of Egas Moniz. Almost seventy-five, he was close to retiring as chief of clinical neurology at the Santa Marta Hospital.†

---

* He was not able, however, to persuade General Richard Southerland, with the U.S. occupation forces in Japan, to allow Dr. Mizuko Nakata—the first surgeon to perform psychosurgery in that country—to attend the conference, even after reminding Southerland that they had been fellow students in a French class at Yale and explaining that "psychosurgery is changing the madhouse into a psychiatric hospital."

† On 14 March 1939, Moniz was almost killed by one of his patients—not one who was lobotomized, however, as has often been reported. The patient, who was probably paranoid, had been sent to Moniz for diagnosis and treatment. While Moniz was writing a prescription for him, the patient pulled a gun from his pocket and fired five bullets—four entering Moniz's body and one injuring his thumb. Of the four bullets in his body, three were located superficially and

For some years, Moniz had been receiving honors from Portugal and abroad. In 1940, he was elected president of medical science of the Academy of Science of Lisbon, a distinguished old body similar to Great Britain's Royal Society, which had made him an honorary member. The universities of Bordeaux, Toulouse, and Lyon gave him honorary degrees, and he was made a commander of France's Legion of Honor. Moniz received, in 1945, the prestigious Oslo Prize for his role in developing cerebral angiography and, later, the president of Portugal presented him with the Cross of Santiago's Sword and the Order of Isabella.

Although Moniz had not published any new articles on psychosurgery after 1937, he had remained active in his profession, continuing to write articles on cerebral angiography. He was still willing to take risks and, in 1943, at the age of sixty-nine, had attempted to use brain surgery to treat patients with Parkinson's disease. Applying essentially the same surgical techniques used in prefrontal leucotomy, he had directed Almeida Lima to destroy part of the motor area of the cerebral cortex by either injecting absolute alcohol or by an electrical cutting tool. The justification for "treatment" was questionable: it was well known at the time that damage to this area of the cortex can produce severe impairment in voluntary movements. Admitting that the "results were not brilliant," and that, ideally, Parkinson's disease should be treated by attacking the "virus," Moniz concluded that, until we are able to do so, "surgery may offer some help to certain patients."[3] Later, in his memoirs, *Confidências de um Investigador Ciêntífico* (which he was completing in 1948), he wrote that he would have preferred to have published this surgical treatment of Parkinson's disease in a United States medical journal rather than, as he did, in a Swiss journal, because it would have received more attention. Always the diplomat, however, Moniz had felt compelled to publish in a neutral country during the Second World War because of Portugal's non-Allied status.

Psychosurgery continued to be used in Portugal, in spite of some opposition to it. It is not true, as Freeman later wrote, that psychosurgery was completely stopped in that country for political reasons.[4] Moniz had his detractors and those who, with good reason, believed that the benefits of psychosurgery had been exaggerated, but prefrontal leucotomy continued to be performed in Portugal; and, regardless of his earlier political activities, Moniz was always a highly respected figure in his native country. The Portuguese government

---

easily removed; but one, having passed through his lung, was lodged in a vertebra. To avoid the risk of injuring his spinal cord, the surgeons decided not to remove the bullet, and it remained there for the rest of his life. He made a rapid recovery for a man of sixty-five and went back to work after a summer convalescing in Avanca. In 1943, at the age of seventy, Moniz had been required to give up his professorship at the University of Lisbon. After his retirement as chief of clinical neurology, he wrote *A Nossa Casa*, a four-hundred-page history of his family, of which he was the only living member.*

generously supported the Lisbon Congress, and the president of the republic, Marshal António Carmona, participated in the ceremonial opening sessions, welcoming the guests and expressing the nation's pride in Moniz's accomplishments. Moniz's failure to publish any articles on psychosurgery had more to do with the fact that others were innovating—developing new surgical procedures and theories—than with any opposition within Portugal.

During the Lisbon Congress, Moniz invited all the participants to his home for a dinner, an experience recalled by Norman Dott, a neurosurgeon at the University of Edinburgh:

> The memory of evening receptions there in 1948 and 1953 conveys one back to an age of grace and elegance. The Senhora's [Moniz's wife] large, faithful domestic retinue, immaculately attired and fortified by some specially selected colleagues for the occasion—the elegance and flavour of the dishes—the aroma of the wines, the respectful, perfect, but amiable service—the glow of rich Portuguese furniture in candle-light—the gleam of plate—the French windows open to the fragrant gardens—the graciousness of host and hostess.[5]

Before the congress, Moniz had arranged for Freeman to be inducted into Lisbon's Academy of Science as a Foreign Member, and personally presented him with the solid gold chain and medal given to new members of that august body. Freeman and Moniz shared much of the spotlight at the congress. They both were in the center of all the group pictures (see figure 11.1), and Freeman gave the second lecture, immediately after Moniz's speech. Speakers at the ceremonial sessions thanked him not only for his effort in organizing the meeting but also for his important contributions in advancing the theory of psychosurgery, in modifying the surgical technique, and in writing such a "persuasive" and "remarkable book." In his own speech, Moniz said that they all were in debt to Freeman for establishing "the solid foundation of the method initiated here, enriching it with improvements." Moniz concluded with this statement: "Leucotomy is at present a Luso-American method. It is due to the renowned Professor that this international assembly takes place in Lisbon. This constitutes further cause for our warmest thanks for this great and unforgettable honour for my country."[6]

It was generally known among the participants at the congress that there was a movement afoot to nominate Moniz for the Nobel Prize. Whether Freeman hoped that he himself would be nominated together with Moniz is not clear. According to one account, Freeman "told his wife that he would never rest until the prize was his"; but without documentation, this statement cannot be accepted as fact.[7]* I talked to several people who knew Freeman

---

* I contacted the author of this account, but was not able to obtain clear evidence of its source.

Figure 11.1.   Walter Freeman and Egas Moniz at the First International Congress of Psychosurgery in Lisbon, August 1948. Almeida Lima is at the extreme right. (Courtesy Ática, S.A.R.L., Editores Livreiros, Lisbon, Portugal.)

well, but they never recalled his having spoken about the prize. We know, however, that, in first nominating Moniz for the prize in 1943, Freeman sent along to the Nobel Prize committee a copy of *Psychosurgery* together with Moniz's monograph,* perhaps hoping to influence the committee to consider the two of them together.[9] The possibility was not completely farfetched: Freeman had advanced the theory of frontal lobe-thalamic interaction as an explanation of how psychosurgery worked and had also been the person most responsible for the acceptance of psychosurgery around the world.

Yet Freeman probably never had a chance to be nominated for the Nobel Prize. However effective he was at developing and improving ideas, he was not an initiator of new ones. Moreover, his habit of alienating people by his bluntness, self-assurance, and combative personality left a trail of opposition, if not enemies. Moniz, in contrast, was a paragon of tact. He made few enemies and had managed to convince nearly everyone that he was a gentle, modest

---

* In the fall of 1943, Freeman had received an invitation signed by Professor Nils Antoni and Folke Henschen—two distinguished Swedish physicians he had met earlier—to submit a nomination for the Nobel Prize in Medicine.[8]

man. Freeman's performance at the congress did not enhance his chances of sharing the nomination. At the first substantive session, devoted to surgical demonstrations, he had antagonized some neurosurgeons by performing two transorbital lobotomies on Portuguese patients selected for the occasion,[10] and his lecture was a rather pedestrian review of what was known of the anatomy of the frontal lobe. Near the end of the congress, when the Brazilian delegation proposed that those in attendance use their influence to nominate Moniz for the Nobel Prize in Medicine, Freeman's name was not mentioned. The motion was quickly adopted by acclamation.

The Nobel Prize committee received many nominations for Moniz during the autumn following the congress.* Some of the nominations apparently were evoked by Moniz himself: Freeman later wrote that, shortly after the Lisbon meeting, he had received "a modestly expressed letter" from Egas Moniz asking whether he [Freeman] could "without embarrassment, suggest his name to the Nobel Committee."[11] Ironically, the Swedish psychiatrist Gösta Rylander was on the selection committee and one of the signers of the award; he had always emphasized that lobotomies produced substantial impairment that was not apparent from superficial examination. With the war so recently over, there had been few nominations that year, and only fifteen nominees were considered worthy of serious consideration for the award in medicine. The committee decided to divide the award between Moniz and Walter Hess, the Swiss physiologist who, in studies of freely moving cats responding to electrical stimulation of the thalamus and the hypothalamus, had revealed much about the neural circuits regulating emotions. Thus, in October 1949, the Nobel Prize Committee announced that Egas Moniz and Walter Hess had both been awarded the Nobel Prize in Physiology and Medicine. An editorial in the *New York Times* applauded the selections:

> Quackish as was their system of feeling bumps on the head and thus judging human capabilities, F. J. Gall and J. C. Spurzheim, founders of phrenology, had the sound conception of a brain in which traits of ability and what we call character were localized. Since their time every nook and cranny of the brain has been poked into, dissected, examined. The brains of animals have been electrically stimulated. . . .
>
> The information that such explorers as Hess, Harvey Cushing, Walter Cannon and others gathered clicked together about 1935, when the International Neurological Congress was held in London. Among those who attended was the Portuguese neurologist, Dr. Egas Moniz. After listening to the papers that were read he decided

---

* For a person to be considered for the 1949 Nobel Prize, nominations had to be received by the committee by 1 February 1949. Immediately after the congress, the Brazilians wrote to the Caroline Institute in Stockholm suggesting Moniz, and the Nobel committee wrote to the University of Lisbon requesting a nomination. António Flores wrote the nominating letter supporting Moniz.

on his return to Lisbon that the time had come to cut worry, phobias and delusions out of the brain. . . .

Surgeons now think no more of operations on the brain than they do of removing an appendix. Hess, Moniz and Cushing before them taught us to look with less awe on the brain. It is just a big organ, with very difficult and complicated functions to perform and no more sacred than the liver.[12]

Moniz's contribution to cerebral angiography was not mentioned in the Nobel committee's award citation. The angiography work was noted in an editorial in the *New England Journal of Medicine,* but its importance was considered dwarfed by the development of lobotomy:

Moniz, however, went into an even more important field than cerebral angiography and in less than a decade was able to announce his radical operation, severing the frontal lobes from the rest of the brain. . . . A new psychiatry may be said to have been born in 1935, when Moniz took his first bold step in the field of psychosurgery.[13]

When, in Lisbon in 1982, I asked Almeida Lima which of Moniz's two discoveries he regarded as the greater, he answered immediately, "Prefrontal leucotomy. It was more original."

Before Watts had moved out of their shared office space, Freeman had suggested to him that they revise their book. The first edition of two thousand copies of *Psychosurgery* had sold rapidly, and the German and Spanish translations—the latter published in Buenos Aires—had also done well. With the idea of updating the book, Freeman had for several years been keeping abreast of everything written about psychosurgery and also corresponded with almost everyone in Europe, South America, and Asia who had significant experience with these operations. He had kept track of all the patients in the original Freeman-Watts prefrontal lobotomy series, pursuing all leads until he had located (or confirmed the death of) every one of them. One patient was traced to Australia with the help of a clue provided by her former husband. Another patient could not be found until he was released from jail, where he had been imprisoned for stealing the "poor box" in church.

Working frantically on the revision of the book, Freeman started writing at four every morning. After a busy day rushing between office, hospital, laboratory, and library, he would return home, eat rapidly, shower, and then work on the book until ten or eleven every evening. He saw little of his family—often refusing to have dinner with them—and, at the end of an evening, was usually so wound up that he needed Nembutal to sleep. As he approached the end of the task, he turned over most of his clinical responsibilities to his assistant Paul Chodoff in order to devote more time to the book.

Watts had a more relaxed attitude toward the project, but Freeman, who was getting increasingly impatient, had no intention of letting him work at his own pace—almost exploding at one point when he discovered that Watts had done nothing on his chapters. Freeman hounded him, at the risk of further straining their relationship:

August 1 came around and Jim went with his family on vacation at Farmington Country Club not far from Charlottesville, Virginia. I followed him one weekend in an effort to get the work finished. I was impatient of delay since I could foresee the opening of the school year with a full schedule. On walking into the dining room I got a black look from his wife, Julia. After allowing Jim and myself a night's rest . . . I took my material out on the porch in the shade, set up a table with a straight chair for the worker and an easy chair for the critic. I was full of the subject and pressed Jim for his part. We worked all morning. Sometimes he dictated while I wrote and sometimes vice versa. When his hand tired I started reading to him one of the new chapters. Presently his head drooped and he was off in a doze. I was proud of the phrasing so I was irritated and must have showed it.[14]

The second edition of the book was published in 1950. The title was changed to *Psychosurgery: In the Treatment of Mental Disorders and Intractable Pain* to acknowledge the application of lobotomy to intractable pain which had developed after the publication of the first edition. The book was extensively revised and about 50 percent longer than the original. There was much more literature to review, and their own prefrontal lobotomy cases had, by that time, exceeded one thousand. The preface brought out into the open Freeman and Watts's disagreement over transorbital lobotomy:

The authors regret to announce that they have been unable to reach an agreement on the subject of transorbital lobotomy. Freeman believes that he has proved the method to be simple, quick, effective, and safe to entrust to the psychiatrist. Watts believes that any procedure involving cutting of brain tissue is a major operation and should remain in the hands of the neurological surgeon. In all other respects the authors are in complete and cordial agreement and they continue as co-heads of the Department of Neurology and Neurological Surgery of George Washington University.[15]

In the text, following Freeman's description of transorbital lobotomy and his results, Watts presented his own views, expressing doubt about how seriously ill some of the patients were:

It is Walter Freeman's opinion that transorbital lobotomy is a minor operation. This is clearly indicated by the fact that of the first ten cases reported four were not disabled [before the surgery] and six had been disabled less than six months. If he is correct in his view, transorbital lobotomy should take precedence over more time consuming psychiatric techniques.

On the other hand, if transorbital lobotomy is a major operation, as I believe it is, it should be performed only rarely in mental cases that are not disabled. In both mental and physical disease, surgical intervention should be reserved for cases in which conservative therapy has been tried and failed or in cases where such treatment is known to be ineffective.

It is my opinion that any procedure involving cutting of brain tissue is a major surgical operation, no matter how quickly or atraumatically one enters the intracranial cavity. Therefore, it follows, logically, that only those who have been schooled in neurosurgical techniques and can handle complications which may arise should perform the operation. If it is a major operation, disability constitutes the indication for transorbital lobotomy.[16]

By the time the second edition of *Psychosurgery* had gone off to the publishers, Freeman had done over four hundred transorbital lobotomies. Not only was Watts concerned about who did the surgery, but he also felt justified in his early suspicions that any operation that could be done so quickly might be done on patients who were not seriously incapacitated. Watts reported in *Psychosurgery* that he had performed six transorbital lobotomies himself, but without encouraging results. He speculated that the type of patient he selected, or his use of conventional anesthesia rather than electroconvulsive shock, might explain why his results were different from Freeman's, but he was clearly skeptical. He left open the possibility, however, that Freeman's modification (the "deep frontal cut") of the transorbital procedure might increase its effectiveness.

For anyone reading the book carefully, it should have become evident that a large price was often paid for the "peace of mind" provided by prefrontal lobotomies:

> It is almost impossible to call upon a person who has undergone operation on the frontal lobes for advice on any important matter. His reaction to situations are direct, hasty, and dependent upon his emotional set at the moment.... There is something childlike in the directness and ingenuousness of people after lobotomy. ... It is hardly fair to the patient who has undergone prefrontal lobotomy to say that he is entirely lacking in insight.... On the other hand, it is rare to have a patient go at all deeply into his introspections. ... They are direct, practical and uninspired.[17]

In this context, Freeman and Watts's comments about the success achieved with different patient populations was most revealing. Proportionally more women than men could be sent home "because of the greater protection afforded them in the home." The authors reported a higher success rate with Jews—probably, they wrote, because of "the greater family solidarity manifested by these people." While college-trained patients did better than those with only a grade-school education, it was reported that those with the "simpler occupations" (clerical and mechanical) were more likely to return to work: "no physician, dentist, artist, musician, or writer has been able to make the grade." Thus, there was much evidence to justify strong reservations about prefrontal lobotomy, and Freeman would soon announce publicly that he would no longer recommend the procedure that he and Watts had developed and that he had promoted so vigorously.

During its peak, between 1949 and 1952, approximately 5,000 lobotomies were performed each year in the United States. One out of three operations was a transorbital lobotomy. Although the transorbital procedure was only rarely performed at Veterans Administration hospitals, Freeman did demonstrate this operation at the V.A. hospital in North Little Rock, Arkansas, from 28 February to 1 March 1949. A few other transorbital operations were performed elsewhere in the V.A. system.[18] Many prefrontal lobotomies, however, were performed by neurosurgeons in V.A. hospitals (by 1949, over 1,200 lobotomies had been done).[19] Indeed, the V.A. played a major role in promoting lobotomy, through its various training programs. The care of neuropsychiatric patients absorbed the largest portion of the available funds; between 1947 and 1949, mental patients occupied over 55 percent of all V.A. beds, and their length of stay in hospitals was more than four times that of the general medical and surgical patients. Whereas many patients in state hospitals had been psychologically abandoned by their families, the relatives of veterans with psychiatric illnesses constituted a pressure group for effective treatment. By the end of 1948, all V.A. hospitals "were making use of modern treatment methods, including electric shock, insulin therapy, prefrontal leucotomy, individual and group psychotherapy, and intensified medical treatment."[20] Also, the V.A.'s neuropsychiatry division, in cooperation with over thirty-eight medical schools, sponsored training programs in psychiatry, and over 385 residents in the program were taught the new somatic therapies.

Freeman began to invest ever more energy and time in transorbital lobotomy. Probably no one had more firsthand experience with the conditions in state mental hospitals around the country than he had. Although transorbital lobotomies comprised 30 percent of the psychosurgery performed during the peak years, in state hospitals they were more than half the total. Virtually all of the transorbital lobotomies in the United States can be traced to the influence of one man—Walter Freeman. In 1951 alone, he visited hospitals in seventeen states as well as provincial hospitals in Canada. He also gave demonstrations in San Juan, Puerto Rico, and Willemstad, Curaçao. On one five-week summer trip that year, he drove 11,000 miles with a station wagon loaded, in addition to camping equipment, with an electroconvulsive shock box, a dictaphone, and a file cabinet filled with patient records, photographs, and correspondence; his surgical instruments were in his pocket (figure 11.2). He recorded the number of operations he performed at each stop:

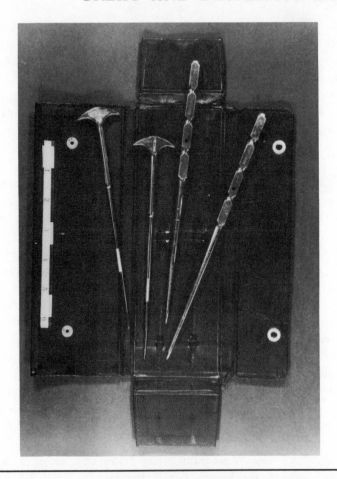

Figure 11.2.   The leucotomes and the orbitoclasts in the case that Freeman carried in the pocket of his suit jacket. (Photograph by Jonathan Williams.)

29 June, Little Rock, Arkansas, 4 patients
30 June, Rusk, Texas, 10 patients
1 July, Terrell, Texas, 7 patients
2 July, Wichita Falls, Texas, 3 patients
9 July, Patton, California, 5 patients
14 July, Berkeley, California, 3 patients
17 July, Mendocino, California, 9 patients
19 July, Fort Steilacoom, Washington, 7 patients
28 July, Sedro Woolley, Washington, 8 patients
4 August, Yankton, South Dakota, 5 patients
6 August, Lincoln, Nebraska, 8 patients
7 August, St. Joseph, Missouri, 1 patient

8 August, Cherokee, Iowa, 25 patients
9 August, Independence, Iowa, 16 patients.[21]

On most days, he left the campsite before dawn and drove four to five hours to arrive at the hospital by midmorning. After meeting with the staff to discuss the patients, and a quick lunch, he would start to operate. The atmosphere was often tense, as some of the patients were agitated or violent, and the operations did not always go well. On the 1951 trip, there were four deaths from the surgery. The experience at the Cherokee State Hospital in Iowa was the worst: three of the twenty-five patients died either during or shortly after surgery. Two deaths were caused by hemorrhage, and one by an accident that caught Freeman completely by surprise. Normally it took a certain force to drive the leucotome through the bony orbit, but in this patient, the bone was so thin that the leucotome went through with hardly a touch from the mallet. When Freeman, as was his custom, stepped back to photograph the leucotome in place,* it sank a full two inches farther into the brain up to the hilt of the handle. This patient never regained consciousness; and when the brain was later autopsied, the tip was found to have gone far beyond the frontal lobes, extending deep into the midbrain: indeed, the tip had penetrated the basal ganglia, skirting the hypothalamus and ending near the red nucleus.[22] Freeman would not allow himself to be emotionally drained by guilt when things went wrong. He had the attitude of a general conducting a military campaign: losses would have to be incurred on the way to ultimate victory.

On another occasion, when visiting the Sykesville State Hospital in Maryland, he attempted to operate on a seriously disturbed woman. She had persecutory delusions and hallucinations and attacked people violently without warning. He had tried to operate on her a year earlier, but her orbital bone was so thick he could not drive the leucotome into her brain, and was forced to give up. This time he managed to get the leucotome through the bone, but the instrument broke into three pieces when he attempted to force up its handle to make the deep frontal cut. One fragment lacerated the woman's eyeball and detached her retina. A neurosurgeon had to be called to remove a piece of metal in her brain, but the eye could not be saved. Because the woman was still very violent, Freeman tried to complete the transorbital lobotomy, but the surgeon objected. On a later visit to Sykesville, Freeman operated on this woman a third time, using the much sturdier orbitoclast,

---

* Freeman took a profile view of most patients with the leucotome in place. It is not clear what value these pictures had, other than to identify patients later or to prove, if necessary, that the surgery had been done. Photographs of patients at other times were used to illustrate their improvement: thus, a woman with tangled hair and an obviously wild, frightened, or agitated expression would be contrasted with her well-groomed and, if not smiling, calm expression after the lobotomy.

and finally succeeded in completing the operation; he reported, a week later, that the patient appeared "calm and accessible."

By 1951, Freeman was taking even less than ten minutes for an operation. Rather than doing the two sides of the brain successively, he would use two leucotomes, forcing one through each orbit before stepping behind the patient to make the deep frontal cut on both sides at the same time. He would grasp the two handles—one in each hand—and simultaneously force them upward and laterally. After a long surgical session, his hands were often tender from the pressure—belying the morning newspapers' description of the "delicate brain operations" he had performed at the local hospital. Freeman remarked, in his unpublished autobiography, that after a long surgical session his hands felt as though he had been working out on parallel bars.

Gradually Freeman began to spend an increasing amount of time on the trips checking on the condition of former patients. He would write to patients who lived close to his planned route, for example:

Dear Robert,*

I am taking a trip and expect to be in your neighborhood around August 2. Please let me know on the enclosed card where I can reach you by telephone. If you have a doctor, please let me have his name, address and telephone number.[23]

He wrote Christmas cards to former patients or their relatives and received a great many replies. Freeman kept these letters on file by a case number that denoted whether they were in the standard (LOB) or transorbital lobotomy (TOL) series. A Christmas card from the wife of LOB case 152 brought good news:

Dear Dr. Freeman,

Earl is just fine, working every day, getting a little roll around the waist as he gets older. Our son is in his second year of Medical School. Needless to say we are very proud of him. We do wish you would stop to see us if you were ever to come to the mid-west. Earl and I are both very grateful to you for making our lives happy. The best wishes for the coming year.

Not all the letters, however, were so comforting. A letter written by the mother of a patient lobotomized eight months earlier stated that her daughter had died of a heart attack. Freeman did not think the operation had been successful and was suspicious of the cause of death. An inquiry to the local Bureau of Vital Statistics confirmed his suspicion: the cause of death was

* The names of all patients have been changed, and any identifying information has been omitted. Copies of the letters are in Freeman's patient correspondence file, Himmelfarb Health Science Library, George Washington University.

suicide by an overdose of phenobarbital, which he had prescribed to control her seizures. He recorded the case as a failure.

Before starting off on one of his trips, Freeman would contact as many as 250 former patients. Some of the patients in the transorbital lobotomy series did not even remember his name—not surprising, as the typical patient had spent little time with him and was usually quite disturbed at the time of the lobotomy. Moreover, the electroconvulsive shock erased much of a patient's memory of the events surrounding the operation.

Freeman paid out of his own pocket most of the expenses he incurred while traveling around the country. He usually did demonstration operations without any charge, although in Virginia, West Virginia, and Ohio, he was paid $25 for each operation—a modest sum by today's standards, even after adjusting for differences in cost of living. Most of the cost of the follow-up visits to check on former patients was assumed by Freeman, who still connected the project with Keen's tragic death and viewed the expenses as coming from the amount he would have spent on his son's education. Only rarely did he receive a substantial payment, as when Dr. Paul Royal at the Green Gables Sanatarium in Lincoln, Nebraska, arranged for him to receive $200 for each operation. The largest sum he received was $2,500 for a special trip to Chicago—money that helped to defray the expenses of his follow-up studies and of some trips to meetings abroad.[24]

A few other groups in the United States reported success with the transorbital lobotomy procedure. One of these was headed by W. Wilson at the New Jersey State Hospital in Trenton.* Wilson first used the transorbital procedure in November 1949, starting with "the criminally psychotic, the most difficult patients in the State of New Jersey." Later, the operation was performed on other patients, the schizophrenic group being the largest. By May 1951, Wilson reported that they were "progressing at the rate of 8 to 12 operations a week," and that "marked improvement was evident in one half of their chronically ill patients, making it possible to open the doors of many seclusion rooms for the first time." Wilson, as so many others had done earlier, emphasized the economic gains from lobotomy, which freed hospital beds and trained personnel.[25]

Lawrence Kolb, a prominent psychiatrist in the audience, was not convinced by Wilson's presentation. He observed that if all the patients had been hospitalized for at least seven years before the lobotomy, as Wilson implied, the results would indeed be spectacular, since few long-term patients improve, but the way the data were presented it was impossible to know how long most of the patients had been hospitalized. In view of the fact that many

* This is the state hospital where Henry A. Cotton had earlier vigorously pursued his focal infection theory of mental disorders (see pages 37–40).

patients hospitalized for less than two years get better with only conservative treatment, Kolb said he could not agree with Wilson's comment that the statistics "speak for themselves."[26]

Matthew T. Moore from the University of Pennsylvania Graduate School of Medicine also used the transorbital lobotomy procedure and, in 1951, presented the results from more than one hundred operations at the Wernersville State Hospital in Pennsylvania. In the catatonic schizophrenic group, he claimed, over 29 percent of the patients had "recovered" and another 24 percent showed "good improvement."[27] J. S. Wilson of the State Hospital in Evanston, Wyoming, also reported good results. He had apparently learned the transorbital lobotomy procedure from Freeman's descriptions in journal articles and, after practicing on cadavers, did over two hundred operations.[28]

Although the vast majority of transorbital lobotomies were performed at state hospitals, some psychiatrists besides Freeman used this procedure on non-institutionalized private patients. Abraham Gardner, for example, reported in the *American Journal of Psychiatry* that he did more than fifteen transorbital operations on non-institutionalized "psychoneurotic" patients at a small community hospital in Lynn, Massachusetts. Gardner observed that his success justified the use of transorbital lobotomy on psychoneurotic patients who had symptoms of tension, fear, agitation, depression, and "tortured self-concern."[29]

Freeman's 1952 schedule of visits to state hospitals was just as demanding as the 1951 trip (see note to chapter 14, pages 323–25, for Freeman's itinerary during this period). During that same year, he also started his West Virginia Lobotomy Project. After visiting their state hospitals for several years, he persuaded the West Virginia Board of Control to finance a large-scale trial of transorbital lobotomy. With the help of the staff at the hospitals in Lakin, Huntington, Spencer, and Easton, Freeman did 225 transorbital lobotomies during a twelve-day period in August 1952. A year later, after he had reported that 85 percent of the patients were out of the hospital, the project was extended.* A total of 787 transorbital lobotomies were performed on patients from the four state hospitals in West Virginia.[30]

Encouraged by the apparent success of this project, Freeman tried to set up similar programs in other states. In October 1953, he was the main speaker at a psychosurgery workshop sponsored by the New Jersey State Hospital in Marlboro. He talked about the advantages of the transorbital lobotomy procedure for state mental hospitals "where there is a shortage of everything but patients."[31] Reminding his audience that the transorbital approach had been introduced by Fiamberti, a psychiatrist, Freeman said that it was desirable for

---

* Edward Reaser, who interned at St. Elizabeth's Hospital and worked under Freeman in the Blackburn Laboratory, became superintendent at the Huntington State Hospital and helped keep track of the discharged patients.

psychiatrists to perform the surgery, as they knew the patients better than anyone else and were also in the best position to help them and their families afterward. He reported that one-tenth of all patients in residence at the Lakin State Hospital in West Virginia were lobotomized during 1952 and 1953, with a discharge rate of 50 percent, and added that one of the most moving experiences of his professional career occurred when, only one week after performing the surgery, he saw twenty previously very disturbed patients enjoying themselves on the hospital grounds with only one attendant. Before the surgery, he said, these patients had not been out of a locked ward for at least six months, and one had not been out for seven years. In conclusion, however, Freeman made explicit what Watts and others had feared all along: Freeman stated that his West Virginia Project had demonstrated that transorbital lobotomy should be considered not the last resort but "the starting point in effective therapy."[32]

Freeman was almost successful in establishing a similar transorbital lobotomy project in Virginia. He was supported by the superintendents of the Eastern and the Western state hospitals, Dr. Granville Jones and Dr. James Pettis, respectively, both of whom had worked with him in the Blackburn Laboratory of St. Elizabeth's Hospital. The three had obtained the support of Dr. Joseph Barrett, the state hospital commissioner. The plan was to have Freeman serve as a consultant, training psychiatrists, who would then perform transorbital lobotomies under his supervision.

Before the "Virginia Project" was officially approved, Freeman started to train some of the psychiatrists—and then another accident occurred.* It was late in the day, and he had rushed off to the airport before the psychiatrist he was training had finished the last operation. When Freeman arrived back in Washington, he learned that, in his absence, the tip of the leucotome had broken off, and the small piece protruding from the orbit had slipped back into the patient's brain. The neurosurgeon who was summoned had to open up the skull to remove the piece of metal. The patient recovered, but the surgeon kept the leucotome tip as "incriminating evidence" to present at the next meeting of the State Advisory Board.

Freeman requested permission to make a presentation at the meeting of the Advisory Board, partly to protect the young staff member but also to try to save the now-endangered project. During his presentation, Freeman reached into his pocket and pulled out a piece of metal larger than the one the neurosurgeon had exhibited. Freeman explained that his surgical colleague, Jonathan Williams, had successfully removed this piece of metal from a patient,

---

* In his unpublished "History of Psychosurgery," Freeman did not specify whether this incident occurred at the Eastern or the Western state hospital; he stated only that psychiatrists had been operating at both.

and argued that a broken leucotome was not a remarkable event—it had happened four times before—and besides, only one patient had been impaired in any way. Freeman then reviewed the good results he had obtained in over a thousand cases, and concluded that it would be unfortunate if the procedure were not available to the state hospitals in Virginia.

After much argument, the Advisory Board finally agreed that transorbital lobotomy would be permitted in the state hospitals as long as a "physician especially skilled in this type of neurosurgery" was available. The local neurosurgeons had argued that a "specialist in neurosurgical surgery" must be present, but this wording was voted down after someone pointed out that this requirement would amount to a prohibition of transorbital lobotomies, since state hospitals did not have funds to hire neurosurgeons. The final wording in the regulations adopted by the State Hospital Board in Richmond in January 1953 was that "contact neurosurgeons may be retained for transorbital lobotomy operations only to the extent that a hospital's appropriations for such services makes it possible." Many transorbital lobotomies were eventually done in Virginia, but only at the discretion of individual hospital superintendents. This was the arrangement in most states. Freeman once remarked that the superintendents of the state hospitals in Texas were generally the most cooperative.[33]

Freeman had concluded that neither a prefrontal lobotomy, nor a transorbital lobotomy—including the "deep frontal cut" modification—were effective with hallucinating, schizophrenic patients. With Jonathan Williams's help, he had tried temporal-lobe amygdalectomies (see note on page 199), but any apparent improvement rarely lasted. He then speculated that perhaps the frontal and the temporal lobes act together to produce hallucinations, as anatomical pathways between the two areas had been described in monkeys. Willing to try almost anything that had any chance of success, he thought about modifying the transorbital operation so that the leucotome tip could be placed at the posterior border of the prefrontal area, in order to sever the nerve fibers connecting these two brain regions. The problem with this scheme was that the bundle of fibers was very close to some large blood vessels and moving a leucotome around in that area could produce a massive hemorrhage. Freeman decided instead to destroy this nerve bundle by injecting a patient's own blood into this region—a technique that had been used earlier in France and Italy (see page 118). As shown in figure 11.3, after removing the leucotome from the orbit, he inserted a hypodermic needle and injected the blood:

I used a syringeful of blood drawn from the patient's arm and injected into the lower part of the incisions on each side. I used blood because of its viscosity, its availability and the tolerance of the brain to spontaneous hemorrhage. Blood in-

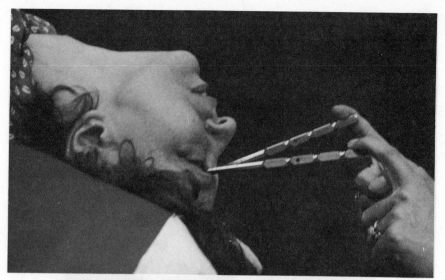

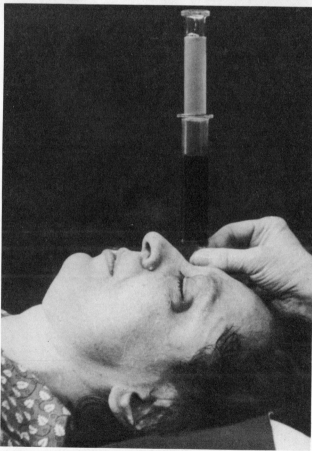

Figure 11.3. (*Top*) Patient with two orbitoclasts—the sturdier leucotomes—in place. (*Bottom*) Injection of patient's own blood after removal of the orbitoclasts; the blood was used to destroy brain tissue in regions where it would have been too dangerous to insert the orbitoclast. (Photograph courtesy of William Beecher Scoville.)

Figure 11.4.   Bookplate used by Egas Moniz; the same design is in tile on the front of his Avanca home, Casa do Marinheiro, now a museum. In the museum as well are the series of scrapbooks, filled mostly with newspaper clippings mentioning his honors and accomplishments, which Moniz had his personal servant, Joaquin, maintain. Moniz was— as Sir Geoffrey Jefferson remarked, with typical English understatement—"not insensitive to praise. He was conscious of great achievement and he would have been scarcely human if he had not valued recognition of his work." (Photograph by the author.)

jections into the brain had been used by French surgeons as a substitute for lobotomy some years before.

It was gratifying to note that following the procedure a number of patients were able to leave the hospital who had not responded to the usual type of transorbital lobotomy. I have not examined any of them but I gather from reports of relatives that they are still hearing voices but are no longer preoccupied by them to the extent of requiring hospitalization. The hazards of such operations with blood injection are greater, a surgical mortality of 6% being recorded, but further trials are desirable. Maybe hot water, as used by [a former classmate] might be better, since heat coagulation near the needle tip might be safer. Injection of blood splits tissues along the course of the fibers rather than cutting across them, but some fibers are undoubtedly destroyed by pressure of the viscous blood. I have tried in vain to get permission to operate upon chronically hallucinated patients, except in a handful of isolated patients.[34]

This procedure did not produce any remarkable success, and Freeman gave up the effort without ever publishing any results.

In July 1953, Freeman went to Lisbon to attend the International Neurological Congress, of which Moniz, going on seventy-nine, was honorary president. (Two years before, he had been asked to be president of Portugal—a ceremonial position—but declined because of his age and poor health.) When the two met, Moniz pressed his cheek against Freeman's and, embracing him with his bandaged hands, said, "Ah, mon cher Freeman, mon maître, you are the great expansionist of leucotomy." Two years later, on 13 December 1955, with his mind still clear and while working on an article about a Portuguese painter, Moniz suffered a sudden, massive, internal hemorrhage. He knew he was dying and quietly acquiesced.

Born of another age, this "Renaissance man" had been instilled from childhood with love of country, an appreciation of the arts and classics, and, above all, a driving ambition to excel and to be recognized (see figures 11.4 and 11.5). Characteristically, Moniz, who had little musical talent, promoted the story that he had written a professionally staged operetta (actually a little at-home show with childhood friends).[35] Though he was determined to be thought of as creative, within neurology he demonstrated little talent for delving into theoretical questions. Moniz made his mark by his ability to select and to solve important applied problems—not through innovation, but through trial and error, dogged persistence, and an uncommon willingness to take risks. Perhaps influenced by the idealized heroics of the great fifteenth-century Portuguese navigators, Moniz identified with and admired other great risk takers as well; in his dressing room in Avanca were three porcelain figures of Winston Churchill, born, like himself, in November 1874. In his memoirs, *Confidências de um Investigador Científico,* which reveal little about his inner thoughts and feelings, he quoted every word of every compliment he ever received from anyone of consequence. Even Walter Freeman, a great admirer of Moniz, was forced to write that "the book will particularly appeal to readers who have praised Egas Moniz, since they will find their words quoted *in extenso.*"[36]

In the fall of 1953, after returning from the Lisbon Neurological Congress, where he had seen Moniz for the last time, Freeman informed the dean of the George Washington University Medical School that he was going to

Figure 11.5.   Statue of Egas Moniz in Lisbon, erected to honor Portugal's only winner of the Nobel Prize. (Photograph by the author.)

move to California. He had been thinking about the move for several years: the wilderness areas attracted him; his brother Norman, a vascular surgeon, was there; and some of his children appeared to be going to settle there. Freeman had hoped he might be appointed professor emeritus at George Washington, but he was only fifty-eight, not yet retirement age.

In June 1954, close to the time of Freeman's scheduled departure from Washington, his colleagues arranged a farewell banquet. Temple Fay,* former head of neurosurgery at Temple University, came down from Philadelphia

---

* Fay had, like Freeman, worked with the University of Pennsylvania pathologist Nathaniel Winkelman. Fay was also a colorful figure in medicine, advocating carbon-dioxide therapy and fluid control (withdrawing spinal fluid and restricting water intake) to treat epilepsy; he is considered the progenitor of the widely known, but controversial [Glen] Doman-[Carl] Delacato treatment of brain-injured children by attempting to "reorganize the nervous system" through a regimen of passive movements.

and reminisced about doing neurology and neurosurgery rotations at the University of Pennsylvania hospitals with Francis Grant and Freeman. Fay talked about Freeman's brilliance and inexhaustible energy and about how for many years Freeman had traveled up from Washington to attend the monthly meeting of the Philadelphia Neurological Society and, in 1943, was elected president of the group—one of the few "out-of-towners" to be so honored. Ted Wipreid, secretary of the Medical Society of the District of Columbia, spoke about Freeman's importance to medicine in Washington and his active role in the society. Freeman had been elected its president in 1949; and, during his term in that office and with his active support, black physicians were admitted to the organization. Watts, too, spoke, saying nothing of his and Freeman's disagreement, but describing their long association during the early days of lobotomy and Freeman's dramatic lectures in the amphitheater of the old George Washington Hospital and his autopsy sessions for students at the morgue of St. Elizabeth's Hospital. Some of Freeman's former students mentioned that he, more than anyone else, had stimulated their interest in the brain, and described his performances at the Saturday clinics as "the best show in town."[37]

Finally, it was Freeman's turn to respond. With hardly any introduction and no explanation that anyone could recall afterward, he cleared his throat and started reading some of his own ribald limericks. His friends and colleagues were shocked, puzzled, and embarrassed. When he finished, they said nothing, but inquired with their eyes, "Why did he do it?" Whatever motivated Freeman to respond this way, it was not an impulsive act: he had brought his folder of limericks with him.* Those who knew him well finally concluded that, after fighting criticism for so many years, he was no longer able to respond appropriately to compliments and affection.

---

* Essentially the same account of Freeman's farewell banquet was given to me by Jonathan Williams and Zigmond Lebensohn, former associates of Freeman who were present at the dinner.

12

Two Patients:
The Effects of Early
Prefrontal Lobotomies

> Is the quieting of the patient a cure? Perhaps
> all it accomplishes is to make things more
> convenient for the people who have to nurse
> them.
>
> —NOLAN LEWIS (1949)

MOST of the early lobotomies were neither clear failures nor outstanding successes. (The data—inadequate though they are—will be reviewed in chapter 13.) Many patients could be sent home to a life that was better than the one they had had in a mental institution, but they were usually not self-sufficient and many were a heavy burden on their families. Other lobotomized patients could not be sent home because either their improvement was not sufficient or no one could assume responsibility for them, but they were able to enjoy more freedom within the institution as they no longer had to be restrained in a locked ward.

Probably no other lobotomized patient has been studied as long or as thoroughly as Carolyn W.* who, since her second lobotomy in 1948, has been examined and tested by experienced psychologists at least twice a year for almost forty years. Her history illustrates that, contrary to a common misconception, lobotomies were not performed exclusively on poor patients.

Born in 1907 into a wealthy Grosse Pointe, Michigan, family, Carolyn enjoyed many of the advantages of growing up in comfortable circumstances. There is no indication in the record that she had any serious psychological problems during her early life. She graduated from a private secondary school and from Vassar College at the usual ages. It is known that when she was in

* Her name and other identifying facts have been changed.

college Carolyn had heated arguments with her mother, usually about missing classes and impulsively going off to New York. It is not clear, however, whether Carolyn's difficulties with her mother were anything more than adolescent rebellious behavior.

After graduating from Vassar, Carolyn attended the University of Pennsylvania Medical School, completing the first two years successfully. Around this time, she fell in love with a young man and married him. Her parents objected strongly to the marriage and managed to have it annulled. Shortly afterward, Carolyn had a "psychotic break," in which she experienced delusions, hallucinations, and mania. She engaged in "stereotyped" behavior such as compulsive picking and tearing at upholstery until it was ruined. At times she was violent and destructive, but most of this behavior was self-directed. She was considered suicidal.

Carolyn was diagnosed as schizophrenic and, in 1935, was brought east and committed to the Hartford Retreat, the private sanatorium in Connecticut later called the Institute of Living. Carolyn received the best medical care available at the time, under the supervision of two experienced psychiatrists, one of whom had a clinical appointment at Yale. She was given all the available therapies for schizophrenia, including insulin coma, electroshock treatment, sedatives (antipsychotic drugs were still in the future), hydrotherapy, and wet packs; when there was a danger that she would injure herself, she was placed in seclusion, sometimes in a straitjacket for restraint. No treatment produced any lasting improvement.

Carolyn's psychiatrists, convinced that they had exhausted all the usual therapies during her almost fourteen years of hospitalization, began to consider prefrontal lobotomy as a possibility. They consulted with C. Charles Burlingame, the psychiatrist-in-chief at the Institute of Living, where these operations had been performed since 1939, and found him sympathetic to the procedure.[1] Carolyn's psychiatrists were encouraged furthermore by the Connecticut Lobotomy Study's early reports that lobotomies had produced beneficial results with schizophrenic patients. As a result, toward the end of 1947, Carolyn was taken to Grace New Haven Hospital in New Haven (now the Yale New Haven Hospital) to undergo a standard Freeman-Watts prefrontal lobotomy. Although the intensity of some of her symptoms was slightly reduced, there was not enough change to warrant her discharge.

During the year after the lobotomy, Carolyn's parents died, leaving half of their substantial estate to her and the other half to her married brother. Following a discussion that included the brother and his wife, permission was obtained to perform a more posterior, "radical" prefrontal lobotomy. Immediately after the second operation, Carolyn became markedly subdued. For a period she was incontinent, unresponsive, and confused, but she was no longer unmanageable or self-destructive. She was sent back home to live with her brother and sister-in-law. By this time, Carolyn was almost forty years old, but she needed a full-time caretaker. Initially, the job was almost like caring for a child; Carolyn had to be brought to the bathroom and essentially toilet-trained. It was also necessary to supervise all her routine activities, even dressing, eating, and bathing. She seemed to be indifferent to pain and once stepped into scalding bath water, severely burning her foot. Like many of the early lobotomized patients, she had bulimia and ate voraciously: it is believed that the frontal lobes normally suppress "lower centers" in the hypothalamus and other areas in the brain that stimulate appetite; following destruction of the frontal lobes, or its exiting fibers, it was not uncommon for patients to gain more than fifty pounds. She also had a tendency to start fires in her room during the first six months after her discharge from the hospital.

One of the senior judges of her home state had agreed to serve as the conservator of Carolyn's large trust fund. It soon became clear to him that the brother and his wife were not going to assume the responsibility of supervising Carolyn's care as they were often away on extended trips. A responsible caretaker was located, and Carolyn moved in with the caretaker's family. A psychologist who is very knowledgeable about brain damage and behavior was given the responsibility of examining Carolyn periodically, monitoring her care, and reporting to the conservator. He supervised Carolyn's care from 1949 until 1981, giving her a battery of tests at least twice a year. After 1981, another noted psychologist experienced with brain-damaged patients assumed this responsibility; he has also regularly tested Carolyn and reported to the present conservator, a senior partner in a prestigious Michigan law firm.

From the beginning, her first psychologist had insisted that Carolyn be given maximal stimulation to prevent further "institutionalization." He encouraged her to read, to take daily walks, and to be given music and art lessons. Carolyn has gradually improved; but in spite of all the stimulation, she is not able to take care of herself and has required full-time caretakers. She is usually able to verbalize what she should do, but there is no guarantee that she will do it. Even though she could state what bus she should take to get to her art teacher, Carolyn took the first bus that came along and had to be rescued at different locations around the city when she was given several opportunities to make the trip alone. She is now always accompanied when she travels.

A similar discrepancy between verbalization and performance was evident on the Wisconsin Card Sorting Test, which is designed to measure "flexibility in thinking." On 64 cards—or 128 in some versions—are printed triangles, stars, crosses, or circles.[2] A given card may have from one to four of any one of these shapes, each colored red, green, yellow, or blue. Thus, a card might picture two red stars or four blue triangles. The subject is asked to stack all the cards in four piles—a task that can be done by using shape, color, or number. After a series of cards has been sorted, the subject is required to switch the criterion by being told that the way he or she has been sorting is "now wrong." The subject who has been using shape can try sorting by color or number. Carolyn can verbalize that the cards can be sorted by color, shape, or number; but even though she has been tested regularly since 1950, she has been completely unable to shift strategies once she has started to sort the cards. This tendency to "perseverate" is characteristic of many people with extensive damage to the frontal lobes, and they often repeat the same response over and over again, even though it is no longer appropriate. In spite of this serious deficit, Carolyn has consistently achieved an IQ of between 115 and 120, in the bright normal range—a score that testifies to the inadequacy of standard intelligence tests for evaluating lobotomized patients. For example, in 1939, Donald Hebb, of McGill University and the Montreal Neurological Institute, described a twenty-nine-year-old man who had a major portion of his abscessed left frontal lobe removed.[3] Four years after the surgery, the former patient made no errors on the Stanford-Binet Superior Adult test, obtaining an IQ of 152.

Recently, Carolyn visited the National Institutes of Health and was examined with a PET scan device—a non-invasive way of visualizing the capacity of different brain regions to utilize various physiologically active substances. The test revealed that in her frontal lobes there were huge nonfunctioning islands of dead brain tissue. Earlier, when given the Luria-Nebraska Test (a battery of tasks used to evaluate brain damage), Carolyn had always performed within the normal range, exhibiting no evidence of "organic impairment."

Carolyn's time sense about past events is also greatly impaired. Although the lobotomies were performed over thirty-five years ago, when asked when the surgery was done she typically replies, "Oh! About five years ago." Yet, in spite of all her deficits, her current psychologist has observed that people meeting her for the first time usually do not detect any abnormality and consider her a proper, affable, well dressed, and somewhat taciturn lady.

Carolyn has been fortunate in having had excellent caretakers and professional supervision and money from a trust fund. The caretakers have virtually adopted her. The present caretaker is the daughter-in-law of the woman who looked after Carolyn for almost twenty-five years. A retired nurse arranges for excursions to museums, restaurants, expositions, and dinner-theater evenings on the average of twice a week. Carolyn swims regularly, and the caretaker and her husband accompany her on several extended trips a year, either within the United States or abroad. Although Carolyn tolerates this traveling, she never expresses much enthusiasm about what she has seen and does not get involved in planning the trips. Her descriptions of her experiences seem as if they were written by a rather dull ten-year-old instead of a woman in her seventies. For example, in a recent letter she wrote of her "nice summer": "It was good weather for swimming and I went in every day it was nice. I went to a Quaker wedding and it was something new." While Carolyn has certain preferences about food and dress, she never seems to initiate any activities. Her room appears to belong to a pre-teenager, with a collection of Teddy bears scattered about. Now an old woman, Carolyn acts essentially like a compliant and not very bright young girl.

Her limitations make it easy to criticize those responsible for the decision to perform the lobotomies. At the time, however, most psychiatrists considered it almost a certainty that any schizophrenic patient hospitalized for more than ten years would only deteriorate further. Carolyn seemed destined to spend her remaining years locked in a mental institution. Today, we might wonder whether some of her limitations were not the result of fourteen years of confinement in a mental institution where more electroconvulsive shock was surely administered than would now be considered safe. It is generally accepted that massive doses of electroshock are likely to produce memory and other cognitive deficits.[4]

Much, of course, has changed since 1948, the year Carolyn had her second lobotomy. Ten years later, she would almost surely not have had a lobotomy but would have been treated with chlorpromazine or one of the other antischizophrenic drugs, which were being used after 1954. Moreover, although some psychosurgery was still being performed in 1958, most people had concluded that these operations were not likely to help chronic schizophrenics. Today, her case might not be considered hopeless: at least some psychiatrists recognize that long-term chronic schizophrenics may occasionally improve even without drugs. At the time of Carolyn's lobotomies, most physicians agreed with the Swiss psychiatrist Eugen Bleuler, who was convinced that schizophrenia was a hopeless disease: "Let us openly say to ourselves and to others that, *at present, we know of no measures which will cure the disease, as such, or even bring it to a halt.*"[5] Manfred Bleuler, Eugen's son, however, has more recently concluded that his father was wrong in his pessimism. Overly influenced by seeing some patients progressively deteriorate, the elder Bleuler did not appreciate that many of the patients not seen "were out for their Sunday walks during his visits" or had been "released and were living at home, recovered."[6]

Based on his experience with many schizophrenic patients followed for more than twenty years, Manfred Bleuler reached a different conclusion:

The condition of many schizophrenics who have been "demented" for many years can still improve appreciably, even decades after the initial onset of their illness. From these data the following conclusions are unequivocal. Schizophrenic "dementia" is by no means always a permanent state. It is accessible to influences from the environment. It is not the "core syndrome" of a hereditary destructive disease process.[7]

Many psychiatrists, however, are still convinced that there is substantial evidence that a predisposition for schizophrenia is genetically determined;[8] and that some of the disorders are progressive, lasting improvement in such cases being therefore rare.

The justification for lobotomizing Carolyn—considering what was known at the time—could be debated endlessly. It may be more useful instead to compare her outcome with that of another patient, a man known in the literature only by the initials J.S. Like Carolyn, he was also given a Freeman-Watts standard lobotomy, but J.S. did not receive the postoperative care available to her. At the time his lobotomy was performed in 1941, J.S. was an inmate in Sing Sing Prison. He was a fifty-two-year-old white man who had been imprisoned for the "carnal abuse" of two seven-year-old boys. He was slightly bent and rather studious in appearance, with gray hair, a deeply lined face, and a slender frame. Five feet ten inches tall, he weighed only 137 pounds. His manner was courteous, almost apologetic.

J.S. had grown up in a middle-class New England Unitarian family and graduated from high school at the normal age. His IQ of 130 on the Stanford-Binet test was significantly above average. After the death of his father, who had earned a reasonably good income as a shoe salesman, J.S. supported his mother by working as a secretary for commercial firms and brokerage houses. Up to the time he was imprisoned, J.S. lived with his mother.

J.S. thought he had been spoiled as a child. He was very fond of his mother even though she was strong-willed and domineering. She had punished him physically for the slightest offense, beating him "with a strap, or birch switch, or with her hands on his bare buttocks."[9] These beatings usually produced an erection and often ended with J.S. ejaculating and his mother kissing him. This behavior continued for about two or three times a week until he was sixteen.

As J.S. grew older, he became aware of the pleasure he derived from beatings—a pleasure that over time became an obsession. He tried self-punishment with razor straps and managed to persuade some intimate friends to punish him. As he was usually not able to persuade people his own age to beat him, he started developing ruses to involve young boys. His strategy was to start talking to young boys and quickly direct the conversation to questions about how their parents punished them. He would tell them that he liked it when he was beaten as a boy, and would make up a game where the boys acted as his parents and would have to punish him. He offered them money to play the game, and several boys came to his home regularly, where he would strip and insist on being beaten severely. Sometimes he would excite the boys by making up stories about sex with girls and then would engage in fellatio.

Shortly before his arrest, he had met two boys on the beach and had talked them into beating him with his leather belt for money. He also played with their penises and induced them to manipulate his until he ejaculated. Two weeks later, while walking with their parents, the boys recognized him. He tried to run away but was caught by the police and eventually sentenced to prison. The year was 1939.

In Sing Sing, J.S. had a relatively easy office job, but he had become increasingly agitated and insecure and compulsively engaged in autoerotic activity. One cause of his agitation was a highly publicized case of an elderly man who had lured a young girl into a vacant house and committed various sexual acts, eventually killing her unintentionally. The man was executed, and J.S. became obsessed with the fear that, after his release, he would have a similar fate.

J.S. came under the care of Ralph S. Banay, a well-known psychiatrist, who was in charge of Sing Sing's psychiatric clinic; he was also founder and director of the Greenmont Sanitorium in Ossining. Banay clearly had much better credentials than most prison psychiatrists.* After observing that J.S. was continually anxious about his obsession, Banay decided that a standard Freeman-Watts prefrontal lobotomy was justified because it was highly unlikely that the prisoner could be helped by conservative treatment. Moreover:

> the compulsive force of the obsession over which he had no control might drive him on to further offenses, thus entailing a great risk to society and himself.
>
> It was expected that by severing the thalamic pathways, the complex formation would lose the emotional depth and driving force.[10]

The operation was performed in 1941 by Leo Davidoff, a noted neurosurgeon. Immediately afterward, J.S. was confused, disoriented, and incontinent. He masturbated and kept placing his finger in his rectum and then in his

---

* Banay had been born in Hungary, where he was a classmate of Ladislas Meduna at the Royal Hungarian University in Budapest and had excellent training, including postgraduate work in Vienna, Munich, and Amsterdam, before coming to the United States in 1927. Before 1941, Banay had worked at the Manhattan State and the Boston State hospitals; he was later affiliated with Columbia and Yale universities. Between 1943 and 1949, Banay directed a large "Deviance Research Program" at Columbia University. At the Greenmont Sanitorium he headed a large drug-addiction treatment program sponsored by the State Narcotics Commission; he often served as a consultant to judges and, in 1955, he testified before Senator Estes Kefauver's Subcommittee on Juvenile Delinquency.

mouth. After several days, however, he regained control of his anal sphincter; and by the tenth postoperative day, the confusion was said to have completely cleared up. Two to three weeks after the operation, all of the undesirable side effects were said to have disappeared entirely. Although J.S. remained under observation in the prison hospital for six months, he was able to resume his secretarial duties throughout most of this period.

Six months after the operation, Banay and Davidoff described the case in an article for the *Journal of Criminal Psychopathology*—a journal read primarily by criminologists and prison psychiatrists. J.S. was described as no longer agitated. His erotic fantasies were gone, and his ability to work was said to be as good as before the operation:

> The initial euphoria gave way to a complacent interest and a tranquil state of mind. . . . His energy output lessened to some degree. . . . There was no sign of increased egotism or conflict with environment. He remained courteous, meek, obliging, and attentive, without undue sensitiveness toward criticism.[11]

The prison psychologist reported that J.S.'s Rorschach responses were the healthiest "in a group of fifty offenders," and J.S. was quoted as saying:

> I feel exceedingly grateful for all that was done for me. My mind is at ease and old temptations do not bother me. It is for this that I give most thanks. I feel free of an old bondage for the first time in my life. I am relaxed now, and can work and read for great lengths of time without fatigue. I enjoy music and games, attend Sunday Bible class and sing in the choir. In general, I feel that the operation was a success.

Banay and Davidoff acknowledged that a lobotomy was not the solution for all sexual offenders, but added that if similar results could be obtained with other patients, this approach "might be a new and important development." They noted that many patients who are a serious risk to themselves and other people have to be penalized as a defensive measure because they do not respond to conservative treatments. Banay and Davidoff concluded that in "persons over 50, whose pathological sexual drive increased with their declining years, whose history consists of persistent criminal sex offense of obsession-compulsion type, accompanied by marked agitation . . . lobotomy might be considered as a choice."[12]

J.S. was released from prison on 10 November 1942, one year after the lobotomy. He reported weekly to his parole officer until May 1944, when he moved to Chicago. There his trail was lost until he was accidentally discovered four years later in the locked ward of the Veterans Administration Hospital at Hines, Illinois. J.S.'s experiences were reconstructed from information obtained from social workers, relatives, former employers, and friends, as well as from a diary and notes found in a large suitcase. In 1948, Banay and Joseph Friedlander published an account of what happened to J.S. after the lobotomy.[13]

Shortly after the lobotomy, J.S. appeared to have undergone rapid improvement, followed by a year-long period of stabilization and then a period of progressive decline. He was not able to hold a job for more than a few weeks before he was fired, because he was "undependable in every way" and did not seem "to know what he was doing." His judgment was childish, he had no insight into his behavior, and he became progressively confused. His memory was very bad, and almost nothing he said about previous events was found, when checked, to be accurate. He was often incontinent, and his buttocks were grossly contaminated with fecal matter. He exuded a profound stench and was covered with bedbug sores. Earlier, the parole officer had given J.S. "a clean slate," reporting that he was "completely free of abnormal desires." In actuality, J.S. continued secretly to practice his perversions in Chicago. In his suitcase were found obscene handwritten propositions on slips of paper, which he had handed out to boys in the parks he frequented.

J.S. was given several psychometric tests of intelligence and personality. He also had a pneumoencephalogram and an electroencephalogram. After weighing all possible explanations of his obviously demented condition—including such organic disorders as Alzheimer's or Pick's disease, cerebral arteriosclerosis, and other conditions that might have explained his progressive decline—Banay and Friedlander concluded that "the lobotomy produced the dementia." They noted that Freeman and Watts had acknowledged that extensive injury to the frontal lobes, such as would be produced by surgery to remove a tumor in this area, could produce "defrontalized dement"—a person "without shame and without ambition, lacking in all the finer qualities that make a man what he is, and debased to the level of the brute."[14] Freeman and Watts had maintained, however, that such a picture does not follow a lobotomy, where presumably the damage was less. After reviewing the evidence from J.S.'s record, Friedlander and Banay were forced to disagree. Noting that several psychiatrists and neurosurgeons had reported cases of patients getting worse, these two physicians concluded that the lobotomy-induced dementia they observed in J.S. might be unrecognized in individuals who, even before a lobotomy, were psychotic.[15] Thus, Banay's initially optimistic report in 1942, which could have encouraged others to treat sexual psychopathy by lobotomy, had to be corrected in 1948, when J.S. was accidentally rediscovered.

While J.S.'s case was atypical in that very few criminals were lobotomized,* it does illustrate two important points about the results of the operation. First,

---

* Three criminally insane patients were lobotomized at the Territorial Hospital for Mental Disorders in Kaneohe, Hawaii;[16] see also the case described on page 191.

the lobotomy did not change the patient's thought processes; it changed only the intensity of his emotional involvement with certain ideas. J.S. still had the same obsessional desires, although—at least for a while—they appeared to have become less intense and preoccupying. Second, a lobotomized patient thrown completely on his or her own resources, without family support, was likely to have great difficulty adjusting. The outcome might have been very different had J.S. had the type of support system that was arranged for Carolyn.

The two patients described in this chapter were not unique. While each is representative of a subset of the very variable effects of a lobotomy, two patients could not possibly convey the total range of results. It has to be understood that the operations varied enormously and, therefore, destroyed different parts of the frontal lobes. Even when surgeons used the same procedure, such as the Freeman-Watts "standard" lobotomy, each surgeon introduced his own variations. Moreover, individual brains are no more alike anatomically than are faces, and even a more precise surgical procedure than lobotomy would destroy functionally different parts of the brain. Also, because blood vessels vary in branching and location, the amount of damage done to them varied greatly. The destruction of large blood vessels could produce "secondary" destruction of a brain region—even at a great distance from the primary surgical site—as a result of deprivation of the oxygen and nutrients normally carried by that vessel. The amount and location of scar tissue was another source of variability. Scar tissue can irritate or excite nerve cells nearby, often "driving" several of them to fire in abnormal synchrony, a condition that may lead to an epileptic seizure.

Variability in results from lobotomy could also be attributed to differences in the patients' personality, education, and general level of intellect. Some patients were much better equipped to adjust to postoperative deficits. While some might learn to use alternative ways of coping, other patients might become completely dependent on their families, while still others might use childishly transparent ways of trying to cover up their deficits. Patients also varied in their mental illnesses. The diagnostic labels used for similar patients varied among both psychiatrists and institutions and were often almost useless for evaluating overall results.

Although, even today, more is unknown than known about the function of the prefrontal brain area, it is generally accepted that specific regions within this area are connected to different parts of the brain and are suspected of having different and, perhaps, opposite functions. Relatively little of this was known when lobotomy was started, and the operations frequently violated functional distinctions.

Parts of the frontal lobes, particularly the ventromedial region, do seem to participate in regulating the intensity of emotional experience. Many of our

current ideas about which parts of the brain play an important role in mod-
ulating the intensity of emotional reactivity are derived from animal experiments. Electrical stimulation of particular brain regions may evoke such visceral responses as increases in heart rate and respiration, pupillary dilation,
and other physiological concomitants of emotions. Destruction of these areas
sometimes produces dramatic changes in the emotional responsiveness of animals. Some clinical observations of people seem to support these ideas. Thus,
a tumor in the hypothalamus appeared to have transformed a kind and con-
siderate schoolteacher into an irritable and explosively aggressive woman who
had to be discharged from her position. As dramatic as some of these experimental and clinical observations may be, our understanding of how all these
different areas function together to regulate emotions is primitive, and our
speculations are undoubtedly wrong in major respects.

Following a lobotomy, many agitated and anxious patients did experience
a striking relief from their most troublesome symptoms. In the best cases,
this led to a normalization of their behavior. In the worst cases, they became
either thoughtlessly impulsive or almost inert, seldom talking and completely
lacking in spontaneity. The latter cases justified the accusation that lobotomies
were producing "zombies." Some patients became careless, indifferent to errors, and generally slovenly. Motivation and drive were often decreased, as
were insight and foresight. Behavior tended to be more concrete, and patients
were less capable of dealing with abstract ideas.

These impairments could vary enormously in magnitude, from tolerable
to devastating levels. In the worst cases, lobotomized patients were not able
to anticipate the consequences of their actions on themselves or on others,
making meaningful interpersonal relationships impossible. Moreover, they
were unable to plan for the future or even for their daily activities. Even
worse, if a major blood vessel in the brain was damaged, but not causing death
during (or shortly after) the surgery—as it did in 5 percent of such cases—
the patient might deteriorate into a completely demented mental state.

In most cases, the impairment was at some intermediate level. Thus, there
was some helpful reduction in the most troublesome symptoms—but at the
price of reducing potential motivation to excel and the capacity for thinking
abstractly, for imagination, for emotional experience, for spontaneity, for
planning ability, for insight, and for social judgment. In the best cases, these
deficits were not great, and, postoperatively, patients could make a good adjustment by compensating for the loss by using remaining abilities resource-
fully, much as some elderly people are capable of doing.

As institutionalized mental patients were rarely demonstrating the capacities
that were impaired by lobotomy, many psychiatrists concluded that what was
lost was not real. Were lobotomies justified? Even considering the lack of

alternatives, my own view is that the risks of lobotomy were too great to justify the surgery except with patients who had been ill for a long time and were highly unlikely to recover. Once lobotomy became routine, however, it gained a momentum of its own—supported, of course, by the factors this history illustrates—and many patients were exposed to an unjustifiable risk when there remained a reasonable chance they might improve. Also, as some of its critics realized, the practice of lobotomy interfered with the search for alternative therapies.

13

Opposition to Lobotomy— And a Brief New Life

> I am sorry to say that even when they [lo-
> botomized patients] improve they are nothing
> to brag about.
> —WINFRED OVERHOLSER (1949)

> I think it should be re-emphasized that by
> psychosurgery an organic brain-defect syn-
> drome has been substituted for the psychoses.
> —JAY HOFFMAN (1949)

THE AWARDING of the 1949 Nobel Prize for the discovery of prefrontal leucotomy created additional interest in psychosurgery. For many, the Prize
bestowed a "seal of approval" on the scientific validity and therapeutic value
of these operations. Ironically, at about the same time, criticisms of lobotomy
began to appear in the popular press and professional journals. Gradually,
over the next few years, the opposition became more effective. The net effect
of these two influences was that the number of operations leveled off for
several years and then, as opposition grew in strength, started to decline. The
introduction of chlorpromazine and other psychoactive drugs in the mid-
1950s provided what the opposition to psychosurgery had always lacked: a
viable alternative for treating the major psychoses. Within a brief span of
time, the use of lobotomy declined precipitately.

During November 1949, the Washington, D.C., Psychiatric Society held
a symposium on lobotomy. After Freeman, Watts, and others had finished
speaking about the effectiveness of different types of lobotomy, Nolan Lewis,
director of the New York State Psychiatric Institute and professor of psychiatry
at Columbia University, began an unusually strong public attack on lobotomy.
The event was fully reported in *Newsweek*. Warning that lobotomy was being
used indiscriminately, Lewis stated that it was not true that these operations

were used only as a "last resort." He asserted that surgeons often operated on a patient without even a preliminary psychiatric examination, and added that "some doctors have shown an utter lack of respect for the human brain." Lewis pointed out that the part of the brain destroyed—that is, the specific type of lobotomy—had no relation to the mental illness being treated. He also questioned the validity of the evidence supporting lobotomy, stating that a great many failures were not reported. Moreover, Lewis questioned whether the "successful cases" should be counted as such:

> Is the quieting of the patient a cure? Perhaps all it accomplishes is to make things more convenient for the people who have to nurse them. . . . The patients become rather childlike. . . . They act like they have been hit over the head with a club and are as dull as blazes. . . . It disturbs me to see the number of zombies that these operations turn out. I would guess that lobotomies going on all over the world have caused more mental invalids than they've cured. . . . I think it should be stopped before we dement too large a section of the population.[1]

Strong words indeed from the director of the institute that was providing most of the staff for the Columbia-Greystone Lobotomy Project! Lewis had not always taken such a strong stand against lobotomy and had even written a somewhat favorable (even if guarded) review of psychosurgery during this period.[2] His criticism on this occasion may have been fueled by the fact that Freeman and Lewis had often clashed when they were both at St. Elizabeth's Hospital.

Watts tried to answer Lewis, but met strong opposition from several psychiatrists. As soon as Watts had finished, Winfred Overholser, the superintendent of St. Elizabeth's Hospital, remarked that there had been about one hundred operations at his institution in six years: "I am sorry to say that even when they improved they are nothing to brag about. . . . These patients have not been completely restored and some of their families are extremely disappointed."[3]

Serious objections to prefrontal lobotomy were increasing and coming from many quarters. In 1949, in an article in the *New England Journal of Medicine,* Jay Hoffman, chief of the V.A. Neuropsychiatric Service, publicly doubted the claims of success following lobotomies on schizophrenic patients:

> The evaluation of the results after prefrontal leukotomy will be greatly influenced by the frame of reference one uses. If the condition of the patient is compared with his condition prior to the onset of his psychosis, all the results must be considered failures. . . . I think it should be re-emphasized that by psychosurgery an organic brain-defect syndrome has been substituted for the psychoses.[4]

This point was supported by two psychologists at McGill University—

H. Enger Rosvold and Mortimer Mishkin—who studied a group of lobotomized Canadian army veterans. As soldiers were routinely given intelligence tests when they were inducted into the army, it was possible to compare their test scores after a lobotomy with those achieved before the psychiatric "breakdown." Rosvold and Mishkin found evidence of a significant decrement in IQ following lobotomy.[5]

The criticisms in the popular media and in the medical and psychological literature gave rise to discussion of the advisability of controlling lobotomy by legislation.* The many criticisms of lobotomy were discussed in the *Stanford Law Review,* but in the end the editors pulled back from recommending any legislation, deciding that the "rare abuses" were not sufficient justification to impose restrictive measures that might stifle beneficial developments in the future. They concluded that "the greater good would be achieved by avoiding legislative fetters."[7]

None of the criticism of lobotomy had any influence on Freeman's transorbital lobotomy project. By this time, he had already performed five hundred transorbital lobotomies and was rapidly extending his activities to more state hospitals. The criticisms of prefrontal lobotomy were used by him to justify transorbital lobotomy, which he considered to have far fewer deleterious side effects. Actually, Freeman had not performed any prefrontal lobotomies with Watts or anyone else for several years. In May 1950, he announced at the American Psychiatric Association Meeting in Detroit that he would no longer recommend prefrontal lobotomy because it produced too many complications. He reported that 47 percent of the patients in the Freeman-Watts series who had two prefrontal lobotomies developed seizures and 15 percent had seizures even after one such lobotomy. In contrast, he said only one-half of 1 percent had seizures after a transorbital lobotomy.[8]

This announcement was Freeman's way of throwing down the gauntlet to advocates of prefrontal lobotomy who criticized the transorbital procedure. In spite of the growing criticism, there were still very many professionals who remained convinced that prefrontal lobotomy was effective; and Freeman's challenge was picked up immediately, not only by neurosurgeons but also by psychiatrists. Paul Hoch of the New York Psychiatric Institute responded by stating that, in his experience, prefrontal lobotomies were more

---

* The Soviet Union declared in 1951 that no more lobotomies would be performed in that country. Up to that time, psychosurgery had been considered to be consistent with their Marxist materialistic philosophy favoring somatic treatments over psychotherapy, especially psychoanalytically oriented treatment which was rejected as "bourgeois idealism." After an autopsy study of the brains of lobotomized patients by Dr. Chevtchenko of Moscow had revealed that the damage was often much more extensive than had been suspected, the Soviet Union decided that psychosurgery was in conflict with Pavlovian theory.[6] The Soviet Union's decision to stop psychosurgery had little influence on public opinion in the United States during the cold war.

effective, and noted that he had seen some of the patients Freeman had operated on as part of the Columbia-Greystone Study.[9] Freeman, however, was convinced that the test was "rigged," as the patients selected for him to operate on were so deteriorated as to be beyond help.

Freeman was pitting transorbital lobotomy performed mainly by psychiatrists against prefrontal lobotomy performed by neurosurgeons. The resulting conflict was inevitable. Almost all neurosurgeons, Watts included, objected to transorbital lobotomy. They were appalled when they watched Freeman adjust the head straps for the electroconvulsive shock and then pick up the instruments with ungloved hands, and they resented the idea of inducing convulsions, which much of their work was aimed at eliminating. They were also concerned about the possibility of hemorrhage and infection. To neurosurgeons, transorbital lobotomy seemed to be a slapdash, hit-or-miss business, and the "assembly line" of operations on twenty or more patients in a row— averaging less than ten minutes each—a vulgar display.

Most of all, neurosurgeons were offended by who was doing the operations—not board-certified neurosurgeons, but psychiatrists. When the neurosurgeon James Poppen made a sarcastic comment about transorbital lobotomy, Freeman, who never failed to meet a challenge head on, retaliated by saying that Poppen's use of aspiration to remove brain tissue was about as elegant as running a "vacuum cleaner over a bathtub of spaghetti." Or when another neurosurgeon mentioned the dangers of a "blind"—in contrast to an "open"—operation, Freeman replied witheringly that "a neurosurgeon may see what he cuts, but he doesn't know what he sees."[10] Poppen retaliated by writing, with obvious reference to Freeman:

> In recent years much has been written about different surgical approaches to the treatment of insanity. I am certain that there will be more to follow. I do hope that in the future we will not be informed initially through the weekly popular magazines. Any procedure which is instituted for such a serious condition should be thoroughly tried and proved to a certain degree before it is advised. Premature information through weekly magazines (not always accurate) has a tendency to give patients or relatives false hopes or impressions.[11]

In the *1948 Yearbook of Neurology, Psychiatry and Neurosurgery,* Percival Bailey wrote:

> The outstanding development of the year is transorbital lobotomy. This procedure is of a nature to distress a surgeon greatly. It is being used in psychiatrists' offices without provision to deal with complications which must occasionally result. Such a blind procedure is unjustifiable, and it is to be feared that its simplicity will cause it to be used too freely and bring disrepute on a potentially very useful measure.[12]

Even as Freeman tried to explain that the results justified the risk and, that besides, the risk had not proved to be great, he knew that neurosurgeons could not accept his methods. Moreover, he understood that no matter how much he "dressed up" the procedure, the neurosurgeons would continue to object until he had moved it into a hospital surgery room where they would then have control. Rather than making any changes, he came to medical meetings armed with statistics showing that the hazard of conventional psychosurgery was greater than that of transorbital lobotomy. These statistics could never override the emotional reaction to his methods or the jurisdictional dispute over who was performing the operation. Referring to Freeman's earlier use of an ice pick, one neurosurgeon declared, "You can't treat the brain like it's a block of ice." Another continually referred to the leucotome as a "stiletto."

On one occasion, hearing that Freeman had been invited to demonstrate transorbital lobotomy at the Tuskegee Veteran's Hospital, the hospital's neurosurgical consultant let it be known that it would take place only "over my dead body."[3] The chairman of the V.A. Lobotomy Advisory Committee supported the neurosurgeon, and the invitation was rescinded. Freeman became even more convinced that he would have to carry on the transorbital project in state hospitals, which seldom had any neurosurgeons to oppose him. The net effect of all these internecine disputes among those who, in general, supported the practice of lobotomy was to accentuate the adverse consequences and to increase the effectiveness of those opposed to the operations.

Although Freeman reported better overall results with transorbital than with prefrontal lobotomy, it is impossible to evaluate these claims today or, for that matter, most of the claims of the supporters of lobotomy. Even if his statistics are accepted at face value, they have no meaning without a truly comparable, unoperated comparison group. The implicit assumption behind the statistical data was that all, or most, of the patients would have progressively deteriorated, requiring institutionalization for the remainder of their lives, if lobotomy had not been performed. If this was really true, then even a 20 percent discharge rate would have been worthwhile, provided the patients discharged were functioning reasonably well. It is not known, however, what the discharge rate would have been in a comparable unoperated group of patients. While the prognosis for deteriorated schizophrenics was poor, it was estimated in the 1930s—even before the introduction of shock treatment— that 41.5 percent of manic-depressive psychotics in state hospitals were "cured," and an additional 27.2 percent were discharged as "improved." According to one early study, 59.7 percent of manic-depressives spent less than a year in hospitals.[14] In their highly regarded textbook on nervous diseases, Smith Ely Jelliffe and William A. White reported in 1935—that is, before shock treatments, psychosurgery, or drugs—that 40 percent of patients suffering from involutional melancholia recover.[15] Taken together, these figures indicate that almost 70 percent of patients with "affective" disorders were eventually discharged even before the major somatic therapies of the mid-1930s had been put into general use.

Moreover, other factors probably distorted most of the estimates of improvement following lobotomy. If, for example, patients were selected for lobotomy who were not as ill as, or who were more likely to improve spontaneously than, the average institutionalized patients, any comparison could be misleading. As Nolan Lewis had argued, there was considerable doubt that all the lobotomized patients were given this treatment as a "last resort." Even Watts had questioned whether all of Freeman's transorbital lobotomy patients were truly "disabled." Furthermore, some lobotomized patients were probably discharged from hospitals primarily because they could be more easily managed at home, rather than because they were otherwise greatly improved.

Later, as a part of the West Virginia Lobotomy Project, which Freeman undertook from 1952 to 1955 (see pages 235–36), he claimed that far fewer of a "roughly comparable" unoperated group of patients improved, but he never described the composition of the control group or how these patients were selected. It is impossible, therefore, to know whether the control group was really comparable to the lobotomized patients.[16] Later, in the 1960s, some carefully controlled studies in Canada failed to find a better improvement rate among patients who had undergone a standard prefrontal lobotomy.[17]

The evaluation of lobotomized patients—as the experimental psychologist Ward Halstead forcefully pointed out in his speech to the American Psychiatric Association in May 1946—was totally inadequate:

> Not a single patient had been adequately studied. For the moral and social responsibility to do this, there has been substituted a phenomenal array of case statistics. Unfortunately, the pyramiding of unknowns is scarcely a pathway to knowledge. . . . In no instance has the psychological test or battery of tests employed ever been shown to be sensitive for frontal lobe function.[18]

Halstead went on to ask how anyone could reasonably expect that the widely used Stanford-Binet test would reveal impairment following a prefrontal lobotomy when the test had been shown to be incapable of detecting deficits even after the much more extensive frontal lobectomy operation, where a patient's impairment was usually obvious to everyone. In a frontal lobectomy, a major portion of the frontal lobes was undercut and removed (see figure 9.4, page 194); whereas in a prefrontal lobotomy, the intent was to restrict the damage to some of the nerve tracts below the surface of the cerebral cortex.

Although Halstead's criticisms were valid, his own experience revealed how difficult it was to measure impairment in lobotomized patients with objective tests. Working at the University of Chicago with the cooperation of neurologists and neurosurgeons—most notably Percival Bailey, Paul Bucy, and A. Earl Walker—Halstead primarily studied neurological patients, most of whom had lobectomies because of brain tumors. However, he also tested a group of lobotomized patients before and after the surgery. Starting with twenty-seven tests, many of which he developed himself, Halstead ended up with ten tests that seemed to be most sensitive to brain damage. Halstead's tests were not like the usual intelligence test: for example, they assessed finger-tapping rate, finger oscillations, flicker fusion (the ability to perceive rapidly flashing lights as separate rather than as continuous, or "fused"), hand steadiness, hearing acuity, peripheral vision, and tactile memory. Although few people accepted Halstead's assertion that his test battery measured "biological intelligence," the results did seem to be sensitive to brain damage. Halstead found that the patients who had frontal lobectomies had the highest "impairment index."* He had to admit, however, that the test performance of the patients in the prefrontal lobotomy group did not reveal any impairment.[19] Halstead attributed the different results following lobectomy and lobotomy to the fact that only in the former did brain damage involve major destruction of the cerebral cortex. Ironically, even his conclusion that frontal lobectomies produce greater impairment than lobectomies in other brain regions was later criticized. Apparently, the major variable in his data was the extent, rather than the location, of brain damage, and the frontal-lobectomized patients had the greatest extent of damage.[20] While Halstead had helped to eliminate some of the subjectivity involved in evaluating brain-damaged patients, he, like most others, was not successful in developing a reliable way of assessing the elusive impairment in personality, motivation, and judgment produced by prefrontal lobotomy.

Earlier, when Earl Walker had attempted to review the results of lobotomies from different institutions, he found great differences in the reported success rates. After examining the evidence, he concluded that the differences in the patient populations, and in the severity of illness of those selected for lobotomy, probably were more important in determining the outcome than were differences in the surgery.[21] There is little doubt that the patients Freeman operated on varied between institutions and over time. He was constantly trying to persuade the staff at state hospitals to provide him with patients who had not been ill too long—or as he put it, before "the psychosis consumed their personalities."[22] During his first visit to Western State Hospital in 1947, the

* Lobectomies can be performed on different lobes, or divisions, of the cerebral cortex: thus, frontal lobectomies could be compared with temporal or parietal lobectomies.

staff had selected only chronic patients, all of whom had been severely deteriorated for many years, and the outcome was poor. A few months later, when Freeman revisited Western, the *Seattle Post-Intelligencer* reported that transorbital lobotomies had been performed for the first time on legally "sane" patients, and quoted William Keller as saying that it was a "brilliant success. . . . I still can't believe it. . . . Every patient on whom the operation has been performed has shown some improvement."[23] Thus, the assumption implicit in much of the statistical evidence supporting lobotomy—that these operations were performed only on otherwise hopeless, deteriorated, and chronically ill mental patients—was clearly incorrect. Over all, the scientific value of most of the prefrontal lobotomy as well as the transorbital lobotomy studies was virtually zero.

At this critical juncture, when criticism of psychosurgery was starting to be effective, new procedures were promoted which, it was claimed, would eliminate the adverse consequences of lobotomy while still providing relief from emotional turmoil. This counter-argument, which helped to sustain interest in psychosurgery, was evident even in some of the criticism published during this period. In 1948, for example, the *Nation* described the danger of creating thousands of people with permanent losses in intellect and personality. The article, a highly emotional attack on lobotomy, spoke of "raping the soul" and of "a pact with the devil in which one may pay too dearly." Nevertheless, the *Nation* left open the possibility that future research might produce refinements in the procedure that would "improve mental conditions by surgery without permanently damaging the personality."[24] A critical editorial in the *New England Journal of Medicine*, although it did not use any flaming prose, essentially took the same position. The editorial pointed to recent evidence suggesting that any benefit produced by lobotomy was accomplished at a price. The alleviation of symptoms was said to come from a "subtraction process" rather than being a cure, and there was usually a "permanent loss to the integrity of the personality" and "vital spark." The editorial concluded, however, with the hope that "refinements of topectomy will avoid some of the disadvantages of lobotomy."[25] Topectomy was being used loosely as a shorthand for the newer operations that restricted damage to relatively discrete brain areas.

John Farquahar Fulton (1899–1960) played an important role throughout the early history of lobotomy; and, during this period, his influence was a major factor in sustaining interest in psychosurgery (see figure 13.1). His prestige and influence were enormous. Born in St. Paul, his father, John Farquahar Fulton, an eminent ophthalmologist, had helped establish the University of Minnesota's medical school. The Fultons were of the same family as Robert Fulton, the inventor of the steamboat. The young John Fulton showed early academic promise, completing high school at the age of sixteen. After serving a brief period in the army during the First World War, he graduated *magna cum laude* from Harvard in 1921. A Rhodes scholar, Fulton went to Oxford University where he first assisted Sir Arthur Shipley, and then worked in Sir Charles Sherrington's laboratory studying frontal lobe regulation of movement. Returning to Harvard, he completed a medical degree in 1927 and spent the following year on a fellowship learning neurosurgical skills from Harvey Cushing at the Peter Bent Brigham Hospital in Boston. At the age of thirty, in 1929, Fulton was appointed professor of physiology at Yale. While he operated on experimental animals, he was never a practicing neurosurgeon; yet he often referred to himself as such, and the lay press at least often assumed that he was. By the late 1940s, however, he was much too busy pursuing other interests to be concerned about the details of research, even that going on in his own Laboratory of Physiology at Yale. Moreover, his health was deteriorating from diabetes (aggravated by a love of good food and drink), and he was by this time mainly devoted to collecting rare medical books and writing articles on the history of medicine.*

Workers in Fulton's laboratory benefited from his broad knowledge of the nervous system, from his contacts with just about every important clinician or scientist working on some aspect of the nervous system, and from his ability to secure the funding that supported their research and often paid their salaries. However, by the 1940s, he was mainly providing "stimulation with minimum direction"—or, as neurosurgeon Earl Walker once put it, people worked in Fulton's laboratory "more with his blessing than with his supervision."[27] Nevertheless, Fulton was often the spokesman for those working in his laboratory, as he received many invitations to give prestigious lectures in the United States and abroad.

His great influence cannot be attributed to the importance of his own laboratory work. Even though he had published well over one hundred research papers, the majority of them on the frontal lobe, they contained no major "discoveries." Indeed, much of his early work at Oxford proved to have been in pursuit of an apparatus artifact.[28] Fulton's influence derived mainly from his important textbooks, from his stewardship of the leading neurophysiological journals, from his wide acquaintance with leading scientists, and from his ability to bridge basic research and clinical concerns.[29] Although he never

---

* For example, Fulton wrote notes on Vesalius, descriptions of early medical libraries, and an analysis of the sixteenth-century poem by Girolamo Fracastoro on the "French disease," *Syphilis sive morbus gallicus*. Helped by considerable wealth, Fulton maintained a large and able secretarial/research staff and a chauffeur who drove him home for lunch. He wrote many articles and monographs on the history of various medical subjects.[26] In 1951, he became the Sterling Professor of the History of Medicine at Yale.

Figure 13.1.   John F. Fulton (1899–1960) in 1953. (From a snapshot by Gregory Zilboorg. Courtesy of John Fulton Papers, Yale University Library.)

practiced medicine, he was highly regarded among clinicians in the United States and abroad. As the head of a large, well-funded laboratory at Yale, he was able to attract, from around the world, a succession of young, energetic neuroscientists who had new ideas and techniques—including, among others, José Delgado, Paul Glees, Carlyle Jacobsen, M. Kennard, Robert Livingston, Haldor Rosvold, Theodore Ruch, J. A. F. Stevenson, and Patrick Wall; the neurosurgeons Paul Bucy and James Watt also worked in Fulton's laboratory. Fulton's influence can hardly be overestimated.

His reputation had also helped secure funding for the Connecticut Cooperative Lobotomy Study and the Yale Veterans Administration Lobotomy Project. He was the recipient of a $100,000 grant from the National Research Council, at a time when this was a much more considerable sum than it is today, and of "generous support" from the Veterans Administration. Lobotomized patients were studied at the Veterans Administration hospitals, the Connecticut State Hospital, the Norwich State Hospital, and the Institute of Living in Hartford. C. Charles Burlingame, the senior psychiatrist at the Institute of Living, had been a helpful contact for Fulton, as the former was both enthusiastic about lobotomy and influential in the Veterans Administration in Connecticut.[30]* (Burlingame, it will be recalled, was consulted in the case of Carolyn W., discussed in chapter 12.)

Fulton never was directly involved in the practice of psychosurgery and was always careful not to take a stand from which he could not withdraw gracefully, but he provided support for psychosurgery at various critical stages during its development. He found the initial reports interesting, deserving of more study, and consistent with the earlier animal studies by Jacobsen and himself. Although he had delegated the work to others and seldom met with those collecting the data, he was responsible for the Yale News Bureau's widely circulated report that the Connecticut Cooperative Lobotomy Project had found that 61 percent of schizophrenics hospitalized more than five years had shown "definite improvement" after a lobotomy.†

When Fulton was invited to England to present the Withering Memorial Lecture at the Birmingham Medical School in 1948, he chose the frontal lobes and lobotomy as his subject. He talked about the localization of different functions within the frontal lobes, arguing that sufficient knowledge was now available to minimize the adverse effects of psychosurgery by a more rational selection of brain targets. And in a passage whose essential message he repeated in many influential lectures and books, he said:

> The results [of lobotomy] are of challenging interest to anyone who concerns himself with the functions of the brain, but the responsibility of those who undertake to interfere with the anatomical structure of the human brain is particularly grave. . . . I utter this word of caution not to discourage those who are performing leucotomies, but in the hope that they will not allow their zeal to outrun their knowledge of function.[32]

Fulton suggested that it was possible to modify the operation based on new information about the brain. Instead of extensive lobotomies, he advised more selective destruction restricted either to the ventromedial region of the frontal lobes or to parts of the limbic system. The limbic system consists of a number of brain structures—located in a ring on the inner margin (*limbus* means edge in Latin) of the cerebral cortex—which James Papez‡ had speculated was a

---

* Describing the Lisbon International Conference of Psychosurgery, Burlingame wrote: "All the reports at the Congress were further evidence that in the hand of the medical profession there had been placed, in the form of psychosurgery, a powerful instrument of therapeutic advance that must not be neglected."[31]

† The psychologist H. Enger Rosvold, who was in charge of much of the testing of the lobotomized patients, has described (personal communication, 17 November 1981) the little time that Fulton gave to this study.

‡ James Papez (1883–1958) had trained in neurology but did not practice medicine. He was a professor at Cornell University, a basic scientist, a teacher, and an eminent neuroanatomist. Late in his life he argued that schizophrenia is caused by a metabolic disorder that produces changes in brain cells; he reported finding such changes in frontal-lobe tissue that had been removed from schizophrenics during prefrontal lobotomy.[33]

circuit regulating emotions. Although the theory was not taken seriously in 1937 when first proposed,[34] by 1949 it had started to make sense as a result of research demonstrating, among other findings, that tame animals could be made savage, and wild animals gentle, following selective damage to different structures within the limbic system.[35] The cingulum, the amygdala, and the hippocampus are prominent brain structures in the limbic system, and Fulton suggested that a more rational lobotomy would be restricted to the cingulum or to the ventromedial portion of the prefrontal brain region. It had been learned in the interim that the ventromedial prefrontal area communicated with the major parts of the limbic system, and the limbic system seemed to be participating in the regulation of emotions.

Fulton's advice was to modify the operation, not to give up psychosurgery. Sir Hugh Cairns, a well-known English neurosurgeon, who had trained Almeida Lima (Moniz's surgeon) and was a personal friend of Fulton, started to perform anterior cingulotomies almost immediately after Fulton's Withering Lectures.* Jacques Le Beau in Paris was similarly influenced by Fulton and also began to perform cingulotomies at the same time. Fulton had based his recommendation on his interpretation of the results of an animal study performed in Seattle by the neurosurgeon Arthur Ward. Where Fulton described the cingulotomized monkeys as "tame," Ward's own description of the monkeys' behavior more clearly reflects their severe impairment:

> There is an obvious change in personality. The monkey loses its preoperative shyness and is less fearful of man. It appears more inquisitive than the normal monkey of the same age. In a large cage with other monkeys of the same size, such an animal shows no grooming behavior or acts of affection towards its companions. In fact, it treats them as it treats inanimate objects and will walk on them, bump into them if they happen to be in the way, and will even sit on them. It will openly eat food in the hand of a companion without being prepared to do battle and appears surprised when it is rebuffed. Such an animal never shows actual hostility to its fellows. It neither fights nor tries to escape when removed from a cage. It acts under all circumstances as though it had lost its "social conscience." . . . It is thus evident that following removal of the anterior limbic area, such monkeys lose some of the social fear and anxiety which normally governs their activity and thus lose the ability to accurately forecast the social repercussions of their own actions.[37]

In numerous lectures and publications during this period, Fulton suggested abandoning the old-style lobotomies and replacing them with psychosurgery restricted to either the ventromedial prefrontal area or the anterior cingulum:

> In the light of the information which we now possess, I believe that the radical

---

* On the evening of 5 September 1933, one of the several times Fulton dined at Hugh Cairns's house, Egas Moniz was the guest of honor.[36]

lobotomy as carried out by Freeman and Watts and by Lyerly should be abandoned in favor of the more restricted lesion and that section of the medial ventral quadrant [orbital area] by electrocoagulation as recommended by Grantham appears to have advantages that make it superior to any other approach. This technique can no doubt be made applicable to the cingulate.[38]

Although Fulton did not hesitate to praise Moniz's monograph and results, there is good reason to believe that he never read it carefully and was surprisingly misinformed not only about this work but even about the details of some of the experiments in his own laboratory. As I noted earlier, he was not aware until 1948 that Moniz had not performed the surgery (see page 148 for Fulton's 1948 letter to Freeman). The same year, in a lecture at the Montreal Neurological Institute, Fulton told his audience—including his host, the neurosurgeon Wilder Penfield—that Moniz's monograph had presented "very encouraging results" based on "fifty patients" many of whom were "deteriorated schizophrenics."[39] Moniz's monograph was, in fact, based on twenty patients, and the results clearly indicated that the poorest outcomes were among the schizophrenic patients.

Fulton introduced his Montreal lecture with a reminder of his own role in the history of psychosurgery:

> The operation of frontal lobotomy was introduced as a result of a brief report made at the International Neurological Congress at London in 1935 by Carlyle Jacobsen and myself on the behavioral changes which developed in two of our chimpanzees, Becky and Lucy, following bilateral ablation of the frontal association areas.

And in the same lecture, he repeated his misleading account of the chimpanzee experiment:

> Prior to the second operation *both* animals showed frustrational behavior, i.e., when unrewarded after having made the wrong choice in the discrimination task or in the delayed reaction procedure, *both* animals had temper tantrums and, if unrewarded many times in succession, signs of experimental neurosis became apparent. Following the second operation the animals seemed devoid of emotional expression.[40] [Italics added]

I have already noted that one of the two chimpanzees actually first began to have temper tantrums after its frontal lobes were damaged (see pages 95–97).*

Fulton's repeated distortion of the chimpanzee experiment came to be so widely accepted that when the Nobel Committee, in their citation of Moniz's

---

* To further complicate these results, during an autopsy on the brain of Becky (the chimpanzee that stopped having temper tantrums after the surgery) Karl Pribram found a large abscess in the frontal lobes (personal communication, 28 August 1984).

award, traced the lines of thought that led to the discovery of prefrontal leucotomy, they perpetuated this error:

> The American physiologist, Fulton, and his collaborators have proved by experiments on anthropoid apes that neuroses caused experimentally disappeared if the frontal lobes were removed and that it was impossible to cause experimental neuroses in animals deprived of their frontal lobes.[41]

Thus, by repeating—in many lectures and books between 1948 and 1952—his personal impression of the results of research done by others, Fulton played a crucial role in modifying psychosurgery and thereby sustained interest in it at a time when criticism of it was on the rise.

14

"A Living Fossil":
The Decline of Lobotomy
and the Final Years
of Walter Freeman

> He [Walter Freeman] has devoted his time,
> energy, and the force of his personality to
> the subject for two decades. He has carried
> out a follow-up program which has never
> been excelled. It has been a "Magnificent
> Obsession."
>
> —JAMES W. WATTS (1957)

BY THE SUMMER OF 1954, Freeman was starting a new life in the San Francisco area, free of academic responsibilities. His transorbital lobotomy project had taken him to more than fifty-five hospitals in twenty-three states,* and uppermost in his mind were his follow-up studies of former patients. This entailed an enormous amount of travel and a herculean effort.

His first drive east began in September, and by the end of October, he had worked his way across the country to Washington, D.C. Using a list of patients that was arranged geographically, he planned his route so that he could visit as many of them as possible along the way. In Washington, he started what would be his pattern of spending several days updating the duplicate patient files he maintained at George Washington University and catching up with the medical literature at the National Library of Medicine.

On later trips, when he included professional meetings in his itinerary, Freeman would call former patients in the area to schedule interviews in his

---

* Freeman had also visited hospitals in Canada, the Caribbean, and South America. A complete list of his itinerary during this period is provided in the end notes to this book.[1]

hotel room. If they failed to show up, he would call again and reschedule the appointment. All of this was done without grant support. Although earlier he believed that objective tests were necessary to assess patients, and had sought the help of trained psychologists—first Thelma Hunt and later Mary Washburn—he came to doubt the value of these tests, owing to the lack of tangible results and to misleading conclusions. He once remarked that he had seen patients "breezing through our tests, better than most of our friends or relatives could do," yet these patients were utterly incapable of "keeping house, holding a job, getting along with their family, or using money wisely."[2]

His interviewing technique was to ask direct questions, and he ran through a list exploring such facets of a patient's life as sexual activity, employment, what he or she did around the house, eating habits, and bowel movements. When possible, Freeman tried to observe patients interacting with others in the household, as he knew answers to questions were not always reliable:

> In Los Angeles one time, I was talking with the mother about the relations between the former patient and his father. I would have been quite misled had not the father come home from work while the interview was going on. The first thing the patient did was to hurl a pillow at his father and then make for the bathroom locking himself in. . . . How this woman was able to put up with this surly, lazy, uncouth son for all the years since the operation is somewhat of a mystery, but she never fails to thank me for what I had done, and with real tears in her eyes.[3]

There was justification for Freeman's rejection of psychological tests, but his personal observations, which he distilled in his publications and lectures to summary statements of percentages of patients "markedly" or "partially" improved or "unchanged," had little scientific value. He left undescribed so much about the patients' lives that his results could not have convinced any skeptic, and his anecdotal accounts of spectacular successes—for example, of a lobotomized physician directing a large clinic—were surely not representative cases.

Shortly after arriving in California, he had contacted Russell Lee, who earlier had implied that he wanted Freeman to join the staff of the Palo Alto Clinic. Reflecting the changing mood, Lee had in the interim decided it would be wiser not to add such a controversial figure to his staff. Freeman took it in stride as other opportunities became available. By January 1955, he figured he had sufficient contacts with local physicians—potential sources of referrals—to open up an office for a private practice on Main Street in Los Altos.*

---

* In 1957, at the age of sixty-two, Freeman decided to join a group of physicians in financing a complex of medical and dental offices in Sunnyvale, California. The space was ready by 1959, and over the years he rented some of the large suite of offices he had purchased to a succession of young neurosurgeons.

Some local physicians invited him to chair a committee to investigate the feasibility of establishing a new hospital in the Palo Alto area. He took the job seriously—even attending a meeting of the American Hospital Association in Atlantic City to get ideas—and helped to shepherd the proposal through a series of meetings with community officials, taxpayer groups, and zoning commissions. Located equally between Los Altos, Mountain View, and Sunnyvale, the El Camino Hospital, as it was called, took seven years to complete. Working with the architect, Freeman took the lead in planning the spacious psychiatric service.*

Freeman was soon quite busy in the Bay Area. He performed transorbital lobotomies at both the Herrick Memorial Hospital in Berkeley with "class one surgical privileges,"† and the Atascadero State Hospital for the Criminal Insane; consulted at the Veterans Administration Hospital in Menlo Park; and taught psychiatric and neurology residents in several nearby towns. In 1958, he became chief of neurology at the small Santa Clara Hospital, resigning in 1966 at the age of seventy-one.

At the November 1957 meeting of the Southern Medical Association in Miami Beach, Freeman presented his "final report" on the original five hundred Freeman-Watts prefrontal lobotomy patients—virtually all of whom he had tracked from ten to twenty years.[5] He dedicated the paper to the memory of Adolf Meyer, who had died in 1950. Meyer had supported lobotomy at the 1936 meeting of the same society, when Freeman and Watts presented their first six cases, but had also stressed the need for scrupulous follow-up studies (see page 144).[6]

Freeman reported that, of the 440 original patients still alive, approximately 75 percent were at home; of these, 45 percent were "usefully occupied in the home or in outside work," and about one-half were employed. Although he presented no data, he noted that the employed group included "business and professional men" who were "carrying on in the pattern they achieved before the onset of the illness." This statement contrasted with Freeman and Watts's comment, in the 1950 edition of *Psychosurgery,* that "no physician, dentist, artist, musician, or writer has been able to make the grade" after a prefrontal lobotomy. Now Freeman said of such patients:

---

* Freeman was drawing on his experience designing the neurology and psychiatry space for the new George Washington University Hospital built after the Second World War. During the 1940s, it was the only general hospital in the D.C. area to have a psychiatric unit. Patients who needed hospitalization for psychiatric problems were—depending on their finances—admitted either to private sanatoriums like Chestnut Lodge or to large public psychiatric institutes. Out of the 4,302 general hospitals in the United States in 1940, only 63 admitted psychiatric patients.[4]

† Even before Freeman was a California resident, Abram Bennett, head of psychiatry at the Herrick Memorial Hospital, had given him surgical privileges. Prior to 1948, Bennett had been chairman of neuropsychiatry at the University of Nebraska Medical School when Freeman had visited.

They may be a bit on the "routine" side, but endowed with enough energy and imagination to make an adequate adjustment in their social lives. . . . There is a surprising number, in addition, who have made progressively better social adjustment in the later years after their post-lobotomy existence.

Most of the failures were in the schizophrenic group, where, he claimed, any subsequent deterioration was the result of further progress of the disease rather than of the lobotomy; he added that, in cases of prolonged illness, "it is safer to operate than to wait." Remember, he said, "a deteriorated schizophrenic looks and acts the same with or without his frontal lobes."[7]

When Freeman finished, Watts, who was in the audience, rose to say that Freeman's follow-up study—his "magnificent obsession"—had never been excelled.[8] But, after thanking his former colleague, Freeman got in the last word by arguing for the superiority of transorbital to prefrontal lobotomy: "There are 2500 transorbital lobotomy patients coming along, and the 5 year results as already published are decidedly superior to those of the major operation."[9]

The five-year results involved over 2,400 transorbital lobotomies, performed in private and state hospitals, which were compared to the 551 patients in the prefrontal lobotomy series.[10] The transorbital lobotomy patients followed most closely were the 628 private patients whom Freeman had operated on between 1946 and 1954. In all respects, he claimed, the record was better following transorbital lobotomy. There was a lower mortality rate, and the incidence of seizures was twenty times less. Moreover, following multiple operations, the incidence of seizures did not rise with transorbital lobotomies but rose to 53 percent after more than one prefrontal lobotomy. Freeman also claimed that the social adjustment after transorbital lobotomy was superior. In the same report, he argued again that lobotomy should be performed early in mental illness. He presented a graph showing that 82 percent of the patients receiving a transorbital lobotomy after less than six months of hospitalization were at home, compared with only 48 percent when the operation was performed after one to two years of hospitalization, and only 26 percent after five to ten years. Apparently, Freeman did not consider an equally likely interpretation: that the patients hospitalized for the shorter times were, on the average, not as sick, and an indeterminate number might have improved without surgery.

After the lecture was over, Paul Royal of the Green Gables psychiatric hospital in Lincoln, Nebraska, rose to support Freeman, stating that, of the eighty-one transorbital lobotomies performed during the previous seven years in his private sanatorium, about 75 percent of the patients were able to return home. Royal added that the patients at Green Gables were generally of higher status than those in state hospitals and that, in his experience, the best results were obtained with college graduates.[11]*

Freeman acknowledged that lobotomy was being eclipsed by the new drugs, whose effect he called "chemical lobotomy," and said that this was a "good thing as far as it concerns chronic hospital patients." But he challenged the strongest advocates of drug treatment to do long-term follow-ups—as he had done; and he noted that "it was nearly time now for a 5 year review of the patients treated by reserpine and chlorpromazine."[13]

Smith, Kline, & French's Thorazine (chlorpromazine) had been approved by the Food and Drug Administration in March 1954. This, and the other drugs that soon followed, had radically decreased the use of lobotomy. Freeman was compelled to justify any further role for the operations. In a 1958 speech given at the V.A. Hospital in North Little Rock, Arkansas, he noted that in some respects—particularly in improving the ward environment—lobotomy and drugs were much the same. He argued, however, that patients treated with drugs do not show the same tendency to become extroverted as they do following a successful lobotomy: only a lobotomy, he insisted, produces a basic change in personality. Moreover, chlorpromazine and reserpine could exacerbate depression to suicidal levels and produce symptoms similar to those seen in Parkinson's disease—symptoms, later called "tardive dyskinesia," that in 1958 were not generally recognized as a side effect of antischizophrenic drugs. Freeman also criticized electroconvulsive shock treatment, noting that some psychiatrists administered as many as four or five shocks daily in an attempt to induce a "therapeutic state of regression, amnesia and disorientation." He proved correct in warning that this "regressive shock therapy" can produce an "organic syndrome" with a persistent amnesia that impairs the intellect.[14]

He also reported that lobotomy had become a safe procedure with a minimum of adverse consequences, and predicted that the operation would take its place alongside drugs and electroshock: the tranquilizing drugs would be used for overactivity and anxiety; electroshock and the "euphoriants"—what are now called "antidepressants"—for depression; and lobotomy for severe obsessions. Some psychiatrists, indeed, still maintain that psychosurgery is particularly effective in treating patients suffering from otherwise intractable obsessional disorders.[15] Freeman was clear, however, that, in his opinion, lobotomy was not useful for psychopathic states, criminality, alcoholism, drug addiction, or mental deficiency. He once remarked that a "peeping Tom" who was lobotomized would stop sneaking up to windows but would instead knock on the front door.[16]

---

* Transorbital lobotomies were also performed in other private hospitals in this country and abroad.[12]

In spite of his almost single-handed effort, the tide had clearly turned, and Freeman was not able to hold it back alone. Lobotomy was not only being replaced by drugs, but the growing popularity of psychoanalysis and psychotherapy made these operations seem archaic. Furthermore, those who still believed in the usefulness of the operations tended not to make them a public issue, while Freeman, in his commitment to being the spokesman for lobotomy, had become almost quixotic. Thus, he became the lightning rod for most of the hostility.

However he tried to make a joke of it, Freeman knew he was becoming a "living fossil," a "dinosaur."[7] It was difficult for him to accept his loss of influence. Without academic affiliation or any office in medical societies, he was no longer even clearly identified with a specialty. Most psychiatrists, committed to psychodynamics, rejected him; and most neurologists, with their new techniques in enzyme biochemistry and electron microscopy, no longer considered him one of their own. Neurosurgeons continued to resent him.

However others regarded him, Freeman increasingly had to consider himself a psychiatrist as most of the patients he saw had psychiatric problems. Not surprisingly, he relied heavily on somatic treatments, using electroconvulsive shock, carbon dioxide therapy, intravenous injections of sodium amytal, and sometimes, as he put it, "direct advice." He advised depressed patients to get "plenty of exercise," as he had become convinced of the therapeutic value of exercise in combating his own depression. Earlier he had written a report about a woman who, following his advice, controlled her depression and hypochondriacal ideas by totally immersing herself in a rigorous program of weight lifting.[18]

Freeman could be open to new ideas other than somatic therapies. He could even envision that such developments as group therapy and psychodrama might be beneficial, but he believed that once the procedures became standardized, they should not be performed by physicians. Similarly, while he understood the fascination psychoanalysts had exploring intrapsychic conflict and the dynamics of interpersonal relations, Freeman thought this approach had little to do with medicine. Acknowledging the brilliance of Freud's speculations, he believed that these had become dogma "to his less gifted followers," and felt strongly that "medically trained people should practice medicine." Whatever value psychotherapy might have, it was a waste of a doctor's training to use a therapy that could be practiced by psychologists, social workers, technicians, clergymen, and, as he wrote, "even bartenders."[19]

Freeman did nothing to decrease his alienation from most of psychiatry; some of his behavior, in fact, exacerbated it. Lecturing to a predominantly psychoanalytic audience at the Langley Porter Clinic in San Francisco, in

January 1961, he picked a topic that could not have been more controversial: lobotomy in children. He questioned the effectiveness of psychotherapy in the cases of six teenagers whom he had lobotomized—one was only twelve years old—and remarked that these children surely "had received adequate psychotherapy"; in fact, one had had three years of it at Langley Porter Clinic.[20]

Freeman had brought along three of the children to demonstrate the outcome. To his prompting, one child kept responding pathetically, "I'm doing the best I can," making Freeman appear unsympathetic. The barrage of hostile criticism that followed his talk was not softened by his replying that the youngsters were "adjusting reasonably well at home," and that some "were even attending school." Apparently anticipating strong criticism, he had brought along a box of five hundred Christmas cards from former patients. Losing his temper completely, he dumped the box on a table, shouting, "How many Christmas cards do you get from your patients?" The meeting was quickly adjourned, but reports of it spread rapidly through the local psychiatric community and contributed to Freeman's dismissal from the Palo Alto Hospital after he suggested that a young schizophrenic boy in that hospital might be helped by transorbital lobotomy.

Freeman's position at the El Camino Hospital, which he had helped found in 1961, became increasingly marginal over the next five years. Few of its psychiatrists approved of his methods, and he ceased actively running the psychiatric service. One night, in 1965, he admitted and administered "emergency" electroconvulsive "shock" treatment (ECT) to a woman the police had found wandering near the hospital. Freeman never explained what about the emergency had prevented him from using drugs. He had seriously breached hospital regulations, which required a two-day waiting period, a head X ray, and an electrocardiogram before administration of ECT. No disciplinary action was taken, but the incident increased a growing suspicion that Freeman needed watching. In 1966, he retired from the staff.

Freeman had continued to perform transorbital lobotomies in the Doctors General Hospital in San José and the Herrick Memorial Hospital in Berkeley, and a few state hospitals he visited. Over the years, however, his surgical privileges became increasingly restricted to the point of de facto prohibition. He performed his last transorbital lobotomy at Herrick Memorial in February 1967, at the age of seventy-two. The patient was one of the ten transorbital lobotomies he had first performed in his office in Washington, D.C., in 1946. She had later moved to California; and, in 1956, Freeman had repeated the operation at Herrick. When, in this, the patient's third transorbital lobotomy, Freeman started to make the deep frontal cut, a blood vessel was ripped open, causing a severe internal hemorrhage. After the woman's death several hours later, his surgical privileges were taken away.[21]

Freeman's health had already begun to decline. In 1963, he developed diabetes, which was controlled with drugs and diet. With a starch-free diet, he lost twenty pounds and looked fragile and older. His eyesight, too, was affected, becoming progressively worse over the ensuing years. He also developed cancer and required surgery several times. While in the hospital following surgery for rectal cancer, he learned that Randy had a melanoma, and knew that his son's days were numbered. In addition, his wife's health had declined, and she was spending much of her time in a nursing home. It was a difficult period, physically and spiritually, and he had to fight depression by keeping his mind busy.

While recuperating from surgery, Freeman read Ernest Jones's biography *The Life and Work of Sigmund Freud* and was struck by the fact that at least eight of Freud's close associates had committed suicide. Examining obituaries in psychiatric journals and tracking down death certificates, Freeman concluded that suicide among psychiatrists was about eight times the frequency among white males. His suicide studies led directly to his writing *The Psychiatrists: Personalities and Patterns,* published in 1968.[22] Not surprisingly, the book reflected his own biological bias in psychiatry. All the people Freeman included among the "great discoverers" in psychiatry—whether psychiatrists or not—had made contributions to biological psychiatry: Julius Wagner-Jauregg, who had introduced malaria-fever therapy; Arthur Loevenhart, whose attempts to induce "cerebral stimulation" led to carbon-dioxide therapy; Ugo Cerletti, the developer of electroconvulsive shock treatment; Ladislas Meduna, who had introduced metrazol shock; Manfred Sakel, the founder of insulin coma treatment; Ivan Pavlov; and, of course, Egas Moniz. The psychoanalytically oriented psychiatrists—Sigmund Freud, Otto Rank, Alfred Adler, Sándor Ferenczi, and Harry Stack Sullivan—were covered in the section called "The Great Theorists."

In the last section, "The Death Instinct," Freeman described the high suicide rate among psychiatrists and concluded that it could not entirely be explained by who selected the field. Psychiatry may actually be "destructive to some of its devotees."[23] His own anti-psychoanalytic views were transparent. Describing one suicide case, Freeman mentions that the neuroanatomist C. Judson Herrick had earlier warned the budding psychiatrist that "psychoanalysis would rob him of his intellectual freedom." Freeman added that "in the final event it [psychoanalysis] may have robbed him of his life."[24] He had made a special effort to determine the cause of death of prominent psychiatrists, especially Harry Stack Sullivan, who had died in Paris under ambiguous circumstances.[25] In his book, Freeman noted that, while the cause of death was apparently heart failure, Sullivan had foreseen with "uncanny accuracy" the time of his death several years before it happened.[26]

He had high hopes for the book and tried to persuade Smith, Kline & French to purchase a large quantity to send to psychiatrists. The pharmaceutical company was not interested. A review in the *American Journal of Psychiatry* dismissed *The Psychiatrists* as "neither good biography nor good history."[27] Freeman's somatic bias was out of step with the time, and his attempt to swim against the stream of contemporary psychodynamic psychiatry was futile. He lost all perspective about the lasting value of the information he was collecting; the task itself gave meaning to his life. He kept driving himself to collect additional information and sometimes would get so wound up that he had difficulty sleeping; but he had learned, after his earlier bout with depression (see pages 138–39), to recognize the danger of missing too much sleep:

> I have skirted the edge on some occasions, when I was close to the breaking point, but I recognized insomnia as my first danger signal and protected myself against it by Nembutal. I have been taking Nembutal every night for nearly thirty years and have found it most helpful. I am dependent on it but do not consider myself addicted to it since I have rarely needed more than three capsules a night.[28]

He had always used the mail extensively to keep track of patients. When he was in Lisbon in 1953 for the International Congress of Neurology, he had sent scenic postcards to many patients, and on a later trip to Europe, he bought two thousand postcards in Vienna, sending most of them to patients with a handwritten "Merry Christmas, Walter Freeman" (see figure 14.1). He always sent cards early in December, hoping that with Christmas approaching patients would be in a mood to respond. He encouraged them to write to him about what they had been doing, and would always send back encouraging, if brief, notes. His patient files are filled with correspondence—for example, these notes to a young lobotomized man in New York City:

> Dear Fred:
>
>   It is good to hear from you even though it happens that you are hospitalized after your hernia operation. I am glad that things are going so well and that you had such good doctors. I guess that you are home by now and from then on will recuperate rapidly. You can expect a certain amount of discomfort till those tissues stretch again, but maybe by the time that you are ready for work the post office job will open up. That would certainly be a fine thing for you.

And two and one-half months later:

> Dear Fred:
>
>   Congratulations on passing your B exam with a 98%. I am sure that you can pass the C exam on August 13, and after that, take your annual leave. I don't think it

Figure 14.1.   Walter Freeman with backpack. Picture was made into a postcard and sent to over fifteen hundred former patients, friends, and colleagues, with the message "Greetings From California. This photo was taken by my son, Dr. Walter J. Freeman, soon after we started on a five day back-pack trip in the High Sierra. A snowstorm in August caught us half way from Kennedy Meadow to Hetch Hetchy. We huddled around the fire all night since we were soaked through and sleep was impossible. I managed to survive somehow, but lost eight pounds when food ran low." (Photograph courtesy of Walter Freeman III.)

will hurt you a bit to study for an hour a day, and I believe that the habit of studying will bring its own rewards and greater confidence.

And to a lobotomized woman, who wrote that she was singing in the church choir:

Dear Judy,

I was delighted to receive the note from you with its quotation from one of the famous hymns. I suspect that it was by Charles Wesley, the brother of the preacher, the great hymnologist of the 18th century. Maybe, someday . . . you will sing it for me, with piano accompaniment.

With the exception of some of the patients in state hospitals, where there was not sufficient money or interest to keep records up to date, Freeman was able to get some information about most of the transorbital lobotomy patients. He was most successful with his private patients and located practically all of them.

When IBM cards became available, Freeman started using them to locate information on patients with particular psychiatric problems or to find records containing useful data about specific behavior after lobotomy. In one article, he argued that lobotomy had proven useful in treating spastic colons and other gastrointestinal disorders, and quipped: "If the trouble's in the head, why work on the belly?" He added that some treatment of psychosomatic disorders made about as much sense as "tinkering with the weathervane in the hope of altering the wind."[29]

In spite of poor health, Freeman decided, at the age of seventy-two, to take off on an extensive cross-country trip to see former patients, to visit family members, and to present papers at several meetings. In February 1968, having transferred his patient files into a Cortez camper-van, he started out (figure 14.2). He worked his way through the southwestern and Gulf states into Florida, where a recurrence of rectal cancer required him to have surgery. While recuperating, he learned that Randy had died, at the age of thirty-two, from a metastatic brain tumor. Never one for ceremonies, and preferring to grieve alone, Freeman did not return home for the funeral. Instead, he continued on through the South, into Ohio, and then east to New Jersey, Washington, D.C., New York, Boston, and Toronto. On the trip back west, he sought out patients in West Virginia, Ohio, Michigan, North Dakota, Idaho, Oregon, and, finally, California. In North Dakota, he searched for a former patient who used to write to him regularly about his farm. The farm had been sold, but Freeman was able to trace the man to Fargo and backtracked the seventy-five miles to interview him there. In Idaho, he located a man who had been given a prefrontal lobotomy in Washington, D.C., thirty years earlier.

It was not long after he had transferred the new follow-up data to IBM cards that he received an invitation from Manuel Velasco-Suarez* to give a talk in Mexico City. Freeman brushed up on his Spanish and, early in 1969,

---

* Manuel Velasco-Suarez, who had done a neurosurgical residency at George Washington University, had great influence on neurology and neurosurgery in Mexico; later he became governor of the state of Chiapas.

Figure 14.2.   Walter Freeman waving from Cortez van, October 1968. (Photograph courtesy of Jonathan Williams.)

took off again in his van—this time on a six-month trip of 22,000 miles, taking him to Mexico City, Texas, Oklahoma, Arkansas, Louisiana, Mississippi, Tennessee, Alabama, on to Washington, D.C., and back to California again. Due to failing eyesight and his habit of driving too fast, Freeman hit an abutment on the trip to Mexico and almost tore the van's door off its hinges. In Mobile, his rectal cancer began acting up once again, and he underwent still another operation. As soon as he was able to walk, he started hunting down former patients in the area. In a prison in Mobile, he located a woman who was one of his transorbital cases. She was serving a term for forging a check. He bought her a carton of cigarettes and decided, after talking to her a while, that the lobotomy had removed her conscience along with her anxiety. Back in California, he calculated that, with the 1968 and 1969 trips, he had added several thousand patient years to the follow-up studies.

During most of Freeman's travels in these years, his wife had remained in a nursing home. She died in April 1970 and was buried in the family plot in the Catholic cemetery in Merced, California. In less than a year, he would announce his retirement from medical practice. He spent much of his time going through the patient files he kept in his apartment in San Francisco. George Washington University had commissioned separate portraits of Watts and himself, and he flew to Washington for the unveiling ceremonies. Watts was portrayed in surgery with several of his former residents looking over his shoulder; Freeman was pictured alone with his pipe in one hand and his other on a skull (figure 14.3).

That year, 1970, about 300 psychosurgical operations were performed in the United States—only about 6 percent of the 5,000 per year performed between 1948 and 1952.[30] Between 1948 and 1952, the annual editions of the *Year Book of Neurology, Psychiatry and Neurosurgery* were filled with summaries of articles about lobotomy and other somatic therapies; by 1970, lobotomy, metrazol, and insulin therapy were rarely mentioned at all. Most of the operations performed in the 1970s were the new procedures using stereotaxic instruments to destroy targets in specific brain areas. No longer was there any "blind" swinging of knives—let alone, ice picks—in the brain.[31] In spite of refinements in the operation, it appeared for a while that psychosurgery would completely disappear. However, toward the end of the 1960s, it had become apparent that the psychoactive drugs were not helping all the mentally ill. In his article "The Frontal Lobe Revisited: The Case for a Second Look," Kenneth Livingston, of the Wellesley Hospital in Toronto, expressed a view shared by the few neurosurgeons still performing psychosurgery:

> In retrospect, even as lobotomy fell from grace clinical evidence consistent with earlier physiological concepts was rapidly accumulating. It had been clinically established that a variety of anatomical lesions of the frontal lobes can produce beneficial change in the affect and behavior of patients suffering from severe psychotic and psychoneurotic disorders. It had been further demonstrated clinically that the areas most effective in producing such change are discrete regions of the medial and orbital frontal cortex. . . . With the impressive array of recently acquired and increasingly sophisticated techniques of both laboratory and clinical investigation, a prospective 'second look' at the frontal lobes seems indicated.[32]

A conference on psychosurgery, to be held in Copenhagen, was planned for August 1970. Walter Freeman, who was almost seventy-five, knew nothing about it until he read an announcement in a journal about the forthcoming "First International Conference on Psychosurgery." Incensed by the slight,

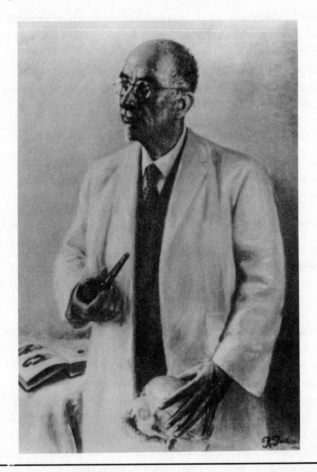

Figure 14.3.   Painting of Walter Freeman by A. Jonniaux. Freeman sent copies to friends and colleagues with the message: "This portrait was presented to George Washington University School of Medicine where I taught for thirty years. I thought you would like to see it." (Photograph courtesy of Walter Freeman III.)

he wrote to the organizers, informing them that the first congress had been held in Lisbon in 1948 and reminding them of his role in organizing it. An apology and an invitation to the now "Second International Conference on Psychosurgery" were sent to him, and he attended, even though a month before he had suffered a concussion from a fall on a backpacking trip.

In Copenhagen, Mogens Fog, professor of neurology at the Kings Hospital, welcomed the guests and complimented them on how far their specialty had advanced. Describing some of the early history of lobotomy, Fog looked at Freeman and said that he could still recall his "astonishment and horror" at the Neurological Congress in Copenhagen in 1939 when Freeman described "the more-or-less blind lobotomies" that he and Moniz were performing at that time.[33]

Freeman gave a paper ostensibly reviewing his long-term follow-up study of 415 lobotomized schizophrenic patients, but his main purpose was to point to his unique role in the history of psychosurgery:

> I have continued in contact with the patients or their relatives, doctors, or hospitals in about 95 per cent of cases. This has been accomplished by means of annual Christmas greetings, as well as by numerous trips around the United States, Canada, and Mexico visiting patients in their homes or by telephone, keeping abreast with changes of name through marriage and more particularly with changes of addresses. On these visits I have often added names of relatives and physicians to whom I could write in case my greetings were returned by the post office as undeliverable. The follow-up has been a rewarding experience. There is no more grateful patient or family than one in which psychosurgery has apparently arrested the downward course of schizophrenia. There are instances of dissatisfaction, of course, and in some cases I have been met with refusal . . . or even denial that an operation had ever been performed, but these are rare.[34]

After reminding the audience that transorbital lobotomy had "proved unacceptable to [his] surgical colleagues," so that he was forced to proceed on his own after Watts and he had parted company, Freeman presented the now-tiresome statistics purporting to show that transorbital lobotomy produced better results than prefrontal lobotomy. He also presented several thumbnail descriptions of highly successful cases: a psychiatrist who, after a transorbital lobotomy, had become chief of clinical service in his own private clinic; another lobotomized physician who had subsequently married, had children, established a ten-man medical clinic, and flew his own plane; a violinist in a metropolitan symphony orchestra; and a college graduate who had become chief construction engineer for a major corporation.

After Freeman finished, there was a long and awkward silence. No one asked a question. There was little interest in any of the "old-style lobotomies" and certainly not in transorbital lobotomy. In spite of this reaction, Freeman was recognized as a pioneer in the field and, at the end of the meeting, was nominated to be honorary president of the Third International Congress of Psychosurgery scheduled for Cambridge, England, in 1972.

Returning to San Francisco early in 1971, Freeman busied himself extracting more information from his follow-up studies. Using the IBM cards, he pulled out information on the suicide rate of lobotomized patients and made this the subject of a lecture he gave in Mexico City at the Fifth World Congress of Psychiatry in December 1971. Plagued by recurrent bouts with cancer and by limited vision, he attended the congress with Zigmond Lebensohn, a former resident who is now a prominent psychoanalyst in Washington, D.C. After the meeting, Freeman flew to Chiapas, where he was entertained by its governor, Manuel Velasco-Suarez. Now needing the cane he had formerly used occasionally as a sporty affectation, Freeman climbed the difficult steps of the Mayan Temple at Palenque with what Lebensohn described as an "uncomplaining stoicism" bordering "on the heroic."[35]

Back in San Francisco, Freeman wrote a note to the *American Journal of Psychiatry* announcing the forthcoming Third International Congress of Psychosurgery and questioning whether lobotomy should stay in a "state of limbo."[36] His health had been steadily declining. He was barely able to finish a paper on the sexual behavior and fertility of lobotomized patients, which he planned to present to the Society of Biological Psychiatry. At the time of the meeting, however, he was unable to get out of bed. The paper was read for him and later published posthumously.[37] In May, he drifted into a coma and, with his four surviving children gathered around, died on 31 May 1972. Three months later, at the Third International Congress of Psychosurgery, Freeman's death was virtually ignored. Almeida Lima and Keiji Sano of Tokyo were made honorary presidents.[38]

Nearly twenty years earlier, before he left Washington, D.C., Freeman had bound together some of his journal articles and presented one copy to the National Library of Medicine and another to St. Elizabeth's Hospital. In the latter, he wrote:

> To St. Elizabeth's Hospital, where I worked 1924–1933, to find some answers to the problems of mental disorders; and where more problems arose than were ever answered.

This inscription, speaking to his unremitting efforts in the years following, might serve as his epitaph.

15

Psychosurgery in the
1970s and 1980s

Tiny electrodes are implanted in the brain
and used to destroy a very small number of
cells in a precisely determined area. As a surgical technique, it has three great advantages
over lobectomy: it requires much less of an
opening in the surfaces of the brain than lobectomy does; it destroys less than one-tenth
as much brain tissue; and once the electrodes
have been inserted in the brain, they can be
left without harm to the patient until the
surgeon is sure which brain cells are firing
abnormally and causing the symptoms of seizures and violence.

—VERNON H. MARK AND
FRANK R. ERVIN (1970)

AT THE END of the 1960s, few people were aware that psychosurgery was still
being performed. The subject was rarely mentioned in the popular media or
even in the medical literature. Yet psychosurgery did not completely fade
away. In 1971, over seventy neurosurgeons in the United States had performed
some psychosurgery. Only six neurosurgeons, however, accounted for most
of the three hundred operations performed that year.

Almost no standard prefrontal lobotomies or transorbital lobotomies were
performed. The operations, as noted earlier, were mainly restricted, "targeted"
psychosurgery, using stereotaxic instruments for directing electrodes into brain
targets. Relatively small areas—compared with the early lobotomies—were
destroyed, usually with radio-frequency waves. Several neurosurgeons experimented with other techniques for destroying brain targets, either by exposing
neural tissue to small amounts of radioactive elements with a short half life,

such as cobalt or yttrium, or by freezing, using cryoprobes.* There was even
an attempt—not yet realized—to use focused ultrasonic beams in a "non-
invasive" procedure.[2] X rays and on-line video monitors were routinely used
to increase the accuracy in placing electrodes into the preselected brain targets.

The new procedures were referred to by such names as "functional neurosurgery," "stereotaxic surgery for behavior disorders," "stereotaxic trac-
totomy," "psychiatric neurosurgery," or simply "psychosurgery"; the term
lobotomy was consciously avoided. By this time, psychosurgery was being used
primarily to treat exaggerated emotional states that did not respond to drugs—
that is, patients who were mostly very depressed, agitated, anxious, obsessed,
or suffering from intractable pain. Schizophrenics were generally treated with
drugs, although a few neurosurgeons believed psychosurgery was useful in
schizophrenia with a significant emotional component ("strong affective
loading" or "prominent tension").

In spite of its limited use, there were signs of a growing interest in psychosurgery. Twenty-two years had elapsed between the First International
Conference on Psychosurgery in Lisbon in 1948 and the Second Conference
in Copenhagen in 1970; but a Third Congress was held in Cambridge, England,
in 1972; a fourth, in Madrid in 1975; and a fifth, in Boston in 1978. Articles
began to appear in the psychiatric literature suggesting that new knowledge
about the brain, combined with the development of better surgical techniques,
justified reconsideration of "psychiatric surgery" for patients who did not
respond to other treatment.[3]

The public became aware of psychosurgery, largely as a result of a controversy stirred up by the 1970 book *Violence and the Brain*.[4] The authors, the
neurosurgeon Vernon Mark and the neuropsychiatrist Frank Ervin, had implied that much of the violence prevalent in our society was caused by brain
pathology, and that neurosurgery could eliminate a significant amount of it.
They described patients with a "dyscontrol syndrome," who were prone to
sudden unprovoked outbursts of "episodic violence," which were claimed to
be triggered by abnormal electrical discharges in the temporal lobe of their
brains. This syndrome was thought to be a variant of temporal lobe epilepsy
in which automatic acts are committed during a period of unconsciousness
or reduced consciousness. Patients with this type of disorder may have the
electroencephalographic pattern of an epileptic but do not necessarily have
any history of convulsions. Mark and Ervin argued, as can be seen in the
epigraph to this chapter, that elegant techniques were now available for lo-

---

* A cryoprobe ("cooling probe," from the Greek *kryo*, meaning "icy cold") physically resembles
an electrode and can be similarly inserted into brain targets with a stereotaxic instrument. The
temperature at the tip of the cryoprobe can be gradually lowered. Activity of nerve cells near the
tip is suppressed at below normal temperatures; the cells are destroyed at below freezing temperatures.

cating the brain "triggers" of these violent outbursts and removing them.

As is usually the case, there was some basis for Mark and Ervin's claim. There is no doubt that the cases, as they described them, do exist, having been well described in the medical literature at least as early as the 1870s.* However, the amount of violence that could be attributed to this cause had been very much exaggerated. Also exaggerated was Mark and Ervin's claims about the effectiveness of the surgery in reducing violence by destroying "a small number of cells in a precisely determined area."[6]

Moreover, Mark and Ervin's recommendation that "we need to develop an 'early warning test' of limbic brain function to detect those having a low threshold for impulsive violence" seemed to suggest a dangerous precedent. Harvard's professor of neurosurgery William Sweet wrote, in his foreword to the book, that "knowledge gained about emotional brain function in violent persons with brain disease can be applied to combat the violence triggering mechanisms in the brains of the nondiseased." Regardless of how it was intended, this statement was interpreted by many readers as suggesting a pacification program with frightening political ramifications.

This interpretation seemed to be strengthened when it was discovered that Mark, Sweet, and Ervin had written a letter to the *Journal of the American Medical Association* after the race riots in Detroit in July 1967. The letter, entitled "The Role of Brain Disease in Riots and Urban Violence," noted that a small proportion of slum dwellers take part in riots and only "a subfraction of these rioters have indulged in arson, sniping, and assault." Asking if there was "something peculiar about the violent slum dweller that differentiates him from his peaceful neighbor," the authors implicitly answered the question by arguing that there is convincing evidence that "a focal brain lesion can play a significant role in the violent and assaultive behavior of thoroughly studied patients." Reviewing reports claiming that persons committing violent crimes have a high incidence of abnormal electrical brain waves, the letter recommended intensive studies of individuals "with low violence thresholds before they contribute to further tragedies."[7] Thus, Mark, Sweet, and Ervin had unwittingly left themselves open to the charge of recommending preventive brain surgery as a means of coping with a desperate social problem; they were thus forced to attempt to clarify the issue by explaining that they always meant to restrict "psychiatric neurosurgery to cases where the primary cause of the violence is brain dysfunction."[8]

---

* In England, Henry Maudsley wrote in 1874 that "the most dangerous cases with which those who take care of insane persons have to do are those of persons suffering from epileptic mania." Sometimes, he wrote, "an attack of furious and destructive mania supervenes, marked by blind and reckless violence." Such outbreaks, Maudsley observed, "may occur in a person who has the epileptic neurosis, without there ever having been an attack of actual epilepsy either in the form of epileptic vertigo or epileptic convulsions."[5]

The fear that plans were brewing to use brain surgery as a means of social control were intensified by a front-page article in the *Washington Post* in 1972 reporting that psychosurgery had been performed on three convicts in California's Vacaville Prison.[9] Several months later, a California neurosurgeon known to be performing a major portion of psychosurgery in the United States was quoted as saying:

> The person convicted of a violent crime should have the chance for a corrective operation. . . . Each violent young criminal incarcerated from 20 years to life costs taxpayers perhaps $100,000. For roughly $6000, society can provide medical treatment which will transform him into a responsible well-adjusted citizen.[10]

This concern with what was called the "captive patient: a forgotten man,"[11] was followed by a heated dispute, widely covered in the press, over a study funded by the state of Michigan to evaluate psychosurgery on violent sexual offenders incarcerated in institutes for the criminally insane.[12] It was also revealed that some prison administrators and officials within the Law Enforcement Assistants Agency, the arm of the Justice Department responsible for prisons, were becoming increasingly interested in biological causes of violent behavior. The widely read journal *Science* asked whether psychosurgery was "legitimate therapy or laundered lobotomy," and it was reported that Mark, Ervin, and Sweet, "in 1970, through various mysterious maneuvers that no one seems to be able to explain, persuaded Congress to direct the National Institute of Mental Health (NIMH) to award them a $500,000 grant to carry on their work."[13]

The suggestion of a relationship between brain pathology and race riots and rumors that plans were under way to perform psychosurgery in prisons— where minorities are disproportionately represented—made the topic a civil-rights issue. Spokesmen for minority groups became involved, and the emotional level of the argument escalated rapidly. It was not long before psychosurgery was being debated in the popular press, in congressional hearings, in symposiums at scientific and professional meetings, in conferences on science, medicine, and ethics, and in legal circles.[14] Most of the debate, however, was centered on the potential danger of using psychosurgery to treat violence, even though a relatively small proportion of these operations were performed on violent patients. Actually, the majority of the neurosurgeons performing these operations believed that psychosurgery should not be performed on violent persons for fear that the operation might make them less inhibited and more impulsive and, in the case of psychopaths, blunt still "further any sense of moral guilt";[15] but a few neurosurgeons did recommend psychosurgery for criminals.[16]

Throughout this period, sociologists, psychologists, and psychiatrists were raising questions about the nature and causes of mental disorders by arguing that such disorders are produced primarily by experiences within the family and society and, therefore, should not be considered medical "diseases" to be treated by physicians.[7] These issues had been debated, for more than half a century, by the protagonists of the functional and the organic views of mental illness. The debate became intensified by the conflicting interests of subspecialties within psychiatry. Increased funding for biological research in psychiatry was undercutting the position of psychoanalysts in university medical schools. By 1980, the process of selecting a new chairman for a department of psychiatry often precipitated a power struggle between the functionalists—usually psychoanalysts, but sometimes behavior and family therapists—and the biological psychiatrists.

Also fueling the debate were the books and lectures of Thomas Szasz, a psychiatrist who became the spokesman for an "anti-psychiatry" movement. He argued that diagnostic labels in psychiatry are "myths" or "metaphors" used to stigmatize social "deviants," and charged that psychiatry, with its use of involuntary confinement, mind-altering drugs, and psychosurgery, had become a political force, concealing social conflict by calling it "illness" and justifying "coercion as treatment."[8] This controversy was dramatized in the 1975 film based on Ken Kesey's novel *One Flew Over the Cuckoo's Nest.*[19]

In an atmosphere of general distrust of authority, enhanced by evidence of government-sponsored experiments using drugs on unsuspecting subjects, there was legitimate concern about plans to solve social and political problems through some form of "brain control." The coalition of civil-rights, anti-psychiatry, and minority groups opposed to psychosurgery proved to be much more effective politically than the earlier opposition to psychosurgery from within the medical profession had been.

In the mid-1970s, the National Commission for the Protection of Human Subjects of Biomedical and Behavioral Research was established with the mandate to study questionable research practices in the United States. Psychosurgery was one of the first issues to be considered. After listening to testimony on all sides of the controversy and to summaries of two contracted studies of lobotomized patients, the commission issued a report that reduced the intensity of the debate. The report specifically criticized any future use of psychosurgery for "social or institutional control" but, on the basis of the testimony presented, concluded that psychosurgery, as currently practiced, might help some patients who had not responded to other treatment. Some critics felt that the National Commission's conclusion that psychosurgery "can be of significant therapeutic value in the treatment of certain disorders," fell only slightly short of endorsing these operations. Nevertheless, the commission, reaching for a compromise, recommended limiting the use of psychosurgery by requiring prior screening of every proposed operation by an institutional review board. At present, these recommendations serve only as guidelines; they have never been translated into federal legislation or regulations.[20]

In Oregon and California, hearings were held, and legislation restricting the practice of psychosurgery was enacted in 1973 and 1976, respectively. Although there were differences, both states required review at a higher level (outside the institution) to determine whether alternative therapies had actually been exhausted and whether truly informed consent had been obtained. Stricter regulation of psychosurgery on incarcerated patients was also legislated.[21] Following the passage of these laws, only one psychosurgical procedure has been performed in Oregon, and none in California. In 1981, a group of physicians in Oregon tried to get the law rescinded, arguing that the cumbersome review process was an unnecessary restraint on their freedom to practice medicine and on their patients' freedom to choose treatment. In a countermove, the state legislator who had been mainly responsible for the original bill mobilized opposition, and a new law prohibiting psychosurgery was passed in 1982. By this time, as a result of the restricting legislation and the increased threat of malpractice suits engendered by the controversy, the amount of psychosurgery performed in the United States was reduced to fewer than two hundred operations a year.

During the same period when these legal restraints were being imposed, extensions of the studies reported to the National Commission had been producing results that were favorable to psychosurgery. These studies had been undertaken by teams of psychologists, neuroscientists, psychiatrists, neurologists, and social workers, who had no personal stake in reporting success. Two independent research teams reported that the quality of life of 70 percent to 80 percent of the patients significantly improved after psychosurgery. Moreover, the investigators found virtually no evidence of physical, emotional, or intellectual impairment caused by the surgery.[22] The conclusion of one study was:

> On the basis of five years of research in which 85 patients who underwent cingulotomy [one form of psychosurgery] were examined, it can be concluded first of all that no patients showed the severe adverse effects that had been associated with prefrontal lobotomy. In fact, there is so far no evidence of any lasting neurological or behavioral deficits after cingulotomy. In addition, this surgical procedure was followed by improvement in the condition of many of the patients, though not all of them.[23]

And of the other:

> Twenty-seven of the cases [more than 50 percent] were adjudged to have improved markedly. This we determined on the basis of subjective relief of symptoms as reported by the patient and a spouse, family member, or close friend. The assessment was supported by the results of independent psychiatric and psychosocial evaluations. This very favorable outcome was not accompanied by detectable neurological deficits.[24]

Encouraged by such reports of favorable results, several psychiatrists and neurosurgeons, both here and abroad, have become willing to reconsider psychosurgery.

At the time of this writing, it is possible to detect a small increase in interest in psychosurgery around the world. While opposition to psychosurgery has been worldwide, there is at the same time a recognition that, in spite of the accomplishments of the ever-increasing variety of psychoactive drugs, some desperate patients still do not respond to any available treatment. Encouraged by some of the recent evidence, a few psychiatrists at institutes where psychosurgery has not been used for over a decade have shown a willingness to undertake a few trial operations. Regardless of the outcome, the existence of alternative treatments and the opposition to psychosurgery make it unlikely that there will be any substantial increase in psychosurgery.

Yet, limited though the use of current psychosurgical techniques may be and out of date though lobotomy is, its history has significant lessons for us today, in both the field of mental health and the wider one of medicine itself. Implicit throughout this book, these lessons are highlighted in the next and final chapter.

16

Final Reflections

The pioneer has his place in medical research.
... On the other hand it is important in
medicine to recognize fully the responsibility
with regard to those who follow voluntarily,
that is physicians; to those who follow
blindly, that is lay people; and to those who
are forced to follow, that is patients.
—OSKAR DIETHELM (1939)

The disadvantage of men not knowing the
past is that they do not know the present.
History is a hill or high point of vantage,
from which alone men see the town in which
they live or the age in which they are living.
—G. K. CHESTERTON (1933)

EARLY IN the nineteenth century, Samuel Taylor Coleridge wrote that history
"gives us a lantern on the stern, which shines only on the waves behind us."[1]
It is indeed all too easy to look back on the past and wonder complacently
how prefrontal lobotomy could have been so readily accepted. Oblivious to
the similarities between past events and present practices, we draw no lessons—
as though, for a brief period, psychiatry, responding to unique circumstances
and the influence of a few zealous, if not malevolent, physicians, had run
amuck.

Although it may be comforting to believe that the forces responsible for
the wide acceptance of prefrontal lobotomy and the other somatic therapies
described in this book are all behind us, they are, in truth, part of the very
bone and marrow of the practice not only of psychiatry but of all medicine.
There are today no fewer desperate patients and desperate families; premature
reports of spectacular cures with minimal risks are still accepted uncritically;
the popular media promote innovative therapies even more enthusiastically;
economics is no less an influence on the selection of treatment; conflicts persist

within and between specialties; and ambitious physicians still strive for "fame and name."

As I write, for example, three French physicians are claiming, with much fanfare, that they may have found a way of arresting the so-far fatal acquired immune deficiency syndrome (AIDS). Whatever will come of this claim, it is abundantly obvious that the proposed treatment, which uses a potentially dangerous drug, had not been tested adequately: only two patients had been treated for only a week. Although, in this instance, the premature publicity has been criticized—a criticism generated, I believe, primarily because of the fierce competition in this field between researchers in France and the United States[2]—most medical "breakthroughs" are described with little hint of the unreliability of the reports. Indeed, while increased popular interest in advances in science and medicine has led to even broader and more rapid coverage by the media, the quality of the reporting has, for the most part, not improved over the years.

Professional science writers are, as Jay Winsten of the Harvard School of Public Health has found, under constant pressure to make their stories "newsworthy." Even writers affiliated with the prestigious news organizations report that they have to "tiptoe along the boundary of truth" in order to demonstrate to their editors that their subjects are "ground breaking" and of "vital importance."[3] As a result, the implications of the most preliminary findings are frequently exaggerated.

Thus, writers and researchers, in a symbiotic relationship, may promote the most preliminary findings—as happened in the fall of 1984 after some researchers attempted to treat Alzheimer's disease by infusing a drug through a "chemical pump" implanted in the abdomen of patients. The researchers had accomplished little more than to demonstrate that the "pump"—which was not new—could be implanted in humans. Information about the effect of the infused drugs on the four Alzheimer patients was based solely on information obtained from relatives interviewed briefly by the researchers, who always knew—contrary to sound scientific methodology—when the patients were receiving the drugs. Nevertheless, as a result of the press conference held at a New England medical center, the possible "breakthrough" was reported on prime-time television shows—NBC's "Nightly News" and "Today Show," the "CBS Morning News," ABC's "Good Morning America," PBS's "MacNeil-Lehrer Report," and the Cable News Network. Many magazines—among them Newsweek, McCall's, Family Circle, People, and Forbes—published articles describing the treatment. The United Press story was run in the Boston Globe under the headline "Researchers Describe Possible Alzheimer's Cure."[4] Never in the days of lobotomy had news of a medical breakthrough reached so many people in so short a time.

During the period of its most rapid growth, prefrontal lobotomy, as I have documented, was promoted as a way to relieve overcrowded and underfunded state mental hospitals. The influence of economic factors is no less important today, but it has a contemporary face. Hospitals are ever on the lookout for the solution to a new problem: that is, for procedures that will generate funds needed to support their huge overheads. The coronary artery bypass experience provides a striking illustration of the magnitude of both the economic influence and—not unrelatedly—of how, once surgeons and hospitals are "tooled up," a new procedure tends to be applied ever more broadly in areas where its effectiveness is questionable. Such extensions may occur with minimal consideration of the risks involved.

From the outset, the enthusiasm for the coronary artery bypass operation has been almost boundless. *Life* reported in 1971 that the operation "has saved two thousand lives," and that "the odds of dying during the bypass surgery are most likely no higher than the chance of having yet another heart attack within the year." The same year, *Newsweek* described it as "a major breakthrough in the conquest of the nation's No. 1 killer," and *Time* commented that nearly 250,000 Americans a year could benefit.[5] After the first bypass operation had been performed in 1964, the number of operations performed annually soon soared: 1,000 in 1968; 2,000 in 1970; 17,000 in 1972; 60,000 in 1975; 180,000 in 1982; and, proving *Time* right, 250,000 in 1985. In total, about 1,000,000 patients have had a coronary bypass operation, and, in an amazingly short period, it became the most frequently performed elective surgery in the United States.[6]

The expense of a single operation in 1985 ranged between $10,000 and $25,000, with the principal surgeon receiving between $3,500 and $4,200. The total cost of these operations was about $1 billion in 1977 and over $1.5 billion by 1985. Together with the cost of all the supportive industry producing supplies and equipment, the money involved annually has been estimated to be $5 billion.[7] For some hospitals and for some surgeons, the operation was clearly an economic windfall.

The coronary artery bypass operation was first performed for angina pectoris on desperate patients who were often suffering from excruciating pain caused by lack of oxygen to heart muscles. Undoubtedly, many of these patients experienced dramatic relief from the operation, and lives were saved. However, once many surgeons had learned the procedure and "bypass units" had been established in hospitals, the operations were increasingly performed to prolong the life of patients who could have been treated by nonsurgical methods. A ten-year study released in 1983 by the National Heart, Lung and Blood Institute found no increase in quality of life or survival rate of surgical patients compared with a matched group of patients treated more conservatively.[8] The study also concluded that at least one out of seven of the operations could have been safely postponed, if not avoided.*

The coronary artery bypass operation is not without risks, however, although they have been given little attention in the media. Much higher at the outset, the mortality rate during the operation is now approximately 1 to 4 percent, depending on patient population and the skill and experience of the surgeon.[11] An estimated 27 percent of patients undergoing the procedure experience serious postoperative physical complications, and about one-third of the patients are troubled postoperatively by depression. Furthermore, those who hoped the operation would make it possible for them to return to work are almost always disappointed.[12] After reviewing the rapid increase in coronary artery bypass operations, two cardiologists concluded that "it becomes difficult not to wonder if there are not other goals such as self-aggrandizement, keeping the beds full, or supporting further personal or institutional expansion."[13] The bypass operation is not controversial and is unquestionably a medical advance, but it clearly has been promoted, in part, for reasons other than the welfare of patients.

"Applause and fame is a thing that physicians much desire," as John Bunyan observed over three centuries ago.[14] I need not document further that physicians are no less influenced by the lure of "fame and name" than are the rest of us. It is not a question of fraudulent reporting of results—although that has occurred—but rather that the very desire for success can distort the facts, subtly and pervasively, often without one's being aware of it. In the great majority of cases—where results are presented prematurely, where success is overestimated and dangers underestimated, where there are biases in the selection of patients, and where failures are explained away as exceptions—the physicians responsible have been genuinely convinced of the validity of their conclusions. Self-deception is by its very nature difficult to guard against—almost impossible when fueled by unbridled ambition.

Patients for whom current therapies are not effective will always be with us, suffering conditions that seem to justify risk, yet their treatment may be determined by factors other than their own welfare. What, then, is the answer? It is much easier to criticize past mistakes than to identify in advance which therapeutic risks should be taken. My personal views depart from the solution most commonly offered. I do not believe, as many well-meaning people have suggested, that there are any scientific or ethical principles that will enable us to make the necessary predictions and decisions. While the quality of the

---

* Similarly, a 1985 report concluded that 30 percent of the cardiac pacemakers surgically implanted in patients were unnecessary. In some hospitals, up to 75 percent were considered unjustified.[9] Several recent reports, however, claim a possible prolongation of life in a small subset of coronary artery bypass patients.[10]

scientific rationale advanced to justify any innovative therapy must be considered, it cannot be the decisive factor. Not only will knowledgeable people differ in their evaluation of any hypothesis, but, even more important, too many medical treatments have proven useful even though the scientific arguments used to justify them initially were later proven false. If we were to reject treatments whose effects we did not understand, even aspirin would not have been acceptable during most of the period of its use. Despite the enormous amount of research on the biochemical action of psychoactive drugs, there is no agreed-upon explanation of how different drugs help anxious, depressed, manic, obsessed, and schizophrenic patients. Psychoactive drugs are used extensively because they are effective and convenient, not because of any compelling scientific rationale. Indeed, the first psychoactive drugs were discovered accidentally,* and only afterward were theories developed of how they might work. Similarly, electroconvulsive treatment is considered by approximately 70 percent of psychiatrists to be effective in alleviating otherwise intractable depression, despite the rejection of the original rationale justifying its use and no agreed-upon explanation today of how it works.[16] In medicine, we have to be most influenced by the quality of the empirical evidence that a treatment does more good than harm.

Discussion of ethical concerns can help to clarify issues and to sensitize physicians and others to questionable practices, but it cannot, in my opinion, provide rules of conduct that apply, unambiguously, to future developments. Most physicians who, after the fact, can be seen to have caused great harm would, had they been questioned at the time, have claimed the noblest motive—that is, alleviation of the suffering of desperate patients—and most would have been able to argue effectively that they had violated no ethical principle. Often, ethical judgments are made retrospectively. A risk with a successful outcome is usually not questioned. Where no effective therapy exists for desperately ill patients, equally ethical and knowledgeable physicians may disagree about treatment. Some risks have to be taken, and some harm will inevitably ensue along with some benefit.

Was, then, the irreparable harm done to many patients by prefrontal lobotomy unavoidable? Must great harm often follow as part of the natural course of progress in medicine? I believe the answer to both questions is no. While it is unrealistic to believe we can formulate any principles for determining with certainty what will prove therapeutically effective, we can develop ways of limiting the harm done until the answer is obtained. In my view, it is essential that we minimize the harm caused by premature claims of cures,

---

* Chlorpromazine was inadvertently discovered following experimentation with antimalarial compounds and drugs to prevent surgical shock,[15] and antidepressant drugs were discovered by their unexpected mood-elevating effects on tubercular patients.

by unbridled ambition, and by uncritically enthusiastic promotion. This end can be accomplished only by establishing procedures for testing innovative therapies before they are broadly used.

Physicians attempting to develop new therapies cannot be expected to regulate themselves. Needed is a way of regulating trials of innovative therapies to test the reliability and validity of initial claims and to provide reasonable estimates of risk. Such procedures already exist in some areas of medicine. The Food and Drug Administration has established steps for approving the marketing and clinical testing of new drugs: namely, animal experimentation to estimate toxicity and other dangers and to provide initial evidence of effectiveness, and then limited clinical trials with an approved experimental methodology which, where feasible, includes randomized control groups. Although these procedures primarily regulate drug companies, they do restrict the availability of untested drugs to the physician.

No safeguards are infallible, and the FDA procedure has been criticized as too conservative—preventing useful drugs from becoming available to desperate patients—and as not conservative enough in protecting patients from potentially dangerous drugs. Despite their shortcomings, the FDA procedures do help to reduce much of the harm that would surely result if there were no controls.

In surgery, where *laissez-faire* conditions have been the rule, some controls are clearly needed now and even more in the future when an increasing number of "high-tech" procedures will be proposed. There are, of course, important differences in the evaluation of innovative drugs and innovative surgical procedures: the latter is more problematic, varying with the skill of the surgeon, which may critically determine the outcome.* This need not be an insurmountable problem: special training can be required of surgeons once the value and safety of an innovative procedure have been adequately established by controlled trials. In order not to hamper progress, there would have to be some tolerance to allow for improvement of technique over a reasonable time, as in surgery it is common for mortality rates to be extremely high initially. There should also be some provision to permit simultaneous testing of a few competing procedures. On the other hand there should be limits to the amount of testing that is permitted, and the widespread adoption of inadequately tested procedures should be prevented.

It goes beyond the scope of this book to propose specific regulations for limiting innovative surgery. There are complex issues involved, and I do not mean to trivialize them. I am confident, however, that formal regulation of

---

* Another difference is that it is neither practical nor ethical to perform "sham operations," whereas "placebos" are used in "double-blind" drug tests.

surgical and other innovative therapies now uncontrolled will ultimately prove preferable to the current random regulation by malpractice suits, political action, and economic pressures. It is to be expected that any controls will be resisted—with the arguments that they will hamper progress and deprive desperate patients of help—and that randomized control studies are unethical in that they deprive some patients of the benefit of treatment. My own view is that the negative effect of reasonable regulation is exaggerated, especially when compared with the cost of uncontrolled experimentation. Furthermore, regulations that lead to improved scientific methodology usually facilitate progress by rapidly eliminating unpromising avenues of exploration. Improved methodology will also reduce the number of patients who undergo useless, and sometimes harmful, treatments.

Despite the shortcomings of the FDA regulations, few would favor abandoning all control of drug treatment. Other areas of experimental medicine must also be subject to more controlled testing procedures. It is my hope, finally, that this history of prefrontal lobotomy and the other "great and desperate cures" will stimulate discussion of these issues, which are so central to medical progress and patient welfare.

Notes

Chapter 1. The Treatment of Mental Illness: Organic versus Functional Approaches

1. E. Kraepelin, *Psychiatrie. Ein Lehrbuch für Studierende und Arzte,* 4 vols. (Leipzig: J. A. Barth, 1909–22).

2. C. Lombroso, *Criminal Man* (1911); E. Kretschmer, *Physique and Character* (1925); W. H. Sheldon (with the collaboration of S. S. Stevens and W. B. Tucker), *The Varieties of Temperament* (New York: Hafner, 1940); Walter Freeman, "Constitutional Factors in Mental Disorders," *Medical Annals of the District of Columbia* 5(1936): 287–336.

3. J. Breuer and S. Freud, *Studien über Hysterie* (Leipzig and Vienna: Franz Deuticke, 1895), translated by A. Brill (New York and Washington, D.C.: Nervous and Mental Disease Publishing Co., 1936).

4. S. Freud, *The Aetiology of Hysteria,* in Philip Rieff, ed., *Early Psychoanalytic Writings* (New York: Collier Books, 1963), pp. 177–87.

5. R. W. Clark, *Freud the Man and the Cause* (New York: Random House, 1980), p. 134.

6. S. Freud, "Psychoanalysis," *Encyclopaedia Brittanica,* 14th ed.

7. L. E. Hinsie, "The Treatment of Schizophrenia," *Psychiatric Quarterly* 3(1929): 5–39; L. E. Hinsie, *Concepts and Problems of Psychotherapy* (New York: Columbia University Press, 1937).

8. For an excellent account of this period, see G. N. Grob, *Mental Illness and American Society 1875–1940* (Princeton: Princeton University Press, 1983).

9. The title of the speech was changed when published; see E. C. Spitzka, "Reform of the Scientific Study of Psychiatry," *Journal of Nervous and Mental Disease* 5(1878): 201–29.

10. Ibid.

11. Ibid.

12. This dispute is described in B. E. Blustein, "A Hollow Square of Psychological Science: American Neurologists and Psychiatrists in Conflict," in A. Scull, ed., *Madhouses, Mad-Doctors, and Madmen: The Social History of Psychiatry in the Victorian Era* (Philadelphia: University of Pennsylvania Press, 1981), pp. 241–70.

13. *New York Times,* 14 November 1879, p. 3, col. 1.

14. H. Maudsley, *Responsibility in Mental Disease,* 2nd ed. (London: Kegan, Paul, 1874; New York: D. Appleton, 1874), p. 15.

15. J. C. Bucknell and D. H. Tuke, *A Manual of Psychological Medicine Containing the Nosology, Aetiology, Statistics, Description, Diagnosis, Pathology, and Treatment of Insanity,* 4th ed. (London: J. A. Churchill, 1879).

The numbers in brackets refer to the original complete citation of that reference in each chapter.

16. "The City Lunatic Asylum," letter to the editor, *New York Times,* 17 November 1879, p. 3, col. 4.

17. Editorial, *New York Times,* 18 November 1879, p. 4, col. 6.

18. S. W. Mitchell, "Address Before the Fiftieth Annual Meeting of the American Medico-Psychological Association," *Journal of Nervous and Mental Disease* 21(1894): 413–37.

19. W. Freeman, *The Psychiatrists* (New York: Grune & Stratton, 1968), p. 1. See also P. Starr, *The Social Transformation of American Medicine* (New York: Basic Books, 1982), p. 345.

20. Mitchell "Address" [18].

21. W. Channing, *American Journal of Insanity* 51(1894): 171–81.

22. A. Meyer, "A Short Sketch of the Problems of Psychiatry," *American Journal of Insanity* 53(1896–97).

23. I. S. Wechsler, *The Neuroses* (Philadelphia: W. B. Saunders, 1929); I. S. Wechsler, *A Textbook of Clinical Neurology* (Philadelphia: W. B. Saunders, 1927–53).

24. D. E. Denny-Brown, "The Changing Pattern of Neurological Medicine," *New England Journal of Medicine* 246(29 May 1952): 839–46.

25. H. Cushing, "Psychiatrists, Neurologists and the Neurosurgeon," *Yale Journal of Biology and Medicine* 7(1934–35): 191–207.

26. C. D. Aring, "The Place of Neurology in the Medical Firmament," *Journal of Medical Education* 21(1946): 220–22.

27. H. Cushing, "The Special Field of Neurological Surgery After Another Interval," *Archives of Neurology and Psychiatry* 4(1920): 603–37.

28. C. Pilcher, "Neurosurgery Comes of Age," *Journal of Neurosurgery* 5(1948): 507–13.

29. P. C. Bailey, "The Present State of American Neurology," *Journal of Neuropathology and Experimental Neurology* 1(1942): 111–17.

30. P. C. Bucy, "Surgical Neurology and Biology," *Journal of the American Medical Association* 115(1940): 261–63.

31. P. C. Bailey, "The Practice of Neurology in the United States of America," *Journal of Medical Education* 21(1946): 281–92.

32. Pilcher, "Neurosurgery Comes of Age" [28].

33. Cited in Bailey, "Practice of Neurology" [31].

34. Ibid.

35. H. A. Riley, "Training of the Neurologist: Neurologia irredenta," *Archives of Neurology and Psychiatry* 29(1933): 862–71.

36. C. P. Symonds, "The Neurological Approach to Mental Disorder," *Proceedings of the Royal Society of Medicine* 34(1941): 289–302.

37. Bailey, "Practice of Neurology" [31].

38. W. L. Russell, "The Presidential Address: The Place of the American Psychiatric Association in Modern Psychiatric Organization and Progress," *American Journal of Psychiatry* 12(1932): 1–18.

39. J. V. May, "Presidential Address: The Establishment of Psychiatric Standards by the Association," *American Journal of Psychiatry* 13(1933): 1–15.

40. W. A. White, "Proceedings of the American Psychiatric Association, May 30, 1933," *American Journal of Psychiatry* 13(1933): 1–15.

41. W. Freeman, F. C. Ebaugh, and D. A. Boyd, "The Founding of the American Board of Psychiatry and Neurology, Inc.," *American Journal of Psychiatry* 115(1959): 769–78.

42. S. Freud, *The Question of Lay Analysis,* translated by Nancy Procter-Gregg (New York: W. W. Norton, 1950).

Chapter 2. Bizarre Illnesses, Bizarre Treatment

1. C. Beers, *A Mind That Found Itself: An Autobiography* (New York: Longmans, Green, 1908).

2. *Nation* 86(1908): 265–66.

3. G. M. Beard, *A Practical Treatise on Nervous Exhaustion (Neurasthenia), Its Symptoms, Nature, Sequence, and Treatment* (New York: Wm. Wood, 1880), p. 115.

4. C. Rosenberg, "The Place of George M. Beard in Nineteenth Century Psychiatry," *Bulletin of Historical Medicine* 36(1962): 245–59.

5. G. M. Beard and A. D. Rockwell, *Medical and Surgical Uses of Electricity*, 8th ed. (New York: Wm. Wood, 1892).

6. M. Prince, *Nature of Mind and Human Automatisms* (Philadelphia: Lippincott, 1885).

7. M. Prince, "Neuroses," in H. R. Bigelow and G. B. Massey, eds. (Philadelphia: F. A. Davis, 1901), pp. D106–D150.

8. R. Hutchings, "Organization of a Physical Therapy Department in a State Hospital," *Psychiatric Quarterly* 3(1929): 203.

9. M. Rosenthal, *A Clinical Treatise on the Diseases of the Nervous System*, vol. II, translated by L. Putzel (New York: Wm. Wood, 1879), pp. 51–52.

10. S. W. Mitchell, *Fat and Blood*, 3rd. ed. rev. (Philadelphia: Lippincott, 1884).

11. G. M. Beard, "The Influence of the Mind in the Causation and Cure of Disease—The Potency of Definite Expectation," *Journal of Nervous and Mental Diseases* 4(1877) 429–34, and discussion that follows.

12. North Carolina Charitable, Penal, and Correctional Institutions Biennial Report 1930–1932, p. 48.

13. New York State Department of Mental Hygiene, annual report, 1939–40, p. 52; H. M. Pollock, "Trends in the Outcome of General Paresis," *Psychiatric Quarterly* 9(1935): 194–211; E. M. Furbush, "General Paralysis in State Hospitals for Mental Disease," *Mental Hygiene* 7(1923): 565–78.

14. M. Nonne, *Syphilis und Nervensystem* (Berlin, 1902).

15. T. H. Kellogg, *A Textbook of Mental Diseases* (New York, 1897), p. 657.

16. F. Nissl, quoted by A. Meyer, "A Few New Trends in Modern Psychiatry," *Psychological Bulletin* 1(1904): 217.

17. R. von Krafft-Ebing, *Die Aetiologie der progressiven Paralyse*, vol. II (1897), p. 657.

18. J. W. Moore, "The Syphilis-General Paralysis Question," *Review of Neurology and Psychiatry* 8(1910): 259–71.

19. A. von Wasserman, Neisser, and Bruck, "Ein serodiagnostische Reaktion bei Syphilis," *Dent. Med. Wochenschr.* 32(1906): 745.

20. J. Wagner-Jauregg, "The Treatment of General Paresis by Inoculation of Malaria," *Journal of Nervous and Mental Disease* 55(1922): 369–75.

21. Cited in E. Jones, *The Life and Work of Sigmund Freud*, L. Trilling and S. Marcus, eds. (New York: Basic Books, 1961), p. 394.

22. W. Breuetsch, "Julius Wagner-Jauregg, M.D., Eminent Psychiatrist and Originator of the Malaria Treatment of Dementia Paralytica," *Archives of Neurology and Psychiatry* 44(1940): 1319–22.

23. Ibid.

24. A. E. Bennett, "Evaluation of Artificial Fever Therapy for Neuropsychiatric Disorders," *Archives of Neurology and Psychiatry* 40(1938): 1141–55; C. A. Neyman, *Artificial Fever* (Springfield, Ill.: Charles C Thomas, 1938).

25. S. E. Jelliffe and W. A. White, *Diseases of the Nervous System. A Textbook of Neurology*, 4th ed. rev. (Philadelphia: Lea & Febriger, 1923).

26. K. Menninger, *The Human Mind* (New York: Alfred A. Knopf, 1945), pp. 42–47.

27. B. Thom, "Tertiary Syphilis Psychosis Other Than Paresis," *American Journal of Insanity* 78(1921): 503.

28. A. P. Noyes, *Modern Clinical Psychiatry* (Philadelphia: W. B. Saunders, 1934; subsequent revised editions, 1939 and 1949); E. A. Strecker, F. G. Ebaugh, and J. R. Ewalt, *Practical Clinical Psychiatry* (Philadelphia: Blakiston, 1st. ed. 1925, 6th ed. 1947); see also L. E. Hinsie, *Concepts and Problems of Psychotherapy* (New York: Columbia University Press, 1937). Quote from Strecker, Ebaugh, and Ewalt.

29. A. Meyer, *Psychobiology. A Science of Man,* Eunice E. Winters and Anna M. Bowers, eds. (Springfield, Ill.: Charles C Thomas, 1957), p. 157.

30. M. Partridge, *Prefrontal Leucotomy: A Survey of 300 Cases Followed over 1½–3 Years* (Springfield, Ill.: Charles C Thomas, 1950).

31. L. Berman, *The Glands Regulating Personality* (New York: Macmillan, 1921; 2nd ed., 1935); C. W. Sawyer, "The Effect of a Previously Performed Thyroid Operation on Involutional Psychosis," *Ohio State Medical Journal* 35(1939): 848–49. Also, many books on depression and endocrine function have been published in the 1980s.

32. W. Freeman, "Personality and the Endocrines: A Study Based upon 1400 Quantitative Necropsies," *Annals of Internal Medicine* 9(1935): 444–50.

33. J. Klaesi, "Uber die therapeutische Andwendung des Daverschafes mittles Somnifen bei Schizophrenen." *Ztschr. f. d. ges Neurol. u Psychiat.* 74(1922): 557; for a brief history of sleep therapy, see H. D. Palmer and A. L. Paine, "Prolonged Narcosis as Therapy in the Psychoses," *American Journal of Psychiatry* 12(1932): 143–64.

34. W. J. Bleckwenn, "Production of Sleep and Rest in Psychotic Cases," *Archives of Neurology and Psychiatry* 24(1930): 365–72; H. D. Palmer and F. J. Braceland, "Six Years Experience with Narcosis Therapy in Psychiatry," *American Journal of Psychiatry* 94(1934): 37–57.

35. See J. Wortis, *Soviet Psychiatry* (Baltimore: Williams & Wilkins, 1950).

36. A. S. Loevenhart, W. F. Lorenz, and R. M. Waters, "Cerebral Stimulation," *Journal of the American Medical Association* 92(1929): 880–83.

37. W. J. Bleckwenn, "The Use of 'Sodium Amytal' in Catatonia," *Research Publication of the Association of Nervous and Mental Diseases* 10(1931): 224–29.

38. S. E. Jelliffe, discussion of Bleckwenn, "Use of 'Sodium Amytal' " [37].

39. "Big Meeting." *Time,* 22 January 1931, p. 29.

40. A. Meyer, cited in R. S. Carroll, "Asceptic Meningitis in Combatting the Dementia Praecox Problem," *New York State Journal of Medicine* 118(1923): 409–11.

41. Carroll, "Asceptic Meningitis" [40]; H. Lundvall, "Blood Changes in Dementia Praecox and Artificial Leucocytosis in Its Treatment," *American Journal of Clinical Medicine* 22(1915): 115.

42. J. H. Talbott and K. J. Tillotson, "The Effects of Cold on Mental Disorders," *Diseases of the Nervous System* 2(1941): 116–26.

43. H. A. Cotton, "The Etiology and Treatment of the So-Called Functional Psychoses. Summary of the Results Based on the Experience of Four Years," *American Journal of Psychiatry* 2(1922): 157–210.

44. Ibid.

45. Ibid.

46. Ibid.

47. Ibid.

48. Ibid.

49. A. Meyer, "Foreword to H. A. Cotton," *The Defective, Delinquent and Insane: The Relation of Focal Infection to Their Causation, Treatment and Prevention* (Princeton: Princeton University Press, 1921).

50. Cotton, "Etiology and Treatment" [43].

51. Ibid.

52. Ibid.

53. N. Kopeloff and G. H. Kirby, "Focal Infection and Mental Disease," *American Journal of Psychiatry* 3(1923): 149–97.

54. H. A. Cotton, "The Physical Causes of Mental Disorders," *American Mercury* 29(1933): 221–25.

55. H. L. Mencken, *Prejudices,* 2nd series (New York: Alfred A. Knopf, 1920), p. 166.

56. Cotton, "Physical Causes of Mental Disorders" [54].

57. A. Meyer, "Henry A. Cotton. In Memoriam," *American Journal of Physiology* 90(part 2) [1934]: 921–23; E. Strecker, "Henry Andrews Cotton, M.D.," *Transactions of the American Neurological Association* (1934): 218; see also *Journal of Nervous and Mental Disease* 78(1933): 579–80.

58. E. Lamphear, "Lectures on Intracranial Surgery. XI. The Surgical Treatment of Insanity," *Journal of the American Medical Association* 24(1895): 883–86.

59. Ibid.

60. R. Semelaigne, Au. Med. Psych., 1895, cited by L. Puusepp, "Acune considerazioni sugli interventi chirurgici nelle malattie méntali," *Giornale della Accademia di Medicina di Torino* 8(1937): 3–16.

61. See E. Valenstein, "Historical Perspective," in E. Valenstein, ed., *The Psychosurgery Debate* (San Francisco: W. H. Freeman, 1980), pp. 14–19, for a brief history of precursors of psychosurgery.

62. G. Burckhardt, "Über Rindenexcisionen, als Beitrag zur Operativen Therapie der Psychosen," *Allegemaine Zeitschrift für Psychiatrie* 47(1891): 463–548.

63. Puusepp, "Acune considerazioni" [60].

Chapter 3. "Anything That Holds Out Hope Should Be Tried"

1. Some of Sakel's biographical material is described in J. Wortis, "In Memoriam, Manfred Sakel, M.D. 1900–1957," *American Journal of Psychiatry* 115(1958–59): 287–88.

2. M. Sakel, "Neue Behandlungsmethode der Morphensucht," *Deutsch Medizinische Wochenschrift* 56(1930): 1777–78.

3. M. Sakel, "Schizophreniebehandlung mittels Insulin-Hypoglykamie sowie hypoglykamischer Schocks," *Wiener Medizinische Wochenschrift* 84(1934): 1211, et seq. to 85(1935): 179; M. Sakel, *Neue Behandlung der Schizophrenia* (Vienna: Moritz Perles, 1935).

4. J. Wortis, "The History of Insulin Shock Treatment," in *Insulin Treatment in Psychiatry,* ed. M. Rinkel and H. Himwich (New York: Philosophical Library, 1959), pp. 19–44.

5. B. Glueck, "The Hypoglycemic State in the Treatment of Schizophrenia," *Journal of the American Medical Association* 107(1936): 1029–31.

6. D. Cameron, "Greetings," in *Insulin Treatment,* ed. Rinkel and Himwich, pp. xxiii–xxvi [4].

7. M. Sakel, "The Pharmacological Shock Treatment of Schizophrenia," with a foreword by Professor Otto Potzl, authorized translation by Joseph Wortis, *Nervous Mental Disease Monograph,* no. 62(1938); see also J. Wortis, "Early Experiences with Sakel's Hypoglycemia Insulin Treatment of the Psychoses in America," *American Journal of Psychiatry* 94(suppl. [1938]): 307.

8. N. Lewis, quoted in Sakel, "Pharmacological Shock Treatment" [7].

9. F. Kennedy, quoted in Sakel, "Pharmacological Shock Treatment" [7].

10. L. Meduna, quoted in M. Fink, "Meduna and the Origins of Convulsive Therapy," *American Journal of Psychiatry* 141(1984): 1034–41. An autobiography of Meduna

has been published in *Convulsive Therapy* (New York: Raven Press), 1(1985): 43–57, and 2(1985): 121–35.

11. W. Oliver, "Account of the Effects of Camphor in a Case of Insanity," *London Medical Journal* 6(1785): 120–30.

12. L. Meduna, quoted in Fink, "Meduna and the Origins of Convulsive Therapy" [10].

13. L. Meduna, "Versuche über die biologische Beeinflussung des Abaufbaues der Schizophrenia. I. Camphor und Cardiazol Krampfe," *Z. Ges. Neurol. Psychiat.* 152(1935): 235–62.

14. N. Redenour, *Mental Health in the United States: A Fifty-Year History* (Cambridge, Mass.: Harvard University Press, 1961), p. 32.

15. L. Meduna, *Carbon Dioxide Therapy* (Springfield, Ill.: Charles C Thomas, 1950); L. Meduna, "The Convulsive Treatment: A Reappraisal," *Journal of Clinical and Experimental Psychopathology* 15(1954): 219–33.

16. P. Verstraeten, "La thérapeutique convulsivante de la psychose maniaco-depressive," *Am. Med.-Psychol.* 95 II(1937): 654.

17. J. Nyiro, "Beitrag zur Wirking der Krampftherapie der Schizophrenie," *Schweiz Arch. f. Neurol. u. Psychiat.* 40(1937): 180.

18. See H. E. Himwich, F. A. D. Alexander, and B. Lipetz, "Effects of Acute Anoxia by Breathing Nitrogen on the Course of Schizophrenia," *Proceedings of the Society for Experimental Biology and Medicine* 39(1938): 367. For other chemical ways of inducing convulsions used as psychiatric treatment, see L. B. Kalinowsky and P. H. Hoch, *Shock Treatments* (New York: Grune & Stratton, 1949), pp. 102–4.

19. U. Cerletti, "Electroshock Therapy," in F. Martí-Ibáñez, A. A. Sackler, M. D. Sackler, and R. R. Sackler, eds., *The Psychodynamic Therapies in Psychiatry* (New York: Hoeber-Harper, 1956), pp. 258–70.

20. L. Kolb and V. Vogel, "The Use of Shock Therapy in 305 Mental Hospitals," *American Journal of Psychiatry* 99(1942): 90–100.

21. "Insulin for Insanity," *Time*, 25 January 1937, pp. 26, 28; "Death for Sanity," *Time*, 20 November 1939, pp. 39–40; "Shock vs. Insanity," *Newsweek* 12(22 August 1938): 23–24; "Bedside Miracle," *Reader's Digest* 35(November 1939): 73–75. See also *New Republic* 91(1937); *Scientific American* 157(1937); *Science News Letter* 31(1937), 33(1938), 34(1938), 35(1939), 36(1939); *Forum and Century* 99(1938); *Hygeia* 18(1940); and *Science Digest* 8(1940).

22. M. Sakel, "Origin and Nature of Hypoglycemic Therapy of the Psychoses," and discussion, *Archives of Neurology and Psychiatry* 38(1937): 188–203.

23. See discussion following Sakel, "Origin and Nature" [22].

24. S. E. Jelliffe, quoted in Sakel, "Origin and Nature" [22].

25. A. Meyer, quoted in Sakel, "Origin and Nature" [22].

26. A. A. Brill, quoted in Sakel, "Origin and Nature" [22].

27. A. A. Brill to W. A. White, 31 October 1936, cited in G. N. Grob, *Mental Illness and American Society 1875–1940* (Princeton: Princeton University Press, 1983).

28. W. Overholser to C. M. Hinks, 27 October 1936; White to Hinks, 27 October 1936, letters cited in Grob, *Mental Illness and American Society* [27].

29. Sakel, quoted in Wortis, "In Memoriam, Manfred Sakel, M.D." [1].

30. B. Malzberg, "Outcome of Insulin Treatment of One Thousand Patients with Dementia Praecox," *Psychiatric Quarterly* 12(1938): 528–53.

31. B. Malzberg, "A Follow-up Study of Patients with Dementia Praecox Treated with Insulin in the New York Civil State Hospitals," *Mental Hygiene* 23(1939): 641–51.

32. "Death for Sanity," *Time*, 20 November 1939, pp. 39–40.

33. E. D. Bond and J. T. Shurley, *American Journal of Psychiatry* 103(1946): 338–41.

34. W. Mayer-Gross, E. Slater, and M. Roth, *Clinical Psychiatry*, 2nd ed. (London: Cassell, 1960), p. 245.

35. The decline of interest in insulin therapy is charted in S. J. Rachman and G. T. Wilson, *The Effects of Psychological Therapy*, 2nd ed. (New York: Pergamon, 1980), pp. 13–14.

36. Wortis, "In Memoriam, Manfred Sakel, M.D." [1].

37. Premier Congrès de Psychiatrie, Paris, 1950, *Comptes Rendus des Séances* (Paris: Hermann & Cie, 1952).

38. M. Sakel, "Insulin Therapy and Shock Therapies: Ascent of Psychiatry from Scholastic Dialecticism to Empirical Medicine," in *Comptes Rendus des Séances* [37].

39. See comments made at the Swiss Psychiatric Association meeting (30 May 1937) in H. Steck and H. Bovet, "The Results Obtained by Insulin Therapy in Cery from 1929 to 1937," *American Journal of Psychiatry* 94(suppl. [1938]): 166–67. For other uses of insulin in psychiatry prior to Sakel's work, see also Wortis in Rinkel and Himwich, eds., *Insulin Treatment* [4], pp. 19–41.

40. J. Meduna, *American Journal of Psychiatry* 94(suppl. [1938]): 41.

41. M. Sakel, "A Reappraisal," *Journal of Clinical and Experimental Psychopathology* and the *Quarterly Review of Psychiatry and Neurology* 15(1954): 255–316.

42. *New York Times*, 3 December 1957, p. 35.

43. Wortis, "In Memoriam, Manfred Sakel, M.D." [1].

44. *New York Times*, 3 December 1957, p. 35.

45. M. Sakel, *Epilepsy* (New York: Philosophical Library, 1958); M. Sakel, *Schizophrenia* (New York: Philosophical Library, 1958).

46. *Journal of Nervous and Mental Disease* 129(1959): 505–6.

47. L. B. Kalinowsky and P. H. Hoch, *Shock Treatments and Other Somatic Procedures in Psychiatry* (New York: Grune & Stratton, 1949), p. 1.

48. B. Rush, *Medical Inquiries and Observations, Upon the Diseases of the Mind* (Philadelphia: Kimber & Richardson, 1812), p. 181.

49. A. W. Stearns, "Report on Medical Progress: Psychiatry," *New England Journal of Medicine* 220(1938): 709–10.

Chapter 4. A Portuguese Explorer: Egas Moniz

1. E. Moniz, *A Vida Sexual (Fisiologia)*, 1901; *A Vida Sexual (Patologia)*, 1902, 1906; *A Vida Sexual (Fisiologia e Patologia)* (19 editions) (Lisbon: Fereira e Oliveira, 1913–33), later ed. (Lisbon: Casa Ventura Abrantes).

2. E. Moniz, *A Neurologia Na Guerra* (Lisbon: Liviaria Ferreira, 1917).

3. E. Moniz, *Um Ano de Politica* (Lisbon: Brazil Editora, 1920; Rio de Janeiro: Companhia Editora Americana, Livaria Francisco Alves, 1919).

4. E. Moniz, *Clinica Neurologica* (Lisbon: Publicacao da Faculdade de Medicina no Primeiro Centeriario da Regia Escola de Chirurgia de Lisboa, 1925).

5. E. Moniz, *O Padre Faria Na Historia do Hipnotismo* (Lisbon: Publicacao da Faculdade de Medicina, 1925).

6. E. Moniz, ed., *Júlio Dinis E. A. Sua Obra*. 2 vols. (Lisbon: Casa Editora Ventura Abrantes, 1924).

7. E. Moniz, *O Papa Joao, XXI* (Lisbon: Casa Ventura Abrantes, 1929).

8. E. Moniz, *Maurico de Almeida-Escultor (1897–1923) Arquivo do Distrito de Aveiro* (1943).

9. E. Moniz, *Historia das Cartas de Jogar*, Primeira parte do vol. Dr. Jose Henriques da Silva—*Tratado jo Jogo do Boston* (Lisbon: Editorial Atica, 1942).

10. J. Berberich and S. Hirsch, "Die röntgenographische Darstellung der Arterien

um Venen am lebenden Menschen," *Klin. Wschr.* 22(1923): 2226–28.

11. B. Brooks, "Intra-arterial Injection of Sodium Iodide. Preliminary Report," *Journal of the American Medical Association* 82(1924): 1016–19.

12. J. Sicard and J. Forestier, "Méthode radiographique d'exploration de la cavité epidurale par le lipiodol," *Revue Neurologique* (Paris) 28(1921): 1264–66.

13. A. von Knauer, *Munchen med. Wochenschr.* 66(1919): 609.

14. A good account of the early history of neuroradiology was written by J. W. D. Brull, in F. C. Rose and W. F. Bynum, eds., *Historical Aspects of the Neurosciences* (New York: Raven Press, 1982), pp. 255–64. See also A. R. Damasio, "Egas Moniz, Pioneer of Angiography and Leucotomy," *Mt. Sinai Journal of Medicine* 42(1975): 502–13; J. L. Autunes, "Egas Moniz and Cerebral Angiography," *Journal of Neurosurgery* 40(1974): 427–32.

15. E. Haschek and O. T. Lindenthal, "Ein Beitrag zur praktischen Verwerthung der Photographie nach Röntgen," *Wien. Klin. Wschr.* 9(1896): 63–64.

16. E. Moniz, "L'Encephalographie arterielle, son importance dans la localisation des tumeurs cérébrales," *Revue Neurologique* 2(1927): 272.

17. Moniz described these events in his book *Confidências de um Investigador Ciêntífico* (Lisbon: Livraria Ática, 1949). Translated from the Portuguese.

18. "Radiography of Blood-Vessels," *Lancet,* 31 January 1931, p. 249; "Cerebral Angiography," *Lancet,* 18 November 1933, pp. 1157–58.

19. W. Lohr and W. Jacobi, "Die kombinierte Encephalo-Arteriographie," *Archiv. für Klinische Chirurgie* 173(1932): 399–420.

20. Moniz, *Confidências* [17], p. 159.

21. Ibid.

22. M. Saito, K. Kamikawa, and H. Yanagizawa, "A New Method of Blood Vessel Visualization (Arteriography, Veinography; Angiography) in Vivo," *American Journal of Surgery* 10(1930): 225–40.

23. Moniz, *Confidências* [17].

24. E. Moniz, *Diagnostic des tumeurs cérébrales et épreuve de l'encéphalographie artérielle. Préface de Monsieur le Docteur J. Babinski* (Paris: Masson, 1931).

25. E. Moniz, *L'Angiographie cérébrale ses applications et résultats en anatomie, physiologie et clinique* (Paris: Masson, 1934); *Clinica delle angiografia cerebrale* (Turin: I.T.E.R., 1938); *Die Cerebrale Arteriographie und Phlebographie* (Berlin: Jules Springer, 1940); *Trombosis y otras obstruciones de las carotidas* (Barcelona: Salvat, 1941).

26. Records of the Medical Nobel Archives (1933).

27. J. W. D. Bull, "The History of Neuroradiography," in F. C. Rose and W. F. Bynum, eds., *Historical Aspects of the Neurosciences* (New York: Raven Press, 1982), pp. 255–64.

28. W. Freeman to J. Fulton, correspondence file, Yale University Library.

29. J. F. Fulton, *Functional Localization in Relation to Frontal Lobotomy* (New York and London: Oxford University Press, 1949), pp. 63–64.

30. W. Freeman, unpublished "History of Psychosurgery," Himmelfarb Health Science Library, George Washington University.

Chapter 5. The Emperor's New Clothes: Moniz's Theoretical Justification
for Psychosurgery

1. Letter to Barahona Fernandes, cited in F. R. Perino, "Egas Moniz. Founder of Psychosurgery, Creator of Angiography. Nobel Prize Winner (1874–1955)," *Journal of the International College of Surgery* 36(1961): 261–71.

2. E. Moniz, "Essai d'un traitement chirurgical de certaines psychoses," *Bulletin de l'Académie de Médicine* 115(1936): 385–92; English translation in *Journal of Neurosurgery* 21(1964): 1108–14.

3. Moniz, "Essai d'un traitement" [2].

4. E. Moniz, "How I Came to Perform Prefrontal Leucotomy," *Proceedings of the First International Congress of Psychosurgery* (Lisbon: 1948).

5. E. Moniz, *Confidências de um Investigador Cíêntífico* (Lisbon: Livraria Atica, 1949). Translated from the Portuguese.

6. P. Flourens, *Recherches expérimentales sur les propiétés et les functions du système nerveux dans les animaux vertébrés* (Paris: Crevot, 1824).

7. International Medical Congress, 7th session held in London, 2–9 August 1881. Transactions, prepared by Sir William Mac Cormac, assisted by Henry Makins and the undersecretaries of the sections (London, 1881): "The International Medical Congress," *British Medical Journal*, 6 August 1881.

8. E. Anderson and W. Haymaker, "Friedrich Leopold Goltz (1834–1902)," in W. Haymaker, ed., *The Founders of Neurology* (Springfield, Ill.: Charles C Thomas, 1953).

9. International Medical Congress [7].

10. Ibid.

11. W. W. Keen, "Vivisection and Brain Surgery," *Harper's New Monthly Magazine* 87(1893): 128–39; see also B. Jennett, "Pioneering Craniospinal Surgery," in P. I. Bucy, *Neurosurgical Giants: Feet of Clay and Iron* (New York: Elsevier, 1985), pp. 35–37.

12. C. S. Sherrington, "Sir David Ferrier," in *Dictionary of National Biography*, L. Stephen and S. Lee, eds. (London, 1908–9; reprinted 1938); D. McK. Rioch, "Sir David Ferrier," in Haymaker, ed., *Founders* [8].

13. E. Hitzig, 1894, quoted by L. Bianchi, *The Mechanism of the Brain and the Function of the Frontal Lobes*, J. H. MacDonald, trans. (Edinburgh: E. and S. Livingston, 1922), p. 348.

14. P. P. Broca, *Bulletin de la Société d'Anthropologie* (Paris) 2(1861): 301–21.

15. D. Ferrier, *The Function of the Brain* (London: Smith Elder, 1876), pp. 231–32.

16. Bianchi, *Mechanism of the Brain* [13].

17. An excellent historical review of the frontal-lobe literature can be found in G. Rylander, *Personality Changes after Operation on the Frontal Lobes* (London: Oxford University Press, 1939).

18. J. M. Harlow, "Recovery from the Passage of an Iron Bar through the Head," *Massachusetts Medical Society* 2(1868): 327–46. The original report of the accident was described by Harlow in a letter published in the *Boston Medical and Surgical Journal* (now the *New England Journal of Medicine*) 39(13 December 1848): 389–92.

19. Bianchi, *Mechanism of the Brain* [13].

20. E. Feuchtwanger's 1923 monograph of 400 soldiers with gunshot wounds (200 frontal and 200 nonfrontal) as summarized by H.-L. Teuber, "The Riddle of Frontal Lobe Function in Man," in J. M. Warren and K. Akert, eds., *The Frontal Granular Cortex and Behavior* (New York: McGraw-Hill, 1964), pp. 415–16, or in the original German publication: E. Feuchtwanger, *Die Funktionen des Stirnhirns, ihre Pathologie und Psychologie* (Berlin: J. Springer, 1923).

21. K. Kleist, "Die Störungen der Ichleistungen und ihre Lokalisation," *Mon. Schr. f. Psychiat. u. Neurol.* 79(1931).

22. P. Bailey, *Intracranial Tumors* (London: Tindall, 1933).

23. E. Moniz, "Essai d'un traitement chirurgical de certaines psychoses," *Bulletin de l'Académie de Médicine* 115(1936): 385–92.

24. A. Lima, "Egas Moniz 1874–1955," *Journal of Neurology* (Springer-Verlag) 207(1974): 167–70 (German). See also A. Lima, "Egas Moniz 1874–1955," *Surgical Neurology* 1(1973): 247–48.

25. Moniz, "Essai d'un traitement" [2].

26. A full description of Joe A may be found in R. M. Brickner, *The Intellectual Functions of the Frontal Lobes* (New York: Macmillan, 1936). See also R. M. Brickner, "An Interpretation of Function Based on the Study of a Case of Bilateral Frontal Lobectomy," *Proceedings of the Association for Research in Nervous and Mental Disorders* 13(1932): 259–351.

27. Moniz, "Essai d'un traitement" [2].

28. W. F. Freeman, "History of Psychosurgery," unpublished manuscript, Himmelfarb Health Science Library, George Washington School of Medicine.

29. S. S. Ackerly, "Instinctive, Emotional and Mental Changes Following Prefrontal Lobe Extirpation," *American Journal of Psychiatry* 92(1935): 717–29.

30. W. C. Halstead, *Brain and Intelligence: A Quantitative Study of the Frontal Lobes* (Chicago: University of Chicago Press, 1947), p. 141 and case 50 (appendix A). For a description of a patient much of whose prefrontal area was destroyed during early childhood, see S. S. Ackerly, "A Case of Paranatal Bilateral Frontal Lobe Defect Observed for Thirty Years," in J. M. Warren and K. Akert, eds., *The Frontal Granular Cortex and Behavior* (New York: McGraw-Hill, 1964), pp. 192–218.

31. W. Penfield and J. Evans, "The Frontal Lobe in Man: A Clinical Study of Maximum Removals," *Brain* 58(1935): 115–33.

32. A. Damasio, "Egas Moniz, Pioneer of Angiography and Leucotomy," *Mt. Sinai Journal of Medicine* 42(1975): 502–13.

33. B. Fernandes, *Egas Moniz, pioneiro de descobrimentos medicos* (Lisbon: Instituto de Cultura e Lingua Portuguesa, Ministerio da Edcacao, 1983).

34. C. F. Jacobsen, "Studies on Cerebral Function in Primates," *Comparative Psychological Monographs* 13 (3) (1936): 1–60.

35. M. P. Crawford, J. F. Fulton, C. F. Jacobsen, and J. B. Wolf, "Frontal Lobe Ablation in Chimpanzee: A Resume of 'Becky' and 'Lucy'," in *The Frontal Lobes*, Research Publications, Association for Research in Nervous and Mental Disease, vol. 27 (Baltimore: Williams & Wilkins, 1948), pp. 3–58.

36. Moniz, "Essai d'un traitement" [2].

37. S. Ramón y Cajal, *Degeneration and Regeneration of the Nervous System*, trans. by R. M. May (London: Oxford University Press, 1928), p. 750.

38. S. Finger and D. G. Stein, *Brain Damage and Recovery: Research and Clinical Perspectives* (New York: Academic Press, 1982).

39. E. Moniz, *Confidências* [5], p. 316. Translated from the Portuguese.

40. Moniz, "Essai d'un traitement" [2].

41. K. Lashley, *Brain Mechanisms and Intelligence* (Chicago: University of Chicago Press, 1929).

42. Comment made during the discussion of E. Moniz and D. Furtado, "Essai de traitement de la schizophrénie par la leucotomie préfrontale," *Annals Médico-Psychologiques* 95(1937): 298.

43. E. Moniz to W. Freeman, 9 July 1946; a copy of this letter was sent by Freeman to John Fulton and is now in the latter's correspondence file in the Yale University Library.

44. A. Flores, "Opening Remarks," in *Proceedings of the First International Congress of Psychosurgery* (Lisbon: 1948).

45. M. W. Govindaswamy and R. Balakrishna, "Bilateral Prefrontal Leucotomy in Indian Patients," *Lancet* 1(8 April 1944): 466–68.

46. For example, W. Freeman, *The Psychiatrists: Personalities and Patterns* (New York: Grune & Stratton, 1968), p. 54.

47. A. Amaral, *O tratamento cirurgico das doencas mentais* (Lisbon: Livraria Luso-espanhola, 1944).

Chapter 6. "Seven Recoveries, Seven Improvements, and Six Unchanged"

1. Statement included as part of Sobral Cid's discussion of the paper presented in Paris by Diogo Furtado and Egas Moniz, 7 July 1937; published in *Annales Médico-Psychologiques* 95(1937): 298.

2. E. Moniz, cited in A. Lima, "Egas Moniz 1874–1955," *Journal of Neurology* (Springer-Verlag) 207(1974): 167–70.

3. E. Moniz, *Confidências de um Investigador Ciêntífico* (Lisbon: Livraria Ática, 1949). Translated from the Portuguese.

4. E. Moniz, "Essai d'un traitement chirurgical de certaines psychoses," *Bulletin de l'Académie de Médecine* (Paris) 115(1936): 385–92.

5. E. Moniz to W. Freeman, 9 July 1946, in the commemorative *Centenario de Egas Moniz*, vol. 11 (Lisbon, 1978), pp. 428–45.

6. E. Moniz, cited in B. Fernandes, "Egas Moniz and the Pre-frontal Leucotomy," *Anais Portugueses de Psiquiatria* 11(1959): 190–205.

7. A. T. Rasmussen, *Some Trends in Neuroanatomy* (Dubuque, Iowa: W. C. Brown, 1947), p. 78.

8. E. Moniz, "Prefrontal Leucotomy in the Treatment of Mental Disorders," *American Journal of Psychiatry* 93(1937): 1379–85.

9. Sobral Cid, quoted by E. Moniz, letter to W. Freeman, 9 July 1946 [5].

10. E. Moniz to W. Freeman, 9 July 1946 [5].

11. W. Freeman, review of "Tentatives Opératoires dans le Traitement de Certaines Psychoses," *Archives of Neurology and Psychiatry* 36(1936): 1413. The review was published unsigned, but a letter by Freeman to John Fulton, dated 6 August 1936 (Yale University Library, History of Medicine Archives), clearly indicates the review was written by Freeman (see chapter 4, p. 112).

12. W. Freeman to J. Fulton, 6 August 1936 [11].

13. J. Fulton to W. Freeman, 12 September 1936, Yale University Library, History of Medicine Archives.

14. Moniz, "Essai d'un traitement" [4].

15. E. Moniz and A. Lima, "Premiers essais de psychochirurgie—Technique et résultats," *Lisboa Médica* 13(1936): 152.

16. E. Moniz, "Prefrontal Leucotomy in the Treatment of Mental Disorders," *American Journal of Psychiatry* 93(1937): 1379–85; E. Moniz, "I principie fisiopatologici della psicochirurgia," *Giornale di Psiquiatria e di Neuropatologia* 43(1937): 360; E. Moniz, "La technique psycho-chirurgicale," *Archives Franco-Belges de Chirurgie* (1937).

17. E. Moniz, "Psychochirurgie," *Der Nervenarzt* 10(1937): 113.

18. W. Freeman, *The Psychiatrists: Personalities and Patterns* (New York: Grune & Stratton, 1968), p. 55.

19. S. Cobb, "Review of Neuropsychiatry for 1940," *Archives of International Medicine* 66(1940): 1341–54.

20. S. Cid, discussion following E. Moniz and D. Furtado, "Essai de traitement de la Schizophrénie par la leucotomie frontale," *Ann. Med. Psychol.* 95, pt. 2 (July 1937): 298.

21. Moniz, "Prefrontal Leucotomy" [16].

22. Moniz to Freeman, 9 July 1946 [5].

23. E. Moniz, "A memoria do professor Sobral Cid," *Imprensa Medica* 8(1941): 3–5.

24. Freeman, *The Psychiatrists* [18].

25. Moniz, "Memoria" [23].

26. Almeida Lima to W. Freeman (n.d.), cited in Freeman, *The Psychiatrists* [18].

27. D. Furtado, et al., "Personality Changes after Lobotomy," *Psychosurgery. First International Conference* (Lisbon, 1949), pp. 35–49.

28. G. Rylander, "Personality Analysis Before and After Frontal Lobotomy," *Arch. Res. in Nervous and Mental Diseases* 27(1947): 691–705.

29. Furtado, et al., "Personality Changes" [27].

30. D. Furtado, "Results of Leucotomy. A Twelve Year Follow-up," *Psychosurgery. First International Conference* (London, 1949), pp. 171–72.

31. E. Lamphear, "Lectures on Intracranial Surgery. XI. The Surgical Treatment of Insanity," *Journal of the American Medical Association* 24(1895): 883–86.

32. M. Ducosté, "L'Impaludation cérébrale," *Bulletin de l'Académie de Médecine* 107(12 April 1932): 516–18.

33. E. Moniz, "L'Arterio-phliborgraphi comme moyen de determiner la vitesse de la circulation du cerveau, des meninges et des parties molles du crane," *Bulletin de l'Académie de Médecine* 107(12 April 1932): 516–18.

34. E. Mariotti and I. Sciuti, "La Terapia Intracerebrale nelle malattie mentali. Contribute alla neurochirurgia della psicosi," *Rivista di Neurologia* (April 1937), XXI Congresso Della *Societa Italiana di Psichiatria*, pp. 238–39.

35. A. M. Dogliotti, "Ventriculographia directa por la vie transorbitale," *Bollettino e mem. della Societa piemontese di chirurgia* 3(1933): 73–84.

36. A. M. Fiamberti, "Proposta di una technica operatoria modificata e semplificata per gli interventi alla Moniz sui lobi prefrontali in malati di mente," *Rassegna di studi psichiat.* 26 (2) [1937]: 797.

37. H. Chavastelon, cited in Mariotti and Sciuti, "La Terapia Intracerebrale" [34].

38. H. Baruk, *Patients Are People Like Us: The Experience of Half a Century in Neuropsychiatry* (New York: William Morrow, 1978), p. 200.

39. W. Mayo, quoted in H. Woltman et al., Proceedings of Staff Meetings, Mayo Clinic (26 March 1941), vol. 16, p. 200.

40. F. Ody, "Le Traitement de la demence precoce par resection du lobe prefrontal," *Archivo Italiano di Chirurgia* 53(1938): 321–30.

41. D. Bagdasar and J. Constantinesco, "L'Operation d'Egas Moniz dans le traitement de certaines psychoses," *Bulletin et Mémoires de la Société Medical des Hospitaux de Buchareste* 3(1937): 78–84.

42. F. Morel, "The Surgical Treatment of Dementia Praecox," *American Journal of Psychiatry* 94(1938): 309–14.

43. Ody, "Le Traitement" [40].

44. Ibid.

45. Ibid.

46. E. Zamiatin, *We* (New York: Dalton, 1924). Translated by Gregory Zilboorg.

47. E. Zamiatin, *We* (London: Jonathan Cape, 1970). Translated by Bernard G. Guerney.

Chapter 7. "The Cat That Walks by Himself"

1. S. W. Mitchell, G. M. Morehouse, and W. W. Keen, *Gunshot Wounds and Other Injuries of Nerves* (Philadelphia: J. B. Lippincott, 1864).

2. W. W. Keen, "Vivisection and Brain-Surgery," *Harper's New Monthly Magazine* 87(1893): 128–39.

3. W. J. Taylor, "Memoir of William Williams Keen, M.D.," *Transactions of the College of Physicians of Philadelphia* 1(ser. 4 [1934]): 62–68; W. W. Keen, "The Surgical Operations of President Cleveland in 1893," in A. Scott Earle, ed., *Surgery in America*, 2nd ed. (New York: Praeger, 1983), pp. 340–50.

4. N. W. Winkelman, "Observations on the Histopathology of Schizophrenia," *American Journal of Psychiatry* 105(1949): 889–96. Later Winkelman recommended transorbital lobotomies; see M. T. Moore and N. W. Winkelman, "Some Experiences with Transorbital Leucotomy: A Review of Results of 110 Cases," *American Journal of Psychiatry* 107(1951): 801–7.

5. C. Trétiakoff, "Contribution à l'étude de l'anatomie pathologique du locus niger de Soemmering avec quelques déductions relatives à la pathogénie des troubles du tonus musculaire et de la maladie de Parkinson," Paris Thesis 1918–19 (Jouve et Cie, 1918).

6. I. Blackburn, *Illustrations of the Gross Morbid Anatomy of the Brain of the Insane* (Washington, D.C.: U.S. Government Printing Office, 1908).

7. W. Freeman, unpublished autobiography.

8. A. Ferraro, "Histopathological Findings in Two Cases Clinically Diagnosed as Dementia Praecox," *American Journal of Psychiatry* 90(1934): 883; see also *Journal of Neuropathology and Experimental Neurology* 2(1943): 84.

9. W. Freeman, "The Small Heart in Schizophrenics: Biometrical Studies in Psychiatry II," *Bulletin of St. Elizabeth's Hospital* 7(1931): 59.

10. I. S. Cooper, *The Vital Probe: My Life as a Brain Surgeon* (New York: W. W. Norton, 1981), pp. 101–2.

11. Z. Lebensohn, "In Memoriam. Walter Freeman, 1895–1972," *American Journal of Psychiatry* 129(1972): 356–57.

12. W. Freeman, *The Psychiatrists* (New York: Grune & Stratton, 1968), pp. 41–42.

13. A few titles from Freeman's publications during this period are "Biometrical Studies in Psychiatry: The Chances of Death," *American Journal of Psychiatry* 8(1928): 425; "The Psychological Panel in Diagnosis and Prognosis; Correlation of Personality with Susceptibility to Disease, Based on 1400 Necropsies," *Annals of International Medicine* 4(1930): 29; "Deficiency of Catalytic Iron in the Brain in Schizophrenia," *Archives of Neurology and Psychiatry* 24(1930): 300; "Psychochemistry: Some Physico-chemical Factors in Mental Disorders," *Journal of the American Medical Association* 97(1931): 293; "Manic-depressive Psychosis: Constitution of the Cyclothymic," *Association for Research in Nervous and Mental Diseases* 51(1930): 51; W. Freeman, W. W. Eldridge, and R. W. Hall, "Malaria Treatment of Dementia Paralytica: Results in 205 Cases after Five to Eleven Years," *Southern Medical Journal* 27(1934): 122–26; "Weight of Endocrine Glands," *Human Biology* 6(1934): 489–523; "Human Constitution," *Annals of Internal Medicine* 7(1934): 805–11; "Personality and the Endocrines; A Study Based on 1400 Quantitative Necropsies," *Annals of Internal Medicine* 9(1935): 444–50; "Constitutional Factors in Mental Disorders," *Medical Annals of the District of Columbia* 5(1936): 287–344.

14. J. W. Watts, "Psychosurgery: The Story of the 20-Year Follow-up of the Freeman-Watts Lobotomy Series," an unpublished lecture given at the Instituto Nacional de Neurologica, Mexico, D.F. (4 November 1974). Personal communication.

15. "Big Meeting," *Time*, 22 June 1931, p. 29.

16. Printed by Jas. C. Wood, Washington, D.C., 1933, for private distribution by Walter Freeman.

17. W. Freeman, "Danger Signals: On the Advantage of a Nervous Breakdown, or a Few Neurotic Symptoms in Certain Men under Forty Years of Age," *Cincinnati Journal of Medicine* 16(1935): 287–91.

Chapter 8. Releasing Black Butterflies

1. W. Freeman, "History of Psychosurgery," unpublished manuscript, Himmelfarb Health Science Library, George Washington University.
2. W. Freeman and J. Watts, "Prefrontal Lobotomy in Agitated Depression: Report of a Case," *Medical Annals of the District of Columbia* 5–6(1936–1937): 326–28.
3. Freeman, "History of Psychosurgery" [1].
4. Freeman and Watts, "Prefrontal Lobotomy in Agitated Depression" [2].
5. W. Freeman and J. Watts, "Prefrontal Lobotomy in the Treatment of Mental Disorders," *Southern Medical Journal* 30(1937): 23–31.
6. Ibid.
7. A. Meyer, comment after paper by Freeman and Watts, "Prefrontal Lobotomy in Agitated Depression" [2].
8. Freeman, "History of Psychosurgery" [1].
9. See W. Freeman, *The Psychiatrists* (New York: Grune & Stratton, 1968), p. 247.
10. Ibid., p. 57.
11. Freeman, "History of Psychosurgery" [1].
12. L. Davis, following W. Freeman and J. Watts, "Subcortical Prefrontal Lobotomy in the Treatment of Certain Psychoses" (Chicago Neurological Association, 18 February 1937), *Archives of Neurology and Psychiatry* 38(1937): 225–29.
13. H. Paskind, following Freeman and Watts, "Subcortical Prefrontal Lobotomy" [12].
14. L. Pollack, following Freeman and Watts, "Subcortical Prefrontal Lobotomy" [12].
15. "The Surgical Treatment of Certain Psychoses" (unsigned editorial), *New England Journal of Medicine* 215(1936): 1088.
16. J. Fulton to W. Freeman, 16 April 1948, Yale University Library.
17. The same patient is described in Freeman and Watts, "Prefrontal Lobotomy in the Treatment of Mental Disorders" [5] (see case 6); and W. Freeman and J. W. Watts, "The Failures: Follow-up Clinic St. Elizabeth's Hospital," *Digest of Neurology and Psychiatry* 17(1949): 445–46 (see case I).
18. W. Freeman and J. W. Watts, "Interpretation of the Functions of the Frontal Lobes Based upon Observations in Forty-eight Cases of Prefrontal Lobotomy," *Yale Journal of Biology and Medicine* 11(1939): 527–38.
19. J. G. Lyerly, "Prefrontal Lobotomy in Evolutional Melancholia," *Journal of the Florida Medical Association* 25(1938): 225–29 (see also discussion following the presentation); J. G. Lyerly, "Transection of the Deep Association Fibers of the Prefrontal Lobes in Certain Mental Disorders," *Southern Surgeon* 8(1939): 426–34.
20. J. C. Davis, following Lyerly, "Prefrontal Lobotomy" [19].
21. P. L. Dodge, following Lyerly, "Prefrontal Lobotomy" [19].
22. J. G. Love, "Prefrontal Lobotomy in the Treatment of Mental Disease: Surgical Technic," *Proceedings of the Staff Meeting of the Mayo Clinic* 18(1943): 372–73.
23. B. G. Lawrence, "Prefrontal Lobotomy: Result in a Case of Agitated Depression," *Delaware State Medical Journal* 10(1938): 81–85; P. F. Elfeld, "Results of Lobotomies at the Delaware State Hospital," *Delaware State Journal* 14(1942): 81–83; D. E. Wynegar, "The Frontal Lobes, the Prefrontal Lobotomy and the Psychoses," *Delaware State Medical Journal* 14(1942): 96–99.
24. M. Tarumianz, in panel discussion, "Neurosurgical Treatment of Certain Abnormal Mental States," *Journal of the American Medical Association* 117(1947): 521–27.
25. J. F. Fulton, preface to *The Frontal Lobes. Research Publication for the Association*

for Research in Nervous and Mental Disease (Baltimore: Williams & Wilkins, 1948), p. xii.

26. Conversation recorded in Freeman, "History of Psychosurgery" [1].

27. *New York Times,* "Find New Surgery Aids Mental Cases," 21 November 1936, p. 10.

28. *New York Times,* 7 June 1937, pp. 1, 10.

29. Ibid.

30. H. A. Dannecker, "Psychosurgery Cured Me," *Coronet* 12(6 October 1942): 8–12.

31. Freeman, "History of Psychosurgery" [1].

32. W. Kaempffert, "Turning the Mind Inside Out," *Saturday Evening Post* 213(24 May 1941): 18.

33. Freeman, "History of Psychosurgery" [1].

34. Ibid.

35. Kaempffert, "Turning the Mind Inside Out" [32].

36. Ibid.

37. Ibid.

38. Ibid.

39. Ibid.

40. G. W. Gray, "The Attack on Brainstorms," *Harper's Magazine* 183(September 1941): 366–76.

41. J. Watts, "Psychosurgery: The Story of the 20 Year Follow-up of the Freeman and Watts Lobotomy Series," unpublished speech read before the Tenth Anniversary Meeting of the Instituto Nacional de Neurologia, Mexico, D.F., 4 November 1974.

42. Freeman, "History of Psychosurgery" [1].

43. R. Corria, "Frontal Decortication in Oligophrenic Eretics (Children with Aggressive Social Behavior)," *Psychosurgery. First International Conference* (Lisbon, 1949), p. 321.

44. E. Rizzatti, "Chirurgia Della Psicosi," *Giornale di Psychiatria edi Neuropatologia* 67(1939): 125–30.

45. Ibid.

46. Ibid.

47. G. Sai, "Esperienze de neuropsichiatria alla Moniz," *Ospedale Psichiatrico e Sanatorio Neurologico Provincale di Trieste* 33(1937): 257–66.

48. D. Bagdasar and J. Constantinesco, "L'Opération d'Egas Moniz dans le traitement de certaines psychoses," *Bulletins et Mémoires de la Société Médicale des Hospitaux de Bucharest* 3(1935): 78–84.

49. E. Mariotti and M. Sciuti, "La terapia intracerebrale nelle malattie mentali. Contributo alla neurochirurgia delle psicosi," *Revista di Neurologia* (April 1937): 238–39.

50. M. E. Torsegno, "La terapia violenta delle psicosi," *Psichiat. Pesaro* 67(1938): 5–44.

51. A. M. Fiamberti, "Proposta di una technica operatoria modificata e semplificata per gli interventi alla Moniz sui lobi prefrontali in malati di mente," *Rasseg. Studi. Psichiatr.* 26(1937): 979.

52. M. Nakata, cited in S. Itai, "Beobachtungen über die frontallobektomierten Patienten," *Psychiatria et Neurologia Japonica* 46(1942): 25–26; see also S. Hirose, "Past and Present Trends of Psychiatric Surgery in Japan," in E. R. Hitchcock, H. T. Ballantine, Jr., and B. A. Meyerson, eds., *Modern Concepts in Psychiatry Surgery* (Amsterdam: Elsevier/North Holland, 1979), pp. 349–57.

53. H. Baruk, *Psychiatric, Médicale, Physiologique et Experimentale* (Paris: Masson, 1938).

54. E. L. Hutton, G. W. Flemming, and F. E. Fox, "Early Results of Prefrontal Leucotomy," *Lancet* 241(1941): 3–7; J. S. McGregor and J. R. Crumbie, "Surgical Treat-

ment of Mental Disease," *Lancet* 2(1941): 7–8.

55. "Psychosurgery," *Time*, 30 November 1942, pp. 48–49.

56. W. Freeman and J. W. Watts, *Psychosurgery: Intelligence, Emotion and Social Behavior Following Prefrontal Lobotomy for Mental Disorders* (Springfield, Ill.: Charles C Thomas, 1942).

Chapter 9. "Hit Us Like a Bomb": Psychosurgery in the 1940s

1. G. Rylander, "The Renaissance of Psychosurgery," in L. V. Laitinen and K. E. Livingston, eds., *Surgical Approaches in Psychiatry* (Baltimore: University Park Press, 1973), p. 3.

2. W. Freeman and J. Watts, *Psychosurgery: Intelligence, Emotion and Social Behavior Following Prefrontal Lobotomy for Mental Disorders* (Springfield, Ill.: Charles C Thomas, 1942), p. 18.

3. Ibid., preface, p. vii.

4. A. E. Walker, *The Primate Thalamus* (Chicago: University of Chicago Press, 1938); C. J. Herrick, "The Evolution of Cerebral Localization Patterns," *Science* 78(1933): 439–44.

5. Freeman and Watts, *Psychosurgery* [2], p. 27.

6. K. Lashley, "The Thalamus and Emotion," *Psychological Review* 45(1938): 42–61.

7. Freeman and Watts, *Psychosurgery* [2], p. 104.

8. T. Hunt, in Freeman and Watts, *Psychosurgery* [2], p. 154.

9. Freeman and Watts, *Psychosurgery* [2], p. 318.

10. Ibid., p. 18.

11. Ibid., p. vii.

12. M. Robinson and W. Freeman, *Psychosurgery and the Self* (New York: Grune & Stratton, 1954).

13. Freeman and Watts, *Psychosurgery* [2], p. 317.

14. W. Kaempffert, "Science in the News: Psychosurgery," *New York Times*, 11 January 1942, sec. II, p. 7.

15. *Time*, 30 November 1942, p. 48.

16. E. C. Dax, "The History of Prefrontal Leucotomy," in J. S. Smith and L. G. Kiloh, eds., *Psychosurgery and Society* (New York: Pergamon Press, 1977), pp. 19–24.

17. E. L. Hutton, G. W. Flemming, and F. E. Fox, "Early Results of Prefrontal Leucotomy," *Lancet* 241(1941): 3–7; J. S. McGregor and J. R. Crumbie, "Surgical Treatment of Mental Disease," *Lancet* 2(1941): 7–8.

18. M. Partridge, *Pre-frontal Leucotomy: A Survey of 300 Cases Personally Followed Over 1½–3 years* (Springfield, Ill.: Charles C Thomas, 1950), pp. 4–5.

19. K. Walsh, *Neuropsychology: A Clinical Approach* (London: Churchill Livingstone, 1978), p. 142; see also F. W. Willway, "The Technique of Prefrontal Leucotomy," *Journal of Mental Science* 89(April 1943): 192–93.

20. P. T. Rees, ed., "Symposium on Pre-frontal Leucotomy," *Journal of Mental Science* 89(April 1943): 11–201.

21. *Time*, 30 November 1942, p. 48.

22. L. Ziegler, "Bilateral Prefrontal Lobotomy: A Survey," *American Journal of Psychiatry* 100(1943): 178–79.

23. *Time*, 23 December 1946, p. 67.

24. C. C. Limburg, "A Survey of the Use of Psychosurgery with Mental Patients," in N. Bigelow, ed., *Proceedings of the First Research Conference on Psychosurgery* (Bethesda,

Md.: National Institutes of Health, U.S. Public Health Service Publication no. 16, 1949), pp. 165–73.

25. J. L. Poppen, "Technic of Prefrontal Lobotomy," *Journal of Neurosurgery* 5(1948): 514–20.

26. M. Kramer, "The 1951 Survey of the Use of Psychosurgery," in W. Overholser, ed., *Proceedings of the Third Research Conference on Psychosurgery, 1951* (Bethesda, Md.: National Institutes of Health, U.S. Public Health Service Publication no. 221, 1954), pp. 159–68.

27. U.S. Bureau of the Census, "Insane and Feeble Minded in Hospitals" (1904), p. 37.

28. N. Dayton, *New Facts on Mental Disorders* (Springfield, Ill.: Charles C Thomas, 1940), pp. 414–29.

29. N. Allen, "Proceedings of the National Conference of Charities and Correction" (1875), p. 43, cited in G. N. Grob, *Mental Illness and American Society* (Princeton, N.J.: Princeton University Press, 1983), pp. 196–97.

30. J. P. Norman, "State-Hospital Psychiatry: An Evaluation," *Mental Hygiene* 31(July 1947): 436–48.

31. F. L. Wright, Jr., *Out of Sight Out of Mind: A Graphic Picture of Present-Day Institutional Care of the Mentally Ill in America, Based on More Than Two Thousand Eyewitness Reports* (Philadelphia: National Mental Health Foundation, 1947), p. 43.

32. A. Q. Marsel, "Bedlam 1946," *Life,* 6 May 1946 (vol. 20, no. 18), p. 102.

33. "Mental Patients Reported Abused at Queens Center," *New York Times,* 14 May 1984, p. 1.

34. West Virginia State Board of Control Report 13(1939–1943), vol. 45.

35. Ibid.; for a description of the various forms of hydrotherapy and "wet packs" used at the time, see R. Wright, *Hydrotherapy in Hospitals for Mental Disease* (Boston: Tudor Press, 1932).

36. The following are only a few of the many exposés that appeared during this time: A. Deutsch, *The Shame of the States* (New York: Harcourt Brace, 1948); A. Deutsch, *The Mentally Ill in America* (New York: Doubleday, Doran, 1949); "Bedlam, USA," *Life,* May 1946; "The Shame of Our Mental Hospitals," *Reader's Digest,* July 1946; and the movie *The Snake Pit* (1948) based on Mary Jane Ward's 1946 novel by that name.

37. C. H. Jones, "Medical Programs in State Mental Hospitals," *Northwest Medicine* 53(1954): 1217–19. See also W. Menninger, "Facts and Statistics of Significance for Psychiatry," *Bulletin of Menninger Clinic* 12(1948): 1–25.

38. Grob, *Mental Illness* [29].

39. Jones, "Medical Programs" [37].

40. W. Menninger, "Facts and Statistics" [37].

41. Deutsch, *The Shame of the States* [36].

42. R. R. Grinker and E. V. McLean, "Course of Depression Treated by Psychotherapy and Metrazol," *Psychosomatic Medicine* 2(1940): 119–38.

43. Menninger, "Facts and Statistics" [37].

44. Veterans Administration memorandum on lobotomy dated 16 August 1943; "Prefrontal Leukotomy, An Evaluation," *V.A. Technical Bulletin* TB 10–46, 21 May 1948.

45. I. Wilson and E. H. Warland, "Pre-frontal Leucotomy in 1,000 Cases" (Great Britain Board of Control [England and Wales]; London: His Majesty's Stationery Office, 1947); abstract in *Lancet* 252(1947): 584–94.

46. C. C. Burlingame, *Digest of Neurology and Psychiatry* 17(1949): 508.

47. G. C. Tooth and M. P. Newton, "Leucotomy in England and Wales 1942–1954" (Great Britain Ministry of Health Reports on Public Health and Medical Subjects, no. 104; London: Her Majesty's Stationery Office, 1961).

48. Cited in W. Freeman and J. W. Watts, *Psychosurgery: In the Treatment of Mental Disorders and Intractable Pain* (Springfield, Ill.: Charles C Thomas, 1950), p. xxi.

49. D. Kelly, "What's New in Psychosurgery?" in S. Arieti, ed., *New Dimensions in Psychiatry: A World View* (New York: John Wiley, 1975), pp. 114–41.

50. W. Freeman and J. Watts, *Psychosurgery: In the Treatment of Mental Disorders and Intractable Pain,* 2nd ed. (Springfield, Ill.: Charles C Thomas, 1950), p. 575.

51. M. V. Govindaswarmy and B. Balakrishna, "Bilateral Prefrontal Leucotomy in Indian Patients," *Lancet* 1(1944): 466.

52. T. De Lehoczky, "Psychosurgery in Hungary," in *Psychosurgery. First International Conference,* 4–7 August 1948 (Lisbon: Livraria Luso-Espanhola, 1949), p. 309.

53. S. Hirose, "Past and Present Trends of Psychiatric Surgery in Japan," in E. R. Hitchcock, H. T. Ballentine, Jr., and B. A. Myerson, eds., *Modern Concepts in Psychiatric Surgery* (Elsevier/North Holland, 1979), pp. 349–57.

54. H. Fernández-Morán, "Leucotomía e inyecciones en los lóbulos prefrontales por la vía transorbitaria," *Archivos Venezolanos de la Sociedad de Oto-Rino Laringologia, Oftalmologia, Neurologia* 7(1946): 109–45.

55. "Psychosurgery," *Life,* 3 March 1947, p. 93.

56. "Kill or Cure," *Time,* 23 December 1946, pp. 66–67. See also editorial comment, "Popular Psychosurgery," *Psychiatric Quarterly* 20(1946 [suppl.]): 307–10.

57. For example, E. L. Hutton, "Leucotomy and the Super-ego," in *Psychosurgery: First International Conference* (Lisbon: Livraria Luso-Espanhola, 1949), p. 327.

58. H. S. Sullivan, editorial, *Psychiatry,* 6 May 1943, pp. 228–29.

59. W. Freeman, unpublished "History of Psychosurgery," Himmelfarb Health Science Library, George Washington School of Medicine.

60. W. Freeman, *The Psychiatrists: Personalities and Patterns* (New York: Grune & Stratton, 1968), p. 137.

61. K. L. Chatelaine, *Harry Stack Sullivan: The Formative Years* (Washington, D.C.: University Press of America, 1981).

62. The proceedings of this postgraduate course in psychosurgery were published in the *Digest of Neurology and Psychiatry* (Institute of Living) 17(1949): 407–54.

63. D. Bullard, quoted in *Digest* [62], p. 427.

64. D. McK. Rioch, quoted in *Digest* [62], p. 428.

65. W. Overholser, quoted in *Digest* [62], p. 430.

66. J. Watts, quoted in *Digest* [62].

67. S. E. Jelliffe, "Some Observations on Obsessive Tendencies Following Interruption of the Frontal Association Pathways," *Journal of Nervous and Mental Disease* 88(1938): 232–33.

68. A. A. Brill, discussion following Jelliffe, "Some Observations" [67].

69. P. Bucy, "Neurosurgical Treatment of Certain Abnormal Mental States" (panel discussion at Cleveland session), *Journal of the American Medical Association* 117(1941): 517–27.

70. R. Grinker, quoted in "Neurosurgical Treatment" [69].

71. Ibid.

72. Ibid, p. 473.

73. S. Cobb, "Review of Neuropsychiatry for 1940," *Archives of Internal Medicine* 66(1940): 1354.

74. E. A. Stephens, forward, *Genetic Psychology Monographs* 29(1944): 7.

75. D. Hebb, "Man's Frontal Lobes: A Critical Review," *Archives of Neurology and Psychiatry* 54(1945): 10–24 (see p. 21).

76. "Frontal Lobotomy," editorial, *Journal of the American Medical Association* 117(16 August 1941): 534–35.

77. "The Lobotomy Delusion," editorial, *Medical Record,* 15 May 1940: p. 335.

78. P. D. Flood, ed., *The Ethics of Brain Surgery from the Cahiers Laennec* (Cork, Ireland: Mercier Press, 1955; translated from 1954 French edition); see especially the foreword and p. 55. Pope Pius XII, "The Moral Limits of Medical Research and

Treatment," 14 September 1952 (National Catholic Welfare Conference, Washington, D.C.), p. 7, no. 15.

79. H. C. Solomon, introduction, in M. Greenblatt, R. Arnot, and H. C. Solomon, *Studies in Lobotomy* (New York: Grune & Stratton, 1950), pp. 1–6.

80. Ibid.

81. Ibid.

82. Partridge, *Prefrontal Leucotomy* [18].

83. Ibid., p. 471.

84. Ibid., p. 473.

85. "Crime Cure," *Time,* 14 July 1947, p. 53. McDonald, quoted in *Time,* 14 July 1947, p. 53.

86. Y. D. Koskoff and R. Goldhurst, *The "Dark Side of the House"* (New York: Dial Press, 1968); E. E. Mayer, "Prefrontal Lobotomy and the Courts," *Journal of Criminal Law and Criminology* 38(1948): 576–83.

87. "New Brain Operation for Mental Illness," *Life,* 16 August 1948, pp. 57–60.

88. E. A. Spiegel, H. T. Wycis, and H. Freed, "Thalatomy in Mental Disorders," *Psychosurgery. First International Conference* (Lisbon, 1949), pp. 91–95 (report made at conference on 5 August 1948). For use of a stereotaxic instrument in human brain surgery, see *Science* 106(1947): 349–50.

89. W. T. Peyton, H. H. Noran, and E. W. Miller, "Prefrontal Lobectomy (excision of the anterior areas of the cerebrum)," *American Journal of Psychiatry* 104(1948): 513–23.

90. W. Penfield, "Symposium on Gyrectomy. Part I. Bilateral Frontal Gyrectomy and Postoperative Intelligence," *The Research Publications of the Association for Research in Nervous and Mental Disease* 28(1948): 519–34.

91. J. Le Beau, *Psycho-Chirurgie et Fonctions Mentalis* (Paris: Masson, 1954); J. Le Beau and M. de Barros, "Bilateral Removal of Some Frontal Areas," in *Psychosurgery. First International Conference* (Lisbon: Livraria Luso-Espanhola, 1949), pp. 73–76.

92. J. L. Pool, "Topectomy: A Surgical Procedure for the Treatment of Mental Illness," *Journal of Nervous and Mental Disease* 110(1949): 164; F. A. Mettler, ed., *Selective Partial Ablation of the Frontal. A Correlative Study of its Effects on Human Psychotic Subjects* (New York: Paul Hoeber, 1949); F. A. Mettler, ed., *Psychosurgical Problems (The Columbia Greystone Associates, Second Group)* (New York: Blakiston, 1952).

93. "Topectomy—New Light on a Stab in the Dark," editorial comment, *Psychiatric Quarterly* 23(1949): 156–63.

94. W. Freeman, *Journal of Nervous and Mental Disease* 88(1938): 233; see also discussion by L. Alexander following W. Mixter, K. Tillotson, and D. Wies, "Frontal Lobotomy in Two Patients with Agitated Depression," *Archives of Neurology* 44(1940): 236–39 (see p. 239).

95. L. Hofstatter, E. A. Smolik, and A. K. Busch, "Prefrontal Lobotomy in Treatment of Chronic Psychoses (with special reference to section of orbital areas only)," *Archives of Neurology and Psychiatry* 53(1945): 125–30.

96. A. Meyer, E. Beck, and T. McLardy, "Prefrontal Leucotomy: A Neuro-anatomic Report," *Brain* 70(1947): 18–49; F. Reitman, "Orbital Cortex Syndrome Following Leucotomy," *American Journal of Psychiatry* 103(1946): 238–41.

97. Poppen, "Technic of Prefrontal Lobotomy" [25]. See also J. L. Poppen, "Prefrontal Lobotomy. Technic and General Impressions Based on Results in 470 Patients Subjected to This Procedure," *Digest of Neurology and Psychiatry* 16(1948): 403–8.

98. W. C. Scoville, "Selective Cortical Undercutting as a Means of Modifying and Studying Frontal Lobe Function in Man," *Journal of Neurosurgery* 6(1949): 65–73.

99. S. Hirose, "Orbito-ventromedial Undercutting 1957–1963," *American Journal of Psychiatry* 121(1965): 1194–1202.

100. W. Freeman and J. Watts, "Pain of Organic Disease Relieved by Prefrontal Lobotomy," *Lancet* I(29 June 1946): 953–55.

101. J. L. Poppen, "Prefrontal Lobotomy for Intractable Pain: Case Report," *Lahey Clinic Bulletin* 4(January 1946): 205–7.

102. J. E. Scarff, "Unilateral Prefrontal Lobotomy with Relief of Ipsilateral, Contralateral and Bilateral Pain; Preliminary Report," *Journal of Neurosurgery* 5(1948): 288–93.

103. E. G. Grantham, "Frontal Lobotomy for the Relief of Intractable Pain," *Southern Surgeon* 16(1950): 181–90. See also F. J. Ayd, "Value of the Grantham Lobotomy," *Southern Medical Journal* 50(1957): 939–42.

Chapter 10. "A New Psychiatry": Transorbital Lobotomy

1. H. Stevens, "The Use of Prefrontal Lobotomy in Schizophrenia," *Quarterly Review of Psychiatry and Neurology* 3(1948): 106–12.

2. W. Freeman and J. M. Williams, "Human Sonar. The Amygdaloid Nucleus in Relation to Auditory Hallucinations," *Journal of Nervous and Mental Disease* 116(1952): 456–62; W. Freeman and J. M. Williams, "Hallucinations in Braille: Effect of Amygdaloidectomy," *Archives of Neurology and Psychiatry* 70(1953): 630–34.

3. F. R. Ewald, W. Freeman, and J. W. Watts, "Psychosurgery: The Nursing Problem," *American Journal of Nursing* 47(1947): 210–13.

4. W. Freeman, "Transorbital Leucotomy: The Deep Frontal Cut," *Proceedings of the Royal Society of Medicine* 42(1949, suppl.): 8–11.

5. W. Freeman, "Transorbital Lobotomy," in W. Freeman and J. Watts, *Psychosurgery: In the Treatment of Mental Disorders and Intractable Pain,* 2nd ed. (Springfield, Ill.: Charles C Thomas, 1950), p. 52.

6. P. Wegeforth, J. B. Ayer, and C. R. Essick, "The Method of Obtaining Cerebrospinal Fluid by Puncture of the Cisterna Magna (Cistern Puncture)," *American Journal of Medical Science* 147(1919): 789–97.

7. As described by J. W. Watts in an unpublished speech, "Psychosurgery: The Story of the 20-year Follow-up of the Freeman and Watts Lobotomy Series," Instituto Nacional de Neurologia, Mexico, D.F., 4 November 1974.

8. W. Freeman, "History of Psychosurgery," unpublished manuscript, Himmelfarb Health Science Library, George Washington University.

9. Ibid.

10. Freeman, "Transorbital Lobotomy" [5].

11. W. Freeman, "Transorbital Leucotomy," *Lancet* 2(1948): 371–73.

12. Quoted from W. Freeman, "History of Psychosurgery" [8].

13. Ibid.

14. Freeman, "History of Psychosurgery" [8].

15. Watts, "Psychosurgery" [7].

16. Freeman, "History of Psychosurgery" [8].

17. Letter of J. F. Fulton to W. Freeman, 2 October 1947; letter of W. Freeman to J. F. Fulton, 6 October 1947, Yale University Library.

18. H. Fernández-Morán, "Leucotomía e inyecciones en los lóbulos prefrontales por la via transorbitaria," *Archivos Venezolanos de la Sociedad de Oto-Rino-Laringologia, Oftamologia, Neurologia* 1(1946): 109–45.

19. W. Freeman, unpublished autobiography.

20. Ibid.

21. C. H. Jones, "William Nobel Keller (1875–1960)," *American Journal of Psychiatry* 118(1961): 94–96.

22. F. W. Haas and D. B. Williams, "Transorbital Lobotomy: A Preliminary Report of Twenty-Four Cases," *South Dakota Journal of Medicine and Pharmacy* 1(May 1948): 191–92.

23. "At State Hospital—Surgery May Free 9 Mental Patients," *Seattle Post-Intelligencer,* 28 August 1947, p. 1.

24. C. H. Jones and J. B. Shanklin, "Transorbital Lobotomy. Preliminary Report of Forty-One Cases," *Northwest Medicine* (Seattle) 47(1948): 421–27.

25. W. Arnold, *Frances Farmer, Shadowland* (New York: McGraw Hill, 1978); D. Shutts, *Lobotomy: Resort to the Knife* (New York: Van Nostrand Reinhold, 1982), pp. 183–84; and the movie *Frances* (1982).

26. Freeman, "History of Psychosurgery" [8].

27. Ibid.

28. Ibid.

29. See *Proceedings of the Royal Society of Medicine* 42(1949): 8–11; and Freeman and Watts, *Psychosurgery* [5].

30. Freeman, "History of Psychosurgery" [8].

31. *Seattle Post-Intelligencer,* 30 October 1947.

32. "Pierced Brains," *Newsweek,* 7 June 1948, p. 46.

33. *Seattle Post-Intelligencer,* 8 July 1948.

34. W. Freeman, unpublished autobiography.

35. W. Freeman, "Transorbital Leucotomy," *Lancet* 2(1948): 371–73.

36. J. Walsh, "Transorbital Lobotomy: Some Results and Observations," *Lancet* 257(1949): 465–66.

37. Freeman, "History of Psychosurgery" [8].

38. Ibid.

39. Ibid.

40. Letter from J. Fulton to W. Freeman, 3 January 1950, Yale University Library.

41. Freeman, "Transorbital Leucotomy" [35].

42. W. Freeman, "Transorbital Leucotomy: The Deep Frontal Cut," *Proceedings of the Royal Society of Medicine* 42(1949, suppl.): 8–11.

43. Freeman and Watts, *Psychosurgery* [5].

44. C. L. Jackson, cited in W. Freeman, "Theoretical and Clinical Observations of Various Leucotomy Techniques," *Proceedings of the Royal Society of Medicine* 42(1949): 22.

Chapter 11. Lobotomy at Its Peak

1. W. Freeman, "History of Psychosurgery," unpublished manuscript, Himmelfarb Health Science Library, George Washington University.

2. E. Moniz, *A Nossa Casa* (Lisbon: Paulino Ferreira, Filhos, 1950).

3. E. Moniz, "De la thérapeutique chirurgicale dans la maladie de Parkinson et les états similares," *Chirurgie Suiss* 7(1943): 385–405.

4. W. Freeman, *The Psychiatrists: Personalities and Patterns* (New York: Grune & Stratton, 1968), p. 54. For evidence of lobotomy continuing in Portugal into the 1940s, see A. Amaral, *O Tratamento Cirúrgico das Doenças Mentais* (Lisbon: Tipographia A. Mendonça, 1944).

5. N. Dott, "Life and Work of Egas Moniz" (Fortieth Anniversary of the Introduction of Cerebral Angiography; Lisbon, 1967).

6. E. Moniz, "How I Came to Perform Prefrontal Leucotomy," in *Psychosurgery* (First International Conference, 4–7 August 1948; Lisbon, 1949).

7. D. Shutts, *Lobotomy: Resort to the Knife* (New York: Van Nostrand Reinhold, 1982), p. 149.

8. Freeman, "History of Psychosurgery" [1].

9. Freeman, *The Psychiatrists* [4], pp. 54–55.

10. *Proceedings of the First International Conference of Psychosurgery, 4–7 August 1948* (Lisbon: Livraria Luso-Espanhola, 1949). For two reviews of the conference, see C. Burlingame, *American Journal of Psychiatry* 105(1949): 550–51; W. Freeman, *American Journal of Psychiatry* 105(1949): 467–68.

11. Freeman, *The Psychiatrists* [4], p. 54.

12. "Explorers of the Brain," editorial, *New York Times*, 30 October 1949, section E, p. 8.

13. "The Nobel Prize in Medicine," editorial, *New England Journal of Medicine* 241(1949): 1025–26.

14. W. Freeman, unpublished autobiography.

15. W. Freeman and J. W. Watts, *Psychosurgery: In the Treatment of Mental Disorders and Intractable Pain*, 2nd ed. (Springfield, Ill.: Charles C Thomas, 1950), p. x.

16. Ibid., pp. 58–61.

17. Ibid., p. 549.

18. J. Ball, J. C. Klett, and C. J. Gresock, "The Veterans Administration Study of Prefrontal Lobotomy," *Journal of Clinical and Experimental Psychopathology* 29(1959): 205–17.

19. Administrator of Veterans Affairs, "Annual Report for Fiscal Years 1947–1949," United States Government Printing Office, Superintendent of Documents. See also statement by H. Tompkins quoted in "Lobotomy Disappointment," *Newsweek*, 12 December 1949, p. 51.

20. Ibid.

21. Freeman, "History of Psychosurgery" [1].

22. Ibid.

23. Ibid.

24. Ibid.

25. W. W. Wilson et al., "Transorbital Lobotomy in Chronically Disturbed Patients," *American Journal of Psychiatry* 108(1951): 444–49.

26. L. Kolb, discussion of Wilson et al., "Transorbital Lobotomy" [25].

27. M. T. Moore and W. M. Lutz, "Transorbital Leucotomy in a State Hospital Program," *Journal of the American Medical Association* 146(1951): 324–30.

28. Described in Freeman's unpublished "History of Psychosurgery" [1].

29. A. Gardner, "Transorbital Leucotomy in Noninstitutional Cases," *American Journal of Psychiatry* 114(1957): 140–42.

30. W. Freeman et al., "West Virginia Lobotomy Project," *Journal of the American Medical Association* 156(1954): 939–43; W. Freeman, "West Virginia Lobotomy Project: A Sequel," *Journal of the American Medical Association* 181(1962): 1134–35.

31. W. Freeman, "Transorbital Lobotomy in State Mental Hospitals," *Journal of the Medical Society of New Jersey* 51(1954): 148–50.

32. Ibid.

33. As reported in Freeman, "History of Psychosurgery" [1].

34. Ibid.

35. F. Perino, "Egas Moniz. Founder of Psychosurgery, Creator of Angiography, Nobel Prize Winner," *Journal of the International College of Surgeons* 36(1961): 261.

36. W. Freeman, "Review of *Confidências de um Investigador Ciêntífico*," *Archives of Neurology and Psychiatry* 63(1950): 191–93.

37. W. Freeman, unpublished autobiography.

Chapter 12. Two Patients: The Effects of Early Prefrontal Lobotomies

1. C. C. Burlingame, *Digest of Neurology and Psychiatry* 16(1948): 650. The *Digest* was published by the Institute of Living and edited by Burlingame; between the years 1947 and 1951, the *Digest* abstracted many articles on lobotomy and other somatic therapies.

2. E. A. Berg, "A Simple Objective Test for Measuring Flexibility in Thinking," *Journal of General Psychology* 39(1948): 15–22; D. A. Grant and E. A. Berg, "A Behavioral Analysis of Degree of Reinforcement and Ease of Shifting to a New Response in a Weigl-type Card-sorting Problem," *Journal of Experimental Psychology* 38(1948): 404–11.

3. D. Hebb, "Intelligence in Man after Large Removals of Cerebral Tissue: Report of Four Left Frontal Lobe Cases," *Journal of General Psychology* 21(1939): 73–87.

4. S. Corkin, "A Prospective Study of Cingulotomy," in E. Valenstein, ed., *The Psychosurgery Debate* (San Francisco: W. H. Freeman, 1980), pp. 164–204.

5. E. Bleuler, *Dementia Praecox or the Group of Schizophrenics* (New York: International University Press, 1950), p. 472; originally published in German in 1911.

6. M. Bleuler, *The Schizophrenic Disorder: Long-Term Patient and Family Studies* (New Haven: Yale University Press, 1978), p. 413; originally published in German in 1972.

7. Ibid., p. 418.

8. S. Kety, "Heredity and Environment," in J. C. Shershow, ed., *Schizophrenia: Science and Practice* (Cambridge: Harvard University Press, 1978), pp. 47–68; S. E. Nicol and I. I. Gottesman, "Clues to the Genetics and Neurobiology of Schizophrenia," *American Scientist* 71(1983): 398–404.

9. R. S. Banay and L. Davidoff, "Apparent Recovery of a Sex Psychopath after Lobotomy," *Journal of Criminal Psychopathology* 4(1942): 59–66.

10. Ibid.

11. Ibid.

12. Ibid.

13. J. W. Friedlander and R. S. Banay, "Psychosis Following Lobotomy in a Case of Sexual Psychopathy. Report of a Case," *Archives of Neurology and Psychiatry* 59(1948): 302–21.

14. W. Freeman and J. W. Watts, "The Radical Treatment of the Psychoses and Neuroses," *Diseases of the Nervous System* 3(1942): 6.

15. Friedlander and Banay, "Psychosis" [13].

16. S. D. Porteus and R. D. Kepner, "Mental Changes after Bilateral Prefrontal Lobotomy," *Genetic Psychology Monographs* 29(1944): 113.

Chapter 13. Opposition to Lobotomy—And a Brief New Life

1. "Lobotomy Disappointment," *Newsweek,* 12 December 1949, p. 51. See also N. Lewis, *American Journal of Psychiatry* 101(1944): 523.

2. N. Lewis, "General Clinical Psychiatry, Psychosomatic Medicine, Psychotherapy, Group Therapy, and Psychosurgery," *American Journal of Psychiatry* 105(1949–50): 512–17.

3. W. Overholser, cited in "Lobotomy Disappointment" [1].

4. J. L. Hoffman, "Clinical Observations Concerning Schizophrenic Patients

Treated by Prefrontal Leukotomy," *New England Journal of Medicine* 241(1949): 233–36.

5. H. E. Rosvold and M. Mishkin, "Evaluation of the Effects of Prefrontal Lobotomy on Intelligence," *Canadian Journal of Psychology* 4(1950): 122–25.

6. H. Baruk, *Patients Are Like Us: The Experience of Half a Century in Neuropsychiatry* (New York: William Morrow, 1978), p. 203. For a discussion of the Soviet Union and somatic treatments, see J. Wortis, *Soviet Psychiatry* (Baltimore: Williams & Wilkins, 1950), pp. 150–76.

7. Editorial Board, "Lobotomy: Surgery for the Insane," *Stanford Law Review,* 1 April 1949, pp. 463–74.

8. Reported in L. Freeman, "Pioneer Sees Peril in Brain Operations," *New York Times,* 5 May 1950, p. 19.

9. Ibid.

10. W. F. Freeman, "History of Psychosurgery," unpublished manuscript, Himmelfarb Health Science Library, George Washington University.

11. J. L. Poppen, "Techniques and Complications of the Standard Prefrontal Leucotomy," in W. Greenblatt, R. Arnot, and H. C. Solomon, eds., *Studies in Lobotomy* (New York: Grune & Stratton, 1950), p. 67.

12. P. Bailey, *The 1948 Year Book of Neurology, Psychiatry and Neurosurgery* (Chicago: Year Book Publishers, 1948), p. 493.

13. Freeman, "History of Psychosurgery" [10].

14. R. G. Fuller and M. Johnston, "The Duration of Hospital Life for Mental Patients," *Psychiatric Quarterly* 5(1931): 341–52, 552–82.

15. S. E. Jelliffe and W. A. White, *Diseases of the Nervous System* (Philadelphia: Lea & Fibiger, 1935), p. 1092.

16. W. Freeman, "West Virginia Lobotomy Project: A Sequel," *Journal of the American Medical Association* 181(1962): 1134–35; W. Freeman et al., "West Virginia Lobotomy Project," *Journal of the American Medical Association* 156(1954): 939–43.

17. K. G. McKenzie and G. Kaczanowski, "Prefrontal Leucotomy: A Five-year Controlled Study," *Journal de l'Association Médicale Canadienne* 91(1964): 1193–96; Editorial, "Standard Lobotomy: The End of an Era," *Journal de l'Association Médicale Canadienne* 91(1964): 1228–29.

18. W. C. Halstead, H. T. Carmichael, and P. C. Bucy, "Prefrontal Lobotomy. A Preliminary Appraisal of the Behavioral Results," *American Journal of Psychiatry* 103(1946): 217–28.

19. W. C. Halstead, *Brain and Intelligence: A Quantitative Study of the Frontal Lobes* (Chicago: University of Chicago Press, 1947), p. 126.

20. M. J. Meier, "Some Challenges for Clinical Neuropsychology," in R. M. Reitan and L. A. Davison, eds., *Clinical Neuropsychology: Current Status and Applications* (New York: John Wiley, 1974).

21. E. A. Walker, "Psychosurgery: Collective Review," *International Abstract Surgery* 78(1944): 1–11.

22. Freeman, "History of Psychosurgery" [10].

23. "New Brain Surgery Technique Tested on Sane Patients," *Seattle Post-Intelligencer,* 30 October 1947, p. 1.

24. M. Gumbert, "Lobotomy: Savior or Destroyer?" *Nation* 167(1948): 517–18.

25. "What Is a Man Profited?" (editorial), *New England Journal of Medicine* 241(1949): 248–49.

26. For a complete list of John Fulton's 520 publications, see *Journal of History of Medicine and Allied Sciences* 17(1962): 51–71.

27. A. E. Walker, "Fulton, John Farquhar," *Dictionary of Scientific Biography,* vol. V (New York: Charles Scribner, 1972), p. 207.

28. The artifact in Fulton research in Sherrington's laboratory is described in J. C.

Eccles, "Life in Sherrington's Laboratory," *Trends in Neurosciences* 5(April 1982): 110.

29. Fulton played a major role in founding *The Journal of Neurophysiology, The Journal of Neurosurgery,* and *The Journal of the History of Medicine.* Among the many books that he wrote (or edited) were: *Physiology of the Nervous System* (New York: Oxford University Press, 1938); *Harvey Cushing: A Biography* (Springfield, Ill.: Charles C Thomas, 1946); *Textbook of Physiology* (Philadelphia: W. B. Saunders, 1949); *Functional Localization in the Frontal Lobes and Cerebellum* (New York: Oxford University Press, 1949); *Functional Localization in Relation to Frontal Lobotomy* (New York: Oxford University Press, 1949); *Functional Lobotomy and Affective Behavior: A Neurophysiological Analysis* (New York: W. W. Norton, 1951); *The Frontal Lobes and Human Behavior (The Sherrington Lectures)* (Liverpool: University Press, 1952). Obituaries of Fulton appeared in *Bulletin of the History of Medicine* 35(1961): 82–86; *New England Journal of Medicine* 262(1960): 1340–41; *Journal of Neurosurgery* 17(1960): 1119–23; *Journal of Neurophysiology* 23(1960): 346–49; *The Yale Journal of Biology and Medicine* 33(1960): 85–93.

30. "In Memoriam. C. Charles Burlingame, M.D. 1885–1950," *American Journal of Psychiatry* 107(1950): 398–400.

31. C. C. Burlingame, *American Journal of Psychiatry* 105(1949): 550–51.

32. J. F. Fulton, *Functional Localization in Relation to Frontal Lobotomy: The William Withering Memorial Lectures* (New York: Oxford University Press, 1949).

33. J. W. Papez and J. Bateman, "Cytological Changes in Nerve Cells in Dementia Praecox," *Journal of Nervous and Mental Disease* 110(1949): 425–37.

34. J. W. Papez, "A Proposed Mechanism of Emotion," *Archives of Neurology and Psychiatry* 38(1937): 725–43.

35. P. MacLean, "Psychosomatic Disease and the 'Visceral Brain': Recent Developments Bearing on the Papez Theory of Emotion," *Psychosomatic Medicine* 11(1949): 338–53.

36. See J. Fulton diary, Yale University Library.

37. A. A. Ward, Jr., "The Anterior Cingular Gyrus and Personality," *Research Publications—Association for Research in Nervous and Mental Diseases* 27(1948): 438–45. For other generalizations from animal experiments to psychosurgery, see E. S. Valenstein, *Brain Control* (New York: John Wiley, 1973), pp. 326–35.

38. J. F. Fulton, *Frontal Lobotomy and Affective Behavior: A Neurophysiological Analysis* (New York: W. W. Norton, 1951), p. 128.

39. J. F. Fulton, "The Surgical Approach to Mental Disorder" (Alpha Omega Alpha Lecture read at the Montreal Neurological Institute, 8 January 1948), *McGill Medical Journal* 7(1948): 133–45.

40. Ibid.

41. H. Olivecrona, Presentation Speech for the Nobel Prize in Physiology and Medicine, Royal Carolina Institute, 1949.

Chapter 14. "A Living Fossil": The Decline of Lobotomy and the Final Years of Walter Freeman

1. The itinerary of Freeman's visits to state hospitals between 1946 and 1960 as recorded in his "History of Psychosurgery" (unpublished manuscript, Himmelfarb Health Science Library, George Washington School of Medicine):

1946	Yankton, S.D.
1947	Ft. Steilacoom, Wash.; Stockton, Cal.; Yankton, S.D.

1948	January	Huntington, W. Va.
	July	New York City
	October	Berkeley, Cal.; Ft. Steilacoom, Wash.; Yankton, S.D.; Lincoln, Neb.; Greystone Park, N.J.
	November	Paterson, N.J.; Hartford, Conn.
1949	February	Galveston, Tex.; Rusk, Tex.; Little Rock, Ark.; Lincoln, Neb.; Rochester, Minn.; Hastings, Minn.; Columbus, O.
	May	Ogdensburg, N.Y.
	June	Berkeley, Cal.; Sedro Woolley, Wash.; Ft. Steilacoom, Wash.; St. Joseph, Mo.
	October	Sykesville, Md.
	December	Sykesville, Md.
1950	February	Milledgeville, Ga.
	April	Milledgeville, Ga.
	May	Sykesville, Md.
	June	Sykesville, Md.
	July	Huntington, W. Va.; Lakin, W. Va.; Spencer, W. Va.
	August	Sykesville, Md.
	October	Sykesville, Md.
	December	Hartford, Conn.
1951	January	Paterson, N.J.; London, Ont.
	March	Milledgeville, Ga.; Philadelphia, Pa.; Sykesville, Md.
	April	Chicago
	May	Spencer, W. Va.; Cincinnati, O.
	June	Little Rock, Ark.; Rusk, Tex.; Terrell, Tex.; Wichita Falls, Tex.; Patton, Cal.; Berkeley, Cal.; Mendocino, Cal.; Ft. Steilacoom, Wash.; Sedro Woolley, Wash.; Yankton, S.D.
	August	Lincoln, Neb.; St. Joseph, Mo.; Cherokee, Ia.; Independence, Ia.
	November	Williamsburg, Va.
	December	Tampa, Fla.; Willemstad, Curacao; San Juan, P.R.
1952	January	Staunton, Va.; Paterson, N.J.
	February	Williamsburg, Va.
	April	Ciudad Trujillo, Dominican Republic; Staunton, Va.
	June	Lincoln, Neb.; Mendocino, Berkeley, and Patton, Cal.
	July–August	Williamsburg, Va.; Huntington, Spencer, Lakin, Weston, Spencer, Lakin, Huntington, Spencer, Lakin, Spencer (W. Va. Lobotomy Project)
	September	Staunton, Va.
	October	Williamsburg, Va.
	November	Mexico, D. F.; Roanoke, Va.
	December	Berkeley, Cal.; Philadelphia, Pa.; Paterson, N.J.
1953	February	St. Joseph, Mo.; Lincoln, Neb.; San Antonio, Tex.
	May	Patton, Auburn, Berkeley, and Agnews, Cal.; Huntington, W. Va.; Williamsburg, Va.
	June	Central Islip, N.Y.; Marion, Va.
	July	Staunton, Va.; Weston, Spencer, Lakin, Huntington, Spencer, Lakin, Huntington, Weston, Spencer (W. Va. patient follow-up)
	August	Fourth International Neurological Congress, Lisbon

	September	Williamsburg, Va.
	October	Marlboro, N.J.
	November	Staunton, Va.; Athens, O.
	December	Compton, Cal.; Lincoln, Neb.; Central Islip, N.Y.
1954	January	Spencer, W. Va.
	March	Berkeley, Cal.; Huntington, Lakin, Spencer, and Weston, W. Va.
	April	Roanoke, Va.; Williamsburg, Va.; Athens, O.; Youngstown, O.
	June	Marion, Va.; Staunton, Va.
	July	Fresno, Cal.
	August	Berkeley, Cal.
	September	Berkeley, Fresno, and Patton, Cal.
	October	Lincoln, Neb.; Cherokee, Ia.; Evansville, Ind.; Athens, O.; Huntington, Lakin, Spencer, and Weston, W. Va.
	November	Washington, D.C.; Williamsburg, Va.; Petersburg, Va.; Staunton, Va.; Rusk, Tex.; Terrell, Tex.
	December	Patton, Cal.
1955	April	Camarillo, Cal.; Patton, Cal.
	May	Spencer, W. Va.; Lakin, W. Va.; Takoma Park, Md.
	June	Youngstown, O.; Athens, O.; Lincoln, Neb.
	August	Berkeley, Cal.; Athens, O.
	October	Takoma Park, Md.
1956	April	Patton, Cal.; Athens, O.; Huntington, W. Va.
	May	Spencer, W. Va.; Youngstown, O.; Lincoln, Neb.
1957	May	Sedro Woolley, Wash.; Athens, O.
	November	Bogotá, Colombia
1958		Local patients in Cal. only
1959	April	Atascadero, Cal.; Camarillo, Cal.
1960	March	Gilroy, Cal.
	April	Atascadero, Cal.
	June	Athens, O.
	September	Compton, Cal.

2. Freeman, "History of Psychosurgery" [1].

3. Ibid.

4. H. Sleeper, "Psychiatry in the General Hospital Today," *Southern Medical Journal* 51(1958): 312–14.

5. W. Freeman, "Prefrontal Lobotomy: Final Report of 500 Freeman and Watts Patients Followed for 10 to 20 Years," *Southern Medical Journal* 51(1958): 739–45.

6. A Meyer, discussion of W. Freeman and J. Watts, "Prefrontal Lobotomy in the Treatment of Mental Disorders," *Southern Medical Journal* 30(1937): 23.

7. Freeman, "Prefrontal Lobotomy" [5].

8. J. Watts, discussion of Freeman, "Prefrontal Lobotomy" [5].

9. W. Freeman, discussion of Freeman, "Prefrontal Lobotomy" [5].

10. W. Freeman, "Frontal Lobotomy 1936–1956: A Follow-up Study of 3000 Patients from One to Twenty Years," *American Journal of Psychiatry* 113(1957): 877–86.

11. P. Royal, discussion following Freeman, "Frontal Lobotomy 1936–1956" [10].

12. A. Gardner, *American Journal of Psychiatry* 114(1957): 140–42; P. Longo, J. Arruda,

and J. Figueiredo, "Lobotomia transorbitaria en 54 pacientes tratados em hospital privado," *Arq. de Neuro-Psiquiat. S. Paulo* 14(1956): 273–84.

13. Freeman, "Prefrontal Lobotomy" [5].

14. W. Freeman, "Psychosurgery: Present Indications and Future Prospects," *California Medicine* 88(1958): 429–34.

15. E. Valenstein, "Who Receives Psychosurgery," in E. Valenstein, ed., *The Psychosurgery Debate: Scientific, Legal and Ethical Perspectives* (San Francisco: W. F. Freeman, 1980), pp. 89–107.

16. Freeman, "History of Psychosurgery" [1].

17. Ibid.

18. W. Freeman, "Physical Exercise in the Treatment of Hypochondriases," *Postgraduate Medicine* 4(1948): 435–37.

19. Freeman, "History of Psychosurgery" [1].

20. W. Freeman, "Adolescents in Distress: Therapeutic Possibilities of Lobotomy," *Diseases of the Nervous System* 22(1961): 1–4.

21. This incident is described in Freeman, "History of Psychosurgery" [1].

22. W. Freeman, *The Psychiatrists: Personalities and Patterns* (New York: Grune & Stratton, 1968).

23. Ibid., p. 284.

24. Ibid., p. 280.

25. Discussion of Sullivan's death can be found in H. S. Perry, *Psychiatrist of America: The Life of Harry Stack Sullivan* (Cambridge: Harvard University Press, 1982); K. L. Chatelaine, *Harry Stack Sullivan: The Formative Years* (Washington, D.C.: University Press of America, 1981).

26. Freeman, *The Psychiatrists* [22], p. 139.

27. W. Burton, *American Journal of Psychiatry* 125(1968): 715.

28. W. Freeman, unpublished autobiography.

29. W. Freeman, "Lobotomy for Intractable Psychosomatic Disorders," *Medical Annals of the District of Columbia* 31(1962): 1–6. See also S. Cheng, H. Tait, and W. Freeman, "Transorbital Lobotomy versus Electroconvulsive Therapy in the Treatment of Mentally Ill Tuberculosis Patients," *American Journal of Psychiatry* 113(1956): 32–35.

30. E. S. Valenstein, "Extent of Psychosurgery Worldwide," in E. S. Valenstein, ed., *The Psychosurgery Debate* (San Francisco: W. H. Freeman, 1980), pp. 76–86.

31. "Standard Lobotomy: The End of an Era," editorial, *Journal de l'Association Médicale Canadienne* 91(1964): 1228–29.

32. K. E. Livingston, "The Frontal Lobe Revisited: The Case for a Second Look," *Archives of Neurology* 20(1969): 90–95.

33. M. Fog, "Welcome," in E. Hitchcock, L. Laitinen, and K. Vaernet, eds., *Psychosurgery: Proceedings of the Second International Conference of Psychosurgery* (Springfield, Ill.: Charles C Thomas, 1972), pp. xiii–xiv.

34. W. Freeman, "Frontal Lobotomy in Early Schizophrenia: Long Follow-up in 415 Cases," *British Journal of Psychiatry* 119(1971): 621–24.

35. Z. Lebensohn, "In Memoriam. Walter Freeman 1895–1972," *American Journal of Psychiatry* 129(1972): 356–57.

36. W. Freeman, "Lobotomy in Limbo?" *American Journal of Psychiatry* 128(1972): 1315–16.

37. W. Freeman, "Sexual Behavior and Fertility after Frontal Lobotomy," *Biological Psychiatry* 6(1973): 97–104.

38. L. Laitinen and K. Livingston, eds., *Surgical Approaches in Psychiatry* (Proceedings of the Third International Congress of Psychosurgery, 14–18 August 1972, Cambridge, England; Baltimore: University Park Press, 1973).

Chapter 15. Psychosurgery in the 1970s and 1980s

1. J. Donnelly, "Psychosurgery," in H. I. Kaplan, A. M. Freedman, and B. J. Sadock, eds., *Comprehensive Textbook of Psychiatry*, vol. III, 3rd ed. (Baltimore: Williams & Wilkins, 1980), pp. 2342–48; E. S. Valenstein, "Extent of Psychosurgery Worldwide," in E. S. Valenstein, ed., *The Psychosurgery Debate* (San Francisco: W. H. Freeman, 1980), pp. 76–86.

2. E. S. Valenstein, "New Surgical Techniques," in Valenstein, *Psychosurgery Debate* [1], pp. 69–75; W. J. Fry and R. Meyers, "Ultrasonic Method of Modifying Brain Structures," First International Symposium on Stereoencephalotomy, *Confinia Neurologica* 22(1962): 315–27.

3. See K. E. Livingston, "The Frontal Lobes Revisited: The Case for a Second Look," *Archives of Neurology* 20(1969): 90–95; J. Holden and L. Hofstatter, "Prefrontal Lobotomy: Stepping-Stone or Pitfall?" *American Journal of Psychiatry* 127(1970): 591–98.

4. V. H. Mark and F. R. Ervin, *Violence and the Brain* (New York: Harper & Row, 1970). For a discussion of the controversy over a feared revival of interest in psychosurgery, see Valenstein, *Psychosurgery Debate* [1], pp. 39–54.

5. H. Maudsley, *Responsibility in Mental Disease* (New York: Appleton, 1874), pp. 169, 230–35, respectively. See also M. Echeverria, "On Epileptic Insanity," *American Journal of Insanity* 30(July 1873): 1–51.

6. E. S. Valenstein, "Brain Stimulation and the Origin of Violent Behavior," in W. Smith and A. Kling, eds., *Issues in Brain/Behavior Control* (New York: Spectrum, 1976), pp. 33–48.

7. V. H. Mark, W. H. Sweet, and F. R. Ervin, "The Role of Brain Disease in Riots and Urban Violence," *Journal of the American Medical Association* 201(1967): 895. See also V. H. Mark and F. R. Ervin, "Is There a Need to Evaluate the Individuals Producing Human Violence?" *Psychiatric Opinion* 5(1968): 32–34.

8. V. H. Mark and R. Neville, "Brain Surgery in Aggressive Epileptics," *Journal of the American Medical Association* 226(1973): 765–72.

9. L. Aarons, "Brain Surgery Is Tested on Three California Convicts," *Washington Post*, 25 February 1972, p. 1.

10. M. H. Brown, *National Enquirer*, 9 July 1972.

11. M. H. Brown, "The Captive Patient: A Forgotten Man," in Valenstein, *Psychosurgery Debate* [1], pp. 537–45.

12. G. Kaimowitz, "My Case Against Psychosurgery," in Valenstein, *Psychosurgery Debate* [1], pp. 506–19; S. L. Chorover, "Big Brother and Psychotechnology," *Psychology Today* 7(1973): 43–54.

13. C. Holden, "Psychosurgery: Legitimate Therapy or Laundered Lobotomy?" *Science* 179(1973): 1109–12.

14. S. L. Chorover, "Psychosurgery: A Neuropsychological Perspective," *Boston University Law Review* 54(1974): 231–48; S. L. Chorover, "Violence: The Pacification of the Brain," in S. L. Chorover, *From Genesis to Genocide* (Cambridge, Mass.: MIT Press, 1979), pp. 135–74; P. R. Breggin, "The Return of Lobotomy and Psychosurgery," *Congressional Record*, 24 February 1972, E 1602–E 1612; B. J. Mason, "New Threat to Blacks: Brain Surgery to Control Behavior," *Ebony*, February 1973, pp. 63–72; S. L. Chorover, "Big Brothers and Psychotechnology II: The Pacification of the Brain," *Psychology Today* 7(1974): 59–70; *Kaimowitz v. Department of Mental Health*, Civil no. 73–19, 434–AW (Cir. Ct. Wayne Co., Michigan), 10 July 1973; *Boston University Law Review* 54 (whole issue [1974]); Kaimowitz, "My Case Against Psychosurgery" [12];

P. Breggin, "Brain Disabling Therapies," in Valenstein, *Psychosurgery Debate* [1], pp. 467–92.

15. W. B. Scoville, in E. Hitchcock, L. Laitinen, and K. Vaernet, eds., *Psychosurgery* (Springfield, Ill.: Charles C Thomas, 1972), p. 19; W. B. Scoville, "The Effect of Surgical Lesions of the Brain on Psyche and Behavior in Man," in A. Winter, ed., *The Surgical Control of Behavior* (Springfield, Ill.: Charles C Thomas, 1971), p. 55.

16. See discussion in E. S. Valenstein, "Aggressive Behavior," in Valenstein, *Psychosurgery Debate* [1], pp. 94–96.

17. E. S. Valenstein, "Causes and Treatment of Mental Disorders," in Valenstein, *Psychosurgery Debate* [1].

18. T. Szasz, *The Myth of Mental Illness* (New York: Harper & Row, 1961; revised ed., 1974); idem, *Law, Liberty and Psychiatry* (New York: Macmillan, 1963); idem, *Psychiatric Justice* (New York: Collier Books, 1965); idem, *The Manufacture of Madness* (New York: Harper & Row, 1970).

19. K. Kesey, *One Flew Over the Cuckoo's Nest* (New York: Viking Press, 1973); see also B. Ennis, *Prisoners of Psychiatry: Mental Patients, Psychiatrists, and the Law* (New York: Harcourt Brace Jovanovich, 1972).

20. National Commission for the Protection of Human Subjects of Biomedical and Behavioral Research, "Report and Recommendations: Psychosurgery" (Department of Health, Education and Welfare, pub. no. (OS) 77–0002, U.S. Government Printing Office, 1977). For a thoughtful criticism of the report, see S. L. Chorover, "The Psychosurgery Evaluation Studies and Their Impact on the Commission's Report," in Valenstein, *Psychosurgery Debate* [1], pp. 245–63.

21. See articles by F. C. Pizzulli and by R. J. Grimm, Valenstein, *Psychosurgery Debate* [1], pp. 367–96 and 421–38, respectively.

22. H.-L. Teuber, S. H. Corkin, and T. E. Twitchell, "Study of Cingulotomy in Man: A Summary," in W. H. Sweet, S. Obrador, and J. G. Martin-Rodriguez, eds., *Neurosurgical Treatment in Psychiatry, Pain, and Epilepsy* (Baltimore: University Park Press, 1977), pp. 355–62; idem, "A Study of Cingulotomy in Man: Report to the National Commission for the Protection of Human Subjects of Biomedical and Behavioral Research," *Appendix: Psychosurgery* (Department of Health, Education and Welfare, pub. no. (OS) 77–0002, U.S. Government Printing Office, 1977), sec. III, pp. 1–115; A. F. Mirsky and M. H. Orzack, "Final Report on Psychosurgery Pilot Study: Report to the National Commission for the Protection of Human Subjects of Biomedical and Behavioral Research," *Appendix: Psychosurgery* (Department of Health, Education and Welfare, pub. no. (OS) 77–0002, U.S. Government Printing Office, 1977), sec. II.

23. S. Corkin, "A Prospective Study of Cingulotomy," in Valenstein, *Psychosurgery Debate* [1], pp. 164–204.

24. A. F. Mirsky and M. H. Orzack, "Two Retrospective Studies of Pyschosurgery," in Valenstein, *Psychosurgery Debate* [1], pp. 205–44; but see S. L. Chorover, "The Psychosurgery Evaluation Studies," idem, pp. 245–63.

Chapter 16. Final Reflections

1. S. T. Coleridge, letter of 18 December 1831, in T. Allsop, *Letters, Conversations and Recollections of S. T. Coleridge* (London: E. Moxon, 1836).

2. C. Norman, "Patent Dispute Divides AIDS Researchers," *Science* 230(1985): 640–42.

3. J. Winsten, "Science and the Media: The Boundaries of Truth," *Health Affairs* 4(1985): 5–23.

4. Ibid.

5. "Lifeline for a Man with a Dying Heart," *Life* 70(5 February 1971): 51; "Saving the Heart," *Newsweek* 78(26 July 1971): 50; "Old Hearts, New Plumbing," *Time* 97(10 May 1971): 51–52.

6. H. D. McIntosh and J. A. Garcia, "The First Decade of Aortocoronary Bypass Grafting, 1967–1977; A Review," *Circulation* 57(1978): 405–31; M. Millman, *The Unkindest Cut: Life in the Backrooms of Medicine* (New York: William Morrow, 1977), see appendix, pp. 217–52; J. Halperin and R. Levine, *Bypass* (New York: Times Books, 1985).

7. Halperin and Levine, *Bypass* [6], see chap. 12, "The Bottom Line: The Economics of Bypass," pp. 229–60.

8. Ibid.

9. B. Phibbs and H. Marriott, "Complications of Permanent Transvenous Pacing," *New England Journal of Medicine* 312(1985): 1428–32.

10. See E. Passamani et al., "A Randomized Trial of Coronary Artery Bypass Surgery," *New England Journal of Medicine* 312(1985): 1665–71; B. Gersh et al., "Comparison of Coronary Artery Bypass Surgery and Medical Therapy in Patients 65 Years of Age or Older," *New England Journal of Medicine* 313(1985): 217–24.

11. Halperin and Levine, *Bypass* [6].

12. Ibid., see chap. 11, "Cardiac Cripples: The Back-to-Work Problem," pp. 205–28.

13. McIntosh and Garcia, "The First Decade" [6], p. 426.

14. John Bunyan, *The Jerusalem Sinner Saved* (1668).

15. J. Swazey, *Chlorpromazine in Psychiatry: A Study of Therapeutic Innovation* (Cambridge, Mass.: MIT Press, 1974).

16. "Electroconvulsive Therapy," Task Force Report 14 (Washington, D.C.: American Psychiatric Association, 1978); V. Milstein and I. Small, "Electroconvulsive Therapy: Attitudes and Experience: A Survey of Indiana Psychiatrists," *Convulsive Therapy* 1(1985): 89–100.

Index